D1283505

Nuclear Magnetic Resonance Spectroscopy

A Physicochemical View

Nuclear Magnetic Resonance Spectroscopy

A Physicochemical View

Robin K Harris

School of Chemical Sciences
University of East Anglia, England

Pitman

PITMAN BOOKS LIMITED
128 Long Acre, London WC2E 9AN

PITMAN PUBLISHING INC
1020 Plain Street, Marshfield, Massachusetts 02050

Associated Companies
Pitman Publishing Pty Ltd, Melbourne
Pitman Publishing New Zealand Ltd, Wellington
Copp Clark Pitman, Toronto

© Robin K. Harris 1983

First published in Great Britain 1983

Library of Congress Cataloging in Publication Data

Harris, Robin Kingsley.
Nuclear magnetic resonance spectroscopy.

Bibliography: p.
Includes index.
1. Nuclear magnetic resonance spectroscopy.
II. Title
QD96.N8H37 1983 538′.362 82-18917
ISBN 0-273-01684-9

British Library Cataloging in Publication Data

Harris, Robin K.
Nuclear magnetic resonance spectroscopy.
1. Nuclear magnetic resonance spectroscopy
I. Title
538′.362 QC762

ISBN 0-273-01684-9

All rights reserved. No part of this publication may be reproduced, stored in a retrieval system, or transmitted, in any form or by any means, electronic, mechanical, photocopying, recording and/or otherwise, without the prior written permission of the publishers. This book may not be lent, resold or hired out or otherwise disposed of by way of trade in any form of binding or cover other than that in which it is published, without the prior consent of the publishers. This book is sold subject to the Standard Conditions of Sale of Net Books and may not be sold in the UK below the net price.

OLSON LIBRARY
NORTHERN MICHIGAN UNIVERSITY
MARQUETTE, MICHIGAN 49855

Typeset in Northern Ireland at The Universities Press (Belfast) Ltd.

Contents

Preface

Nuclear magnetic resonance spectroscopy has occupied an important place in the armoury of physical techniques available to the chemist for over 25 years. Indeed, it is arguable that it is the single most important tool for obtaining detailed information on chemical systems at the molecular level. In spite of the high cost of NMR spectrometers no moderately-sized chemical laboratory, whether industrial or academic, can be considered to be properly-equipped without at least one such instrument. No university undergraduate chemistry course is complete without a discussion of the basic principles of NMR. The subject has proved to be remarkable for the number of innovations that have appeared and become established as valuable additions. Thus, in the early 1970s the use of high-resolution techniques based on the Fourier transform (FT) principle profoundly modified the practice of NMR and provided a quantum jump in the amount of information accessible by NMR. The FT mode also ushered in an era in which an increasing number of spectrometer operations are computer-controlled, leading also to the ability both to perform new experiments and to enhance the quality of the results. The past five or six years have seen several additional 'quantum jumps' in NMR, for instance (a) multinuclear studies, (b) very-high-field operation, (c) high-resolution work on solids, (d) 'two-dimensional' operation, and (e) spin imaging. These advances owe much to the ever-increasing variety of pulse sequences which can be implemented.

It is clear that many of these advances must be incorporated into undergraduate and graduate teaching. In 1969 Dr Ruth Lynden-Bell and I wrote a book entitled *Nuclear Magnetic Resonance Spectroscopy* (published by Thomas Nelson & Sons Ltd, but now out of print) which took a physicochemical approach to such teaching. Although very little in that book has become outmoded, the material is now wholly inadequate for a proper view of NMR—the FT technique, for instance, was not mentioned. The aim of the present work is to provide an updated text based on the same philosophy of a physicochemical approach. Material from the earlier book has been incorporated (with modifications), particularly in Chapters 1, 2 and 8, but the total amount of text has roughly doubled, since much has been added. The new book is intended to cover NMR adequately for students taking Bachelor's and Master's degrees in chemistry, and should also be valuable in Ph.D. work and for people involved in NMR to a greater or lesser extent in industrial and Government laboratories.

It is, perhaps, obvious that considerable difficulty was encountered in deciding which topics to include and which to exclude. The choice is, to some extent, personal, but the policy has been to avoid mention of topics at a trivial level, such as would have been necessary to keep the size of the book within reasonable bounds if all the conceivably relevant areas were to be mentioned. Therefore several topics have been totally omitted. The principal among these are (a) NMR in liquid crystals and liquid crystal solutions, (b) spin imaging and related topics, and (c) chemically-induced dynamic nuclear polarization. These areas are self-contained and may be considered somewhat away from mainstream chemical usage of NMR. Specifically biochemical applications of NMR have also been largely omitted, in spite of their importance, mainly because it is difficult to develop such aspects without extensive discussion of strictly biochemical principles. Some will doubtless be critical of these decisions, but I defy anyone to produce a set of topics for a book of this size which would not arouse similar criticisms! Some texts on the omitted topics are listed under 'Further reading' at the end of this book (p. 241).

Problems are also encountered in deciding the level of mathematical sophistication used. I have tried to combine a degree of rigour with a desire both to give a physically-meaningful description of NMR behaviour and to minimize the amount of mathematics that is needed. A deliberate decision was taken not to introduce density matrix formalism, in spite of its power. Throughout the book SI units and equations are used. The form of relaxation equations may on occasion seem unfamiliar because of the factors of $(\mu_0/4\pi)$, and many scientists still apparently find Gauss rather than Tesla come first to mind for units of magnetic field, but these problems are not usually of great significance. I believe familiarity with SI is now such that a more detailed discussion of units is not required, but readers should refer if necessary to the booklet by M. L. McGlashan entitled *Physico-Chemical Quantities and Units* (Royal Society of Chemistry Monographs for Teachers, Number 15, 1968). Lists of SI units and of fundamental constants, in tabular form, follow this preface. IUPAC recommendations for NMR have been used throughout the book (see *Pure Appl. Chem.*, **29,** 627 (1972) and **45,** 219 (1976)).

Chapters 1, 2 and 8 cover roughly the same ground as in the corresponding chapters of the earlier book. Chapter 1 is, indeed, a self-contained discussion of the simpler basic principles of NMR, as involved in the continuous wave (CW) mode of operation, and is intended for beginners of the subject. Chapter 2 discusses second-order spectral analysis, and it is largely possible for the student to omit detailed reading of this Chapter if it is not directly relevant to his or her interests. Chapter 3 is largely new, and introduces the Fourier transform principle, together with the concept of relaxation. Chapter 4 continues the discussion of relaxation but concentrates on the dipolar interaction and on double resonance. The fifth chapter covers chemical exchange and the effects arising from quadrupolar interactions, in a form greatly expanded from the earlier book. Chapters 6 and 7 are entirely new. The former deals with NMR of solids, including the high-resolution techniques that are now finding increasing application. The latter discusses a number of special pulse sequences and develops the concepts of two-dimensional NMR. The final chapter is about the factors affecting chemical shifts and coupling constants. Problems are provided at the end of each chapter. A limited number of references, together with suggestions for further reading, are also included.

It is hoped that the whole volume provides a basis of understanding of NMR that will be valuable for chemists of many complexions, but particularly for those who value a physicochemical approach.

University of East Anglia
5 April 1982

Robin Harris

Acknowledgements

I am very grateful to Dr Ruth M. Lynden-Bell who joined me in writing a book on NMR in 1969. In particular, Chapter 2 of that book, which was written by Dr Lynden-Bell, has become, with modifications, Chapter 2 of the present book. Dr Lynden-Bell has also kindly read and commented on the present Chapter 3. I am also grateful to my colleagues Professors Norman Sheppard and Ken Packer who have, over the years, been the source of countless stimulating discussions of NMR. The work of many graduate students and other research workers at the University of East Anglia has also contributed to my enlightenment and has provided a number of the figures in the present book. Dr G. A. Webb has made some useful comments on Chapter 8 at galley-proof stage, which has resulted in improvements, and I thank him for his help. I would also like to thank Dr A. J. Jones of the National NMR Centre, the Australian National University, with whom I spent two months of sabbatical leave in 1981 during which the writing of this book was commenced. Some figures have been taken from the literature, and acknowledgements are given in the appropriate figure captions. A number of people have assisted with typing this book, but the largest share has been carried out by Mrs Janice Hancock, whom I would like to thank for all her work.

List of SI units and related quantities

Abbreviations

Property	*Symbol*	*SI unit*	*Symbol of unit*	*Equivalent SI units*	*Equivalents in other systems*
Length	ℓ, r	metre	m	–	10^2 centimetres
Mass	m	kilogramme	kg	–	10^3 grammes
Time	t	second	s	–	–
Force	F	newton	N	kg m s^{-2}	10^7 dynes
Energy	U	joule	J	kg m^2 s^{-2}	10^7 ergs
Angle	θ	radian	rad	$1/2\pi$ cycles	–
Frequency	ν	hertz	Hz	–	1 cycle per second (c/s or cps)
Electric charge	q	coulomb	C	A s	$2{\cdot}9979 \times 10^9$ e.s.u.
Electric current	i	ampere	A	–	10^{-1} e.m.u.
Magnetic induction field	B	tesla	T	kg s^{-2} A^{-1}	10^4 gauss (e.m.u.)
Magnetic field intensity	H	ampere metre^{-1}	–	m^{-1} A	$4\pi \times 10^{-3}$ oersted
Magnetic dipole moment	μ	ampere metre^2	–	m^2 A	10^3 e.m.u. sec^{-1}
Magnetogyric ratio	γ	radian tesla^{-1} second^{-1}	–	rad kg^{-1} s A	10^{-4} rad gauss^{-1} sec^{-1}
Reduced coupling constant	K	newton ampere^{-2} metre^{-3}	–	kg m^{-2} s^{-2} A^{-2}	10 cm^{-3}

Fundamental constants[a]

Constant	Symbol	Value	Unit: SI	Unit: CGS
Elementary charge	e	1·602192	$\times 10^{-19}$ C	$\times 10^{-20}$ e.m.u.
		4·80325	–	$\times 10^{-10}$ e.s.u.
Electron rest mass	m_e	9·10956	$\times 10^{-31}$ kg	$\times 10^{-28}$ g
Proton rest mass	m_p	1·67261	$\times 10^{-27}$ kg	$\times 10^{-24}$ g
Planck constant	h	6·62620	$\times 10^{-34}$ J s	$\times 10^{-27}$ erg s
	$\hbar = h/2\pi$	1·054592	$\times 10^{-34}$ J s	$\times 10^{-27}$ erg s
Charge-to-mass ratio for electron	e/m_e	1·758803	$\times 10^{11}$ C kg^{-1}	$\times 10^{7}$ e.m.u.
Permeability constant	μ_0	4π	$\times 10^{-7}$ kg m s^{-2} A^{-2}	–
		1	–	$\times 1$ e.m.u.
Free electron Landé splitting factor	g_s	2·00232	–	–
Magnetogyric ratio of proton	γ_p	2·675197[b]	$\times 10^{8}$ rad s^{-1} T^{-1}	$\times 10^{4}$ rad s^{-1} gauss^{-1}
	$\gamma_p/2\pi$	4·25771[b]	$\times 10^{7}$ Hz T^{-1}	$\times 10^{3}$ s^{-1} gauss^{-1}
Bohr magneton	μ_B	9·27410	$\times 10^{-24}$ J T^{-1}	$\times 10^{-21}$ erg gauss^{-1}
Nuclear magneton	μ_N	5·05095	$\times 10^{-27}$ J T^{-1}	$\times 10^{-24}$ erg gauss^{-1}
Boltzmann constant	k	1·38062	$\times 10^{-23}$ J K^{-1}	$\times 10^{-16}$ erg K^{-1}
Avogadro constant	N	6·02217	$\times 10^{23}$ mole^{-1}	$\times 10^{23}$ mole^{-1}

[a] Taken from B. N. Taylor, W. H. Parker & D. N. Langenberg, *Rev. Mod. Phys.* **41,** 375 (1969), but quoted to fewer places

[b] Corrected for the diamagnetic shift of H_2O.

Symbols and abbreviations

These lists contain the symbols and abbreviations most frequently used in this book, but they are not expected to be exhaustive. Some specialized notation is only defined in the relevant chapter. An attempt has been made to standardize usage throughout the book as far as is feasible, but it must be borne in mind that the original research literature certainly is not standardized in this way, and some difficulties may arise from this fact. Trivial use of subscripts, etc. is not always mentioned in the symbols list below. Some of the other symbols used in the text, e.g. for physical constants such as h or π, or for thermodynamic quantities such as H or S, are not included in the list since they are reckoned to follow completely accepted usage. In general, vectors are in bold print, and quantum mechanical operators are indicated by circumflexes.

Symbols

a	hyperfine (electron-nucleus) coupling constant
$\mathbf{B}$	magnetic induction field (magnetic flux density)
$\mathbf{B}_0$	static magnetic field of an NMR spectrometer
$\mathbf{B}_1, \mathbf{B}_2$	r.f. magnetic fields associated with ν_1, ν_2.
$\mathbf{B}_\mathrm{L}$	local magnetic field (components $B_{x\mathrm{L}}$, $B_{y\mathrm{L}}$, $B_{z\mathrm{L}}$)—of random field or dipolar origin
c	coefficient in linear expansion of wave functions
C_X	(i) natural abundance of nuclide X, expressed as a % (ii) spin-rotation coupling constant of nuclide X
D^C	nuclear receptivity relative to that of carbon-13
D^P	nuclear receptivity relative to that of the proton
$\mathbf{E}$	electric field
F	spectral width
$\hat{\mathbf{F}}_X$	nuclear spin operator for a group, G, of nuclei (components $\hat{F}_{Gx}$, $\hat{F}_{Gy}$, $\hat{F}_{Gz}$, $\hat{F}_{G+}$, $\hat{F}_{G-}$)
F_G	magnetic quantum number associated with $\hat{\mathbf{F}}_G$
g	nuclear or electronic g factor (Landé splitting factor)
H_{ij}	element of matrix representation of $\hat{\mathscr{H}}$
$\hat{\mathscr{H}}$	Hamiltonian operator (in energy units)—subscripts indicate the nature of the operator
i	$\sqrt{-1}$
$\hat{\mathbf{I}}_j$	nuclear spin operator for nucleus j (components $\hat{I}_{jx}$, $\hat{I}_{jy}$, $\hat{I}_{jz}$)
$\hat{I}_{j+}, \hat{I}_{j-}$	'raising' and 'lowering' spin operators for nucleus j
I_j	magnetic quantum number associated with $\hat{\mathbf{I}}_j$
I	moment of inertia

${}^nJ$	nuclear spin–spin coupling constant through n bonds (in Hz). Further information may be given by subscripts or in brackets. Normally subscripts are only used for algebraic symbols for nuclei in spectral analysis cases, e.g. J_{AX}. Brackets are used for indicating the species of nuclei coupled, e.g. $J(^{13}C, ^1H)$ or, additionally, the coupling path, e.g. J(POCF)
$J(\omega)$	spectral density at angular frequency ω
${}^nK$	reduced nuclear spin-spin coupling constant (see the notes concerning ${}^nJ$)
m_e	mass of the electron
m_j	eigenvalue of $\hat{I}_{jz}$ (magnetic component quantum number)
m_T	total magnetic quantum number for a spin system (eigenvalue of $\sum_j \hat{I}_{jz}$)
$m_T(X)$	total magnetic quantum number for X-type nuclei
$\mathbf{M}_0$	equilibrium macroscopic magnetization of a spin system in the presence of B_0
M_x, M_y, M_z	components of macroscopic magnetization
M_n	moment of a spectrum (M_2 = second moment, etc.)
n_α, n_β	populations of the α and β spin states
N	total number of nuclei of a given type in the sample
P	(i) angular momentum; (ii) transition probability
q	electric field gradient (principal components q_{xx}, q_{yy}, q_{zz}.)
Q	nuclear quadrupole moment
r	(i) general symbol for distance (ii) general symbol for spin state (as $\langle r|$ or $|r\rangle$)
R	dipolar coupling constant, $(\mu_0/4\pi)\gamma_1\gamma_2(\hbar/2\pi)r^{-3}$
s	general symbol for spin state (as $\langle s|$ or $|s\rangle$)
S	(i) signal height (ii) electron (or, occasionally, nuclear) spin—cf. I
T	temperature
T_c	coalescence temperature for an NMR spectrum
T_1^X	spin–lattice relaxation time of the X nuclei (further subscripts refer to the relaxation mechanism)
T_2^X	spin–spin relaxation time of the X nucleus (further subscripts refer to the relaxation mechanism)
T_2'	inhomogeneity contribution to the dephasing time for M_x or M_y
T_2^*	total dephasing time for M_x or M_y; $(T_2^*)^{-1} = T_2^{-1} + (T_2')^{-1}$
$T_{1\rho}^X$	spin-lattice relaxation time of the X nuclei in the frame of reference rotating with B_1
T_d	pulse delay time (in FT NMR)
T_{ac}	acquisition time (in FT NMR)
T_p	period for repetitive pulses (= interpulse time = $T_{ac} + T_d$ if τ_p is negligible)
u	in-phase (dispersion mode) signal
v	out-of-phase (absorption mode) signal
W_0, W_1, W_2	relaxation rates between energy levels differing by 0, 1 and 2 (respectively) in m_T

α nuclear spin wave function (eigenfunction of $\hat{I}_z$) for a spin-$\frac{1}{2}$ nucleus

α_A^2 s-character of hybrid orbital at atom A

α_E the Ernst angle (for optimum FT sensitivity)

β nuclear spin wave function (eigenfunction of $\hat{I}_z$) for a spin-$\frac{1}{2}$ nucleus

γ_X magnetogyric ratio of nucleus X

δ_X chemical shift (for the resonance) of nucleus of element X (positive when the sample resonates to high frequency of the reference). Usually in ppm. Further information regarding solvent, references or nucleus of interest may be given by superscripts or subscripts or in brackets.

Δ spectral width

Δn population difference between nuclear states (Δn_0 at Boltzmann equilibrium)

$\Delta\delta$ change or difference in δ

$\Delta\nu_{1/2}$ full width (in Hz) of a resonance line at half-height

$\Delta\sigma$ (i) anisotropy in $\sigma(\Delta\sigma = \sigma_{\parallel} - \sigma_{\perp})$
(ii) difference in σ for two different situations

$\Delta\chi$ (i) susceptibility anisotropy ($\Delta\chi = \chi_{\parallel} - \chi_{\perp}$)
(ii) difference in electronegativities

ε_0 permittivity of a vacuum

η (i) nuclear Overhauser enhancement
(ii) asymmetry factor (e.g. in e^2qQ/h)
(iii) viscosity

θ angle—especially for that between a given vector and $\mathbf{B}_0$

$\boldsymbol{\mu}_0$ (i) magnetic dipole moment (component μ_z along $\mathbf{B}_0$)
(ii) electric dipole moment

μ_0 permeability of a vacuum

μ_B Bohr magneton

μ_N nuclear magneton

ν_c carrier frequency of the radiation

ν_j Larmor precession frequency of nucleus j (in Hz)

ν_0 (i) spectrometer operating frequency
(ii) Larmor precession frequency (general, or of bare nucleus)

ν_1 frequency of 'observing' r.f. magnetic field

ν_2 frequency of 'irradiating' r.f. magnetic field

Ξ_X resonance frequency for the nucleus of element X in a magnetic field such that the protons in TMS resonate at exactly 100 MHz

σ_j shielding constant of nucleus j (used sometimes in tensor form). Usually in ppm. Subscripts may alternatively indicate contributions to σ

$\sigma_{\parallel}, \sigma_{\perp}$ components of σ parallel and perpendicular to a molecular symmetry axis

τ time between r.f. pulses (general symbol)

τ_A pre-exchange lifetime of molecular species A

τ_c correlation time, especially for molecular tumbling

τ_d dwell time

τ_e electronic correlation time

τ_{null} recovery time sufficing to give zero signal after a 180° pulse

τ_p pulse duration

τ_{sc}	correlation time for relaxation by the scalar mechanism
τ_{sr}	correlation time for spin-rotation relaxation
$\tau_{\parallel}, \tau_{\perp}$	correlation times for molecular tumbling parallel and perpendicular to a symmetry axis
χ	(i) magnetic susceptibility (ii) electronegativity (iii) nuclear quadrupole coupling constant ($=e^2qQ/h$)
$\omega_j, \omega_0, \omega_1, \omega_2, \omega_c$	as for $\nu_j, \nu_0, \nu_1, \nu_2, \nu_c$ but in rad s^{-1}

Abbreviations

acac	acetylacetonato
ASIS	aromatic solvent-induced shift
CIDNP	chemically-induced dynamic nuclear polarization
CPMG	Carr–Purcell pulse sequence, Meiboom–Gill modification
CW	continuous wave
DD	dipole–dipole (interaction or relaxation mechanism)
DMSO	dimethylsulphoxide
EFG	electric field gradient
en	ethylenediamine
ESR	electron spin resonance
FID	free induction decay
fod	1,1,1,2,2,3,3-heptafluoro-7,7-dimethyl-4,6-octanedionato
FT	Fourier transform
LCAO	linear combination of atomic orbitals
MO	molecular orbital
mol. wt.	molecular weight
NMR	nuclear magnetic resonance
NOE	nuclear Overhauser effect
NQR	nuclear quadrupole resonance
o.d.	outside diameter
ppm	parts per million
QF	quadrupole moment/field gradient (interaction or relaxation mechanism)
r.f.	radiofrequency
r.m.s.	root mean square
SA	shielding anisotropy
S/N	signal-to-noise ratio
SPI	selective population inversion
SPT	selective population transfer
SR	spin–rotation (interaction or relaxation mechanism)
tmhd	2,2,6,6-tetramethylheptane-3,5-dionato (also known as dpm, dipivaloylmethanato)
TMS	tetramethylsilane
UE	unpaired electron relaxation mechanism
WAHUHA	Waugh, Huber and Haeberlen (pulse sequence)

1 The fundamentals

1-1 Introduction

Spectroscopy may be defined as the interaction between matter and electromagnetic radiation such that energy is absorbed or emitted according to the Bohr frequency condition, $\Delta E = h\nu$, where ΔE is the energy difference (normally quantized) between the initial and final states of the matter, h is Planck's constant, and ν is the frequency of the electromagnetic radiation. It proves convenient to sub-divide the subject according to the nature of the energy levels considered. This classification usually correlates with the region of the spectrum (i.e. the range of frequencies) used. Altogether, spectroscopy is the most powerful tool the chemist has for investigating molecular structure and molecular processes. This statement is true regardless of whether the chemist's interest is organic, inorganic, physical, theoretical, or biochemical.

This book sets out to discuss the physicochemical basis of that branch of spectroscopy known as nuclear magnetic resonance (NMR). NMR involves the magnetic energy of nuclei when they are placed in a magnetic field, and the transitions occur in the radio wave region of the spectrum. The first experiments were carried out[1] in 1945, but useful chemical applications became possible only after the discovery of the chemical shift effect[2] in 1949. From modest beginnings the subject developed very rapidly, especially after the appearance of the first commercial NMR spectrometer in 1953. Today the subject has expanded so that it is of equal importance with the older-established vibrational (infrared) and electronic (ultraviolet) branches of spectroscopy. Unfortunately, the cost of NMR instrumentation is rather high.

All forms of spectroscopy give spectra that may be described in terms of the frequency, intensity and shape of spectral lines or bands. These observable properties depend on molecular parameters of the system, which for NMR are found to be the shielding constants and coupling constants of the nuclei, and the lifetimes of the energy levels. It is these parameters which are of fundamental physicochemical importance. The NMR spectroscopist will be concerned in (a) obtaining spectra and developing the techniques used, (b) evaluating the nuclear magnetic parameters from the spectra, (c) interpreting or predicting these parameters using physical or theoretical models, and

(d) applying the knowledge gained to problems of chemical interest such as structure determination, evaluation of kinetic or thermodynamic data, etc. This book will be concerned with all four aspects.

We shall deal primarily with NMR of liquid or solution systems. Studies in the gas phase are uncommon, partly because of the rather low sensitivity of the NMR technique (vapour pressures of the order of one atmosphere are needed) and partly because the resonance lines are usually rather broader (of the order of a few Hz) than for liquids, since collision narrowing does not always occur efficiently for gases at atmospheric pressure. NMR studies of solids are of increasing importance since the development of special techniques allows high resolution to be attained, and this area is discussed in Chapter 6. In the absence of these special techniques the spectra from solids usually contain very broad bands (of the order of tens of kHz), due to the direct effects of the interaction of nuclear spin dipoles in fixed orientations, whilst spectral lines for fluids are very narrow, since molecular tumbling averages the direct effects to zero.

This first chapter deals with the basis of the subject of NMR, explains some of the concepts involved, discusses the simplest types of spectra obtained for liquids or solutions, and sets the scene for the more detailed discussion in the remaining chapters. The treatment of relaxation phenomena is postponed to Chapter 3.

1-2 Quantization of angular momentum

One of the fundamental postulates of modern chemical physics is that the total angular momentum of any isolated particle cannot have any arbitrary magnitude but may only take certain discrete values. Angular momentum is said to be quantized, and its magnitude, P, can be specified in terms of a quantum number, say R, by means of Eqn 1-1:

$$P = \hbar[R(R+1)]^{1/2} \tag{1-1}$$

The constant $\hbar$ is $h/2\pi$, where h is known as Planck's constant; R is either integral or half-integral. Frequently, the various possible modes of motion (rotation, vibration, etc.) that contribute to the total angular momentum may be treated independently, and each assigned a separate quantum number. This quantum number specifies (by means of an equation analogous to 1-1) the angular momentum attributed to the particular mode of motion considered.

The quantization of angular momentum implies quantization of energy. Each energy state of the particle may be specified by the appropriate quantum numbers and has a stationary-state wave function (eigenfunction). In some instances, states with different eigenfunctions may have the same energy, and they are then said to be degenerate. The total energy of the particle may be considered as the sum of individual contributions due to the various possible modes of motion. The total wave function describing the particle may then be written as a *product* of separate wave functions for each type of motion.

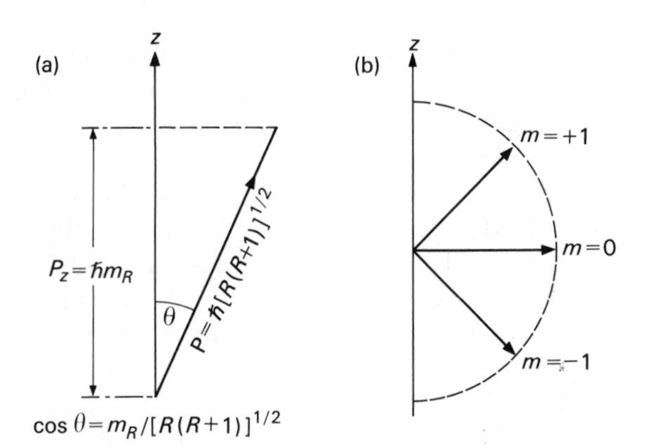

Fig. 1-1 Angular momentum vectors: (a) the general case, showing the use of quantum numbers R and m_R, (b) the possible orientations for an angular momentum vector of quantum number $R = 1$. Note that R and m_R do not specify direction in the xy plane at all.

Angular momentum is actually a vector property, so that for a full description direction has to be specified as well as magnitude. In quantum-mechanical terms this is done by using another quantum number, m_R say, such that the component of angular momentum, P_z, along the direction considered (z), is given by Eqn 1-2:

$$P_z = \hbar m_R \tag{1-2}$$

Quantum numbers specifying vector magnitudes must, of course, be positive. Since the component cannot be greater than the vector itself, $|m_R|$ must be less than or equal to R, but m_R can be either positive or negative. In fact, quantum mechanics shows that the allowed values of m_R are as in Eqn 1-3; there are thus $2R+1$ possible values of m_R.

$$m_R = R, R-1, R-2 \ldots -R \tag{1-3}$$

Figure 1-1 illustrates the general situation for a quantized angular momentum vector.

1-3 Electron and nuclear spin

The particular type of angular momentum pertinent to this book is that of nuclear spin. Transitions between nuclear spin energy levels give rise to the phenomenon of NMR; the corresponding property of electron spin leads to electron spin resonance (ESR). The concept of electron spin is probably familiar to chemistry students at an early stage, since the structure of the periodic table can be explained on the basis of four electronic quantum numbers, n, ℓ, m_ℓ and m_s, the last-mentioned being the directional (component) spin quantum number. The postulate that electrons possessed angular momentum which could be attributed to a spinning motion was first made in 1925, in order to explain both the fine structure of bands observed in atomic spectra, and the effect of a magnetic field on such spectra (the anomalous Zeeman effect). Electron spin also appears as a natural result from relativistic wave mechanics. The spin quantum number of a single electron is $s = \frac{1}{2}$ and the associated angular momentum is $\hbar\sqrt{3}/2$ (see Eqn 1-1). The possible values of the directional quantum number, m_s, are therefore $+\frac{1}{2}$ and $-\frac{1}{2}$. This is reflected in one of the ways of expressing Pauli's exclusion principle, namely that only two electrons ($m_s = \pm\frac{1}{2}$) may occupy an electronic orbital specified by the quantum numbers, n, ℓ, and m_ℓ.

The existence of nuclear spin was first suggested by Pauli in 1924, a year before the idea of electron spin was mooted, in order to explain further splittings in atomic spectra (hyperfine structure). These splittings demonstrate that nuclei of different elements (and different isotopes of the same element) differ in spin angular momentum. This fact may be traced to details of the internal construction of nuclei. The proton has a spin quantum number, I, of $\frac{1}{2}$, as does the neutron. For nuclei other than ^1H the spin angular momenta of the individual nucleons couple (together with their orbital-type angular momenta) to give the observed total. This is characterized by a quantum number, I, which may be integral or half-integral or zero. Excited states of the nucleus with different values of I exist; these are important for the Mössbauer effect, but the energy differences between the possible states are so huge that only the ground state need be considered here. Certain systematic features may be noted as follows:

(a) Nuclei with an odd mass number have half-integral spin.

(b) Nuclei with an even mass number and an even charge number have zero spin (e.g. ^{12}C, ^{16}O, and ^{32}S).

(c) Nuclei with an even mass number but an odd charge number have integral spin.

Table 1-1 lists values of the nuclear spin quantum number of some of the more popular nuclides for NMR. More extended lists of nuclei are given in Appendices 1 and 2. Since NMR depends on the existence of nuclear spin, nuclei with $I = 0$ do not show any direct effects and for most purposes may be ignored. Since the ^{12}C and ^{16}O isotopes account for almost the whole of naturally-occurring carbon and oxygen nuclei, the NMR spectra of many quite complex organic chemicals are remarkably simple and easy to interpret. This fact makes NMR in many cases a more convenient tool for structure determination than infrared or ultraviolet spectroscopy.

The direction of a spin angular momentum vector is conventionally defined such that the spinning body, viewed along the vector, twists in the same sense as a right-handed screw. The same convention is applied to the vector $\boldsymbol{\omega}$ describing the angular velocity of the rotation. The magnitude of the nuclear spin angular momentum is given by the principles of Section 1-2 as

$$P = \hbar[I(I+1)]^{1/2} \tag{1-4}$$

Nuclei with spin quantum number greater than $\frac{1}{2}$ also possess electric quadrupole moments. This results in lifetimes for their nuclear magnetic states in solution which are normally much shorter than those for spin-$\frac{1}{2}$ nuclei. The NMR absorption lines of $I > \frac{1}{2}$ nuclei are therefore generally broad (by virtue of the uncertainty principle) and not so readily studied as those for $I = \frac{1}{2}$ nuclei. For some purposes such nuclei

Table 1-1 Nuclear spin properties for the commonly-studied nuclides[(a)]

Isotope	*Spin*	*Natural abundance* C/%	*Magnetic moment*[(b)] μ/μ_N	*Magnetogyric ratio* $\gamma/10^7$ rad T^{-1} s^{-1}	*NMR frequency* Ξ/MHz	*Usual reference*
electron	$\frac{1}{2}$	—	$-3{\cdot}18392\times10^3$	$-1{\cdot}76084\times10^4$	$[6{\cdot}582\times10^4]$	—
neutron	$\frac{1}{2}$	—	−3·31362	−18·3257	[68·50]	—
^1H	$\frac{1}{2}$	99·985	4·83724	26·7519	100·000000	Me_4Si
^2H	1	0·015	1·2126	4·1066	15·351	Me_4Si-d
^3H	$\frac{1}{2}$	—[(c)]	5·1596	28·535	106·664	Me_4Si-t
^7Li	$\frac{3}{2}$	92·58	4·20394	10·3975	38·866	Li^+aq
^{11}B	$\frac{3}{2}$	80·42	3·4708	8·5843	32·089	$Et_2O.BF_3$
^{13}C	$\frac{1}{2}$	1·108	1·2166	6·7283	25·145004	Me_4Si
^{14}N	1	99·63	0·57099	1·9338	7·228	$MeNO_2$ or
^{15}N	$\frac{1}{2}$	0·37	−0·4903	−2·712	10·136783	$[NO_3]^-$
^{17}O	$\frac{5}{2}$	0·037	−2·2407	−3·6279	13·561	H_2O
^{19}F	$\frac{1}{2}$	100	4·5532	25·181	94·094003	CCl_3F
^{23}Na	$\frac{3}{2}$	100	2·86265	7·08013	26·466	Na^+aq
^{27}Al	$\frac{5}{2}$	100	4·3084	6·9760	26·077	$[Al(H_2O)_6]^{3+}$
^{29}Si	$\frac{1}{2}$	4·70	−0·96174	−5·3188	19·867184	Me_4Si
^{31}P	$\frac{1}{2}$	100	1·9602	10·841	40·480737	85% H_3PO_4
^{59}Co	$\frac{7}{2}$	100	5·234	6·317	23·61	$[Co(CN)_6]^{3-}$
^{77}Se	$\frac{1}{2}$	7·58	0·925	5·12	19·071523	Me_2Se
^{113}Cd	$\frac{1}{2}$	12·26	−1·0768	−5·9550	22·193173	$CdMe_2$
^{119}Sn	$\frac{1}{2}$	8·58	−1·8119	−10·021	37·290662	Me_4Sn
^{195}Pt	$\frac{1}{2}$	33·8	1·043	5·768	21·414376	$[Pt(CN)_6]^{2-}$

(a) For more information see Appendix 1.
(b) See note [3] at the end of the chapter.
(c) Radioactive.

may be regarded as effectively non-magnetic. For most of this book we shall be considering only spin-$\frac{1}{2}$ nuclei, though some of the equations and the majority of the concepts apply equally to spin$>\frac{1}{2}$ nuclei. Some specific properties of such nuclei will be considered in Chapter 5.

1-4 Nuclear magnetic moments

Any motion of a charged body has an associated magnetic field; on a macroscopic scale an electrical current, which is due to motion of electrons along a conductor, produces such a field. Current travelling in a loop has an associated magnetic dipole moment. This phenomenon also occurs on an atomic scale, for whenever electrons or nuclei possess angular momentum there is a magnetic moment. Since angular momentum is quantized on this scale, so are magnetic moments. The magnitude of such moments is most easily discussed from the classical standpoint by considering orbital electronic motion. Suppose an electron is travelling in an orbital at an angular velocity ω. Such motion is equivalent to an electrical current in the opposite direction of magnitude $i = e\omega/2\pi$, where e is the magnitude of the charge on the electron. The orbital angular momentum, denoted P, is $m_e r^2\omega$ where

m_e is the mass of the electron and r its distance from the nucleus. Thus, the current is

$$i = e\omega/2\pi = eP/2\pi m_e r^2 \qquad (1\text{-}5)$$

The magnetic moment μ generated by such motion is given in electromagnetic theory by $\mu = Ai$, where A is the area marked out by the orbital. If it is supposed, for simplicity, that the electron moves in a circular orbit, $A = \pi r^2$, and substitution in Eqn 1-5 gives

$$\boldsymbol{\mu} = -(e/2m_e)\mathbf{P} \qquad (1\text{-}6)$$

This equation illustrates the fact that $\boldsymbol{\mu}$ and $\mathbf{P}$ are anti-parallel vectors for electrons. As has been stated above, on the atomic scale angular momentum is quantized in units of $\hbar$. It follows that electron magnetic moments are quantized in units of $(eh/4\pi m_e)$. This unit, which will be written herein as μ_B, is an important quantity, known as the Bohr magneton. It has a value of $9{\cdot}27410 \times 10^{-24}$ J T^{-1}. When spin motion is involved, this simple approach does not reproduce quantitatively the quantum-mechanical expression relating magnetic moment and angular momentum. It is convenient to introduce a factor g (known as the Landé splitting factor, or simply the g-factor) to take account of this fact. Thus, we write

$$\boldsymbol{\mu} = -g(e/2m_e)\mathbf{P} = -g\mu_B \mathbf{P}/\hbar \qquad (1\text{-}7)$$

It is important to note that the observed values of the g-factor in various situations can be reproduced by a full quantum-mechanical treatment. Equation 1-7 can also be used for coupled spin–orbital motion if appropriate values of g are used.

Analogous equations may be used for the spin motion of nuclei. The constant of proportionality relating μ and P depends on the mass of the particle and on its charge. The appropriate equation for nuclei is therefore Eqn 1-8:

$$\mu \propto (Ze/2m_N)P \qquad (1\text{-}8)$$

where m_N is the mass of the nucleus and Z its charge number. For nuclei it proves to be convenient to use a unit of magnetic moment based on the application of Eqn 1-8 to the proton. This unit is known as the nuclear magneton, $\mu_N = eh/4\pi m_p$, where m_p is the mass of the proton. The magnitude of μ_N is $5{\cdot}05095 \times 10^{-27}$ J T^{-1}. For other nuclei the charge number and the difference in mass from that of the proton are both included in the nuclear g-factor, which also allows for quantum-mechanical effects. Thus

$$\boldsymbol{\mu} = g_N \mu_N \mathbf{P}/\hbar \qquad (1\text{-}9)$$

where g_N is the *nuclear* g-factor. It should be noted that unlike Eqn 1-7, which includes a negative sign explicitly (so that g is positive for electrons), Eqn 1-9 allows the nuclear g-factor to be either positive or negative. It might be thought that, since all nuclei are positively charged, all nuclear g-factors must be positive. However, this is not so, and the reason for this may be traced to the structure of the nucleus. Negative nuclear g-factors imply that the spin magnetic moment is anti-parallel to the angular momentum (as for the electron); positive values indicate that $\boldsymbol{\mu}$ and $\mathbf{P}$ are parallel.

Usually, nuclear magnetic moments are described in terms of magnetogyric ratios, γ, rather than nuclear g-factors. The value of γ is defined as the ratio of the magnetic moment to the angular momentum (taking explicit account of the sign of the relationship):

$$\boldsymbol{\mu} = \gamma \mathbf{P} \qquad (1.10)$$

From Eqns 1-9 and 1-10 it can be seen that

$$\gamma \hbar = g_N \mu_N \qquad (1\text{-}11)$$

Selected values of nuclear magnetic dipole moments and magnetogyric ratios are given in Table 1-1. Use of Eqn 1-4 with Eqns 1-9 and 1-10 gives[3]

$$\mu = g_N \mu_N [I(I+1)]^{1/2} = \gamma \hbar [I(I+1)]^{1/2} \qquad (1\text{-}12)$$

Nuclear g-factors are small numbers (as are electron g-factors) since the effects of charge and mass are partially compensatory. Nuclear magnetic moments are, however, only ca. 10^{-3} times as large as electronic magnetic moments due to the effect of the mass to charge ratio—the value of μ_N/μ_B depends on m_e/m_p.

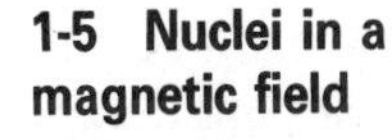

1-5 Nuclei in a magnetic field

In the absence of a magnetic field the energy of an isolated nucleus is independent of the quantum number m_I. This is equivalent to stating that the energy of a magnet is independent of its orientation when no magnetic field is present. However, when a magnetic field of strength B is applied in a direction defined as z, a magnetic moment $\boldsymbol{\mu}$ would have, in classical terms, an energy U given by

$$U = -\boldsymbol{\mu} \cdot \mathbf{B} = -\mu_z B \qquad (1\text{-}13)$$

where μ_z is the component of $\boldsymbol{\mu}$ in the direction z. This component is, in the case of nuclear spin, defined by m_I and, using an equation analogous to Eqn 1-12, is given by

$$\mu_z = g_N \mu_N m_I = \gamma \hbar m_I \qquad (1\text{-}14)$$

Thus Eqn 1-13 becomes

$$U = -\gamma \hbar m_I B \qquad (1\text{-}15)$$

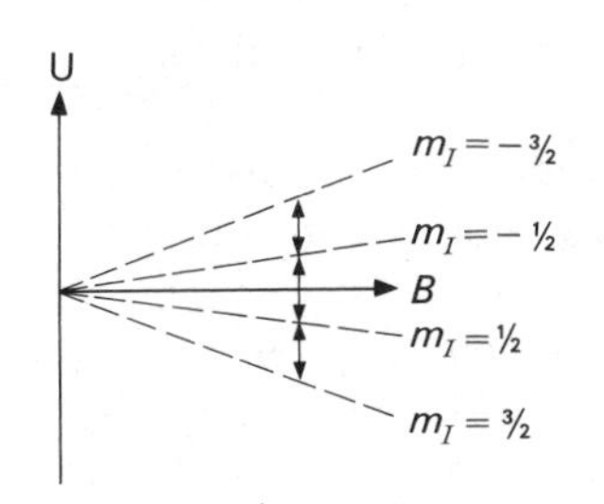

Fig. 1-2 The nuclear spin energy for a single nucleus with $I = \frac{3}{2}$ (such as ^{11}B), plotted as a function of the magnetic field B. The transitions obeying the selection rule $\Delta m_I = \pm 1$ for a particular value of B are indicated; these three transitions are degenerate for an isolated spin.

In this situation there are $(2I+1)$ non-degenerate energy levels (see Fig. 1-2) corresponding to the $2I+1$ values of m_I, each separated by $|\gamma \hbar B|$. It is clear that in principle transitions may be induced between these levels by the use of appropriate electromagnetic radiation. Such transitions will be governed by the Bohr frequency condition, whereby

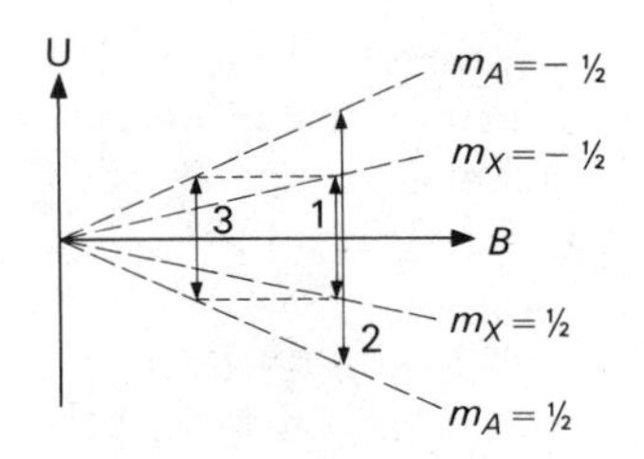

Fig. 1-3 Energy diagrams for two different isolated nuclei of spin $I=\frac{1}{2}$, A and X, illustrating constant frequency and constant field conditions for resonance. The diagram is drawn with $\gamma_A > \gamma_X$ (both positive). If resonance conditions are such that transition 1 is stimulated for nucleus X, then to achieve resonance for A, either the frequency must be increased at constant field (transition 2), or the field must be decreased at constant frequency (transition 3).

energy is transferred when ΔU, the gap between the energy levels considered, is equal to $h\nu$, where ν is the frequency of the radiation.

Thus for NMR transitions of a single nucleus

$$h\nu = |\gamma \hbar B \, \Delta m_I| \tag{1-16}$$

$$\nu = |(\gamma/2\pi) B \, \Delta m_I| \tag{1-17}$$

The selection rule governing such transitions is $\Delta m_I = \pm 1$ (see Chapter 2) so Eqn 1-17 reduces to

$$\nu = |\gamma/2\pi| B \tag{1-18}$$

In the rest of this book the modulus signs will generally be dropped from equations analogous to 1-18. It may be seen from Eqn 1-18 that for $I > \frac{1}{2}$ the $2I$ possible transitions are all degenerate in this simple case. Normally, magnetic fields in the range 1·5–12 T (15–120 kgauss) are used, and from Table 1-1 it can be seen that frequencies of the order of tens or hundreds of MHz are necessary. Such values are in the radiofrequency range of the spectrum.

Since the resonance frequency is proportional to magnetic field, either ν or B may be varied in order to achieve resonance. If it is required to change from observing the resonance of nucleus X to that of nucleus A, where the latter has a higher magnetogyric ratio than the former, then Eqn 1-18 states that either ν may be *increased* at constant B or B may be *decreased* at constant ν (see Fig. 1-3). There are advantages in using the highest value of B consistent with magnetic field homogeneity requirements, so it is normal to alter ν. For example, in a field of 2·35 T protons resonate at ca. 100 MHz whereas ^{11}B nuclei absorb at ca. 32 MHz. In contrast to the case just considered, where field and frequency have opposing effects as far as achieving resonance is concerned, for a given nucleus, *increase* of B requires an *increase* of ν if resonance is to be maintained. Thus protons resonate at ca. 40 MHz in a 1 T field, and at ca. 60 MHz in a 1·5 T field.

1-6 Larmor precession

An alternative approach to NMR, which is frequently of great use, is to consider the classical motion of a magnetic moment in a uniform magnetic field, $\mathbf{B}_0$, under the condition of constant total energy. A torque is exerted on the magnetic moment which tends to align it perpendicular to the field. However, since this torque can only alter the component of angular momentum perpendicular to $\mathbf{B}$ and $\boldsymbol{\mu}$, the net result corresponds to a rotation of the direction of $\boldsymbol{\mu}$ in a cone with its axis along $\mathbf{B}_0$, i.e. the movement of μ traces out a cone about B_0. Such a movement, being analogous to the motion of a gyroscope, is referred to in general as precession (see Fig. 1-4), and in such instances as are considered here it is known as Larmor precession. The induced angular velocity of the precession is given by $\boldsymbol{\omega}_i = -\gamma \mathbf{B}_0$ (Section 3-2). Thus the Larmor frequency, ν_i, is

$$\nu_i = |\gamma/2\pi| \, B_0 \tag{1-19}$$

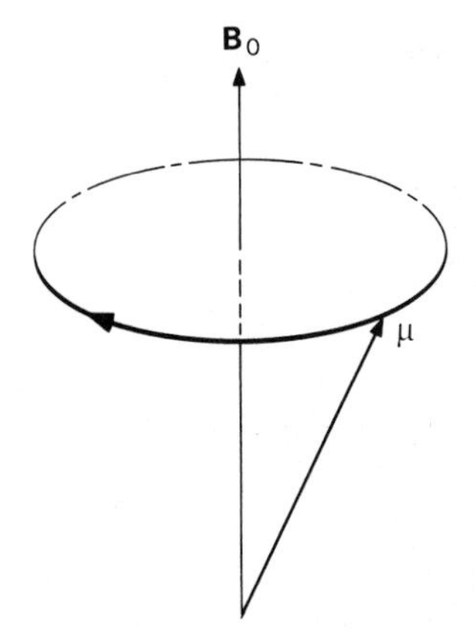

Fig. 1-4 Precession of a magnetic moment $\boldsymbol{\mu}$ about an applied magnetic field $\mathbf{B}_0$. It is assumed that the magnetogyric ratio is positive, as for the proton.

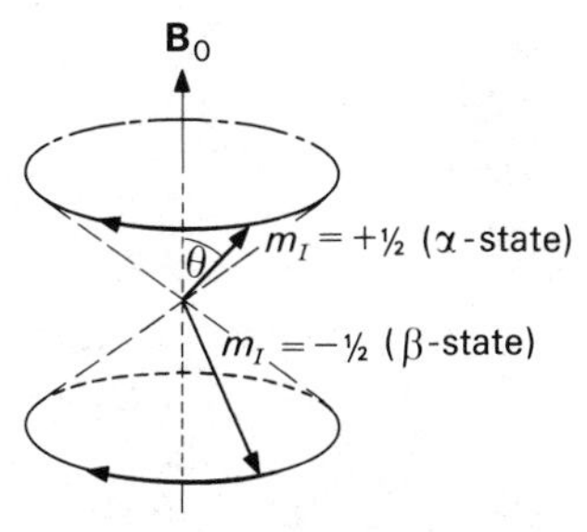

Fig. 1-5 The two precessional cones for a spin-$\frac{1}{2}$ nucleus in a magnetic field, showing typical spin vectors. The angle θ is 54° 44′ in this case.

Quantization of magnetic dipolar energy in the field B_0 can readily be incorporated in this picture; the half-angle of the cone of precession is determined by the quantum number m_I, i.e. the half-angle is given by $\cos\theta = m_I/[I(I+1)]^{1/2}$. Thus nuclei of spin I will be distributed among $2I+1$ precession cones. The spin-$\frac{1}{2}$ case is shown in Fig. 1-5. It should be noticed that the direction and frequency of precession are independent of m_I; when γ is positive the motion about $\mathbf{B}_0$ is left-handed as shown in Fig. 1-5. For a spin-$\frac{1}{2}$ nucleus the two spin states $m_I = +\frac{1}{2}$ and $m_I = -\frac{1}{2}$ are frequently referred to as α and β respectively.

Suppose there is an additional weak magnetic field, $\mathbf{B}_1$, perpendicular to $\mathbf{B}_0$. Such a field will also exert a torque on $\boldsymbol{\mu}$, tending to change the angle θ between $\boldsymbol{\mu}$ and $\mathbf{B}_0$. However, if $\mathbf{B}_1$ is fixed in direction it will alternately try to increase and decrease θ as $\boldsymbol{\mu}$ precesses. Since $\mathbf{B}_1$ is stated to be weak, the net effect will be a slight wobbling in the precession of $\boldsymbol{\mu}$; such an effect is referred to as nutation. Alternatively, the motion of $\boldsymbol{\mu}$ can be described as caused by a resultant field $\mathbf{B}_0+\mathbf{B}_1$. If, on the other hand, $\mathbf{B}_1$ is not fixed in direction, but is rotating about $\mathbf{B}_0$ with the same frequency as the precession of $\boldsymbol{\mu}$ and in the same direction, its orientation with respect to $\boldsymbol{\mu}$ will be constant. Suppose this orientation is such that $\mathbf{B}_1$ is always perpendicular to the plane containing $\mathbf{B}_0$ and $\boldsymbol{\mu}$ as in Fig. 1-6; then the torque exerted on $\boldsymbol{\mu}$ by $\mathbf{B}_1$ will always be *away* from $\mathbf{B}_0$. Consequently, a large effect on $\boldsymbol{\mu}$ is possible (see Fig. 1-6 and Chapter 3). Since changing θ corresponds to changing the energy of $\boldsymbol{\mu}$ in $\mathbf{B}_0$, this condition is described as resonance—the frequency, ν, of the field $\mathbf{B}_1$ required must, by the above discussion, equal the Larmor precession frequency of Eqn 1-19. This statement is clearly in agreement with the quantum-mechanical formula for NMR absorption given by Eqn 1-18. The energy for the change of θ is, of course, derived from the rotating field $\mathbf{B}_1$, which is supplied by radiofrequency electromagnetic radiation.

1-7 The intensity of an NMR signal

So far, only the behaviour of individual nuclear spins has been considered. However, a macroscopic sample contains an ensemble of many such spins, which will be distributed at random around the precessional cones depicted in Fig. 1-5. If the sample consists of many identical molecules, each with one magnetic nucleus of spin $I=\frac{1}{2}$, the populations of the two states depicted in Fig. 1-5 can be derived using Boltzmann's equation. The energy difference between the states (see Eqn 1-15) is $\Delta U = \gamma\hbar B_0$, and at normal temperatures this is much less than the thermal energy kT. Therefore it can be shown (Problem 1-15) that the population difference between the two states is

$$n_\alpha - n_\beta = \Delta n_0 = N\,\Delta U/2kT \qquad (1\text{-}20)$$

where N is the total number of nuclei in the relevant sample, and the

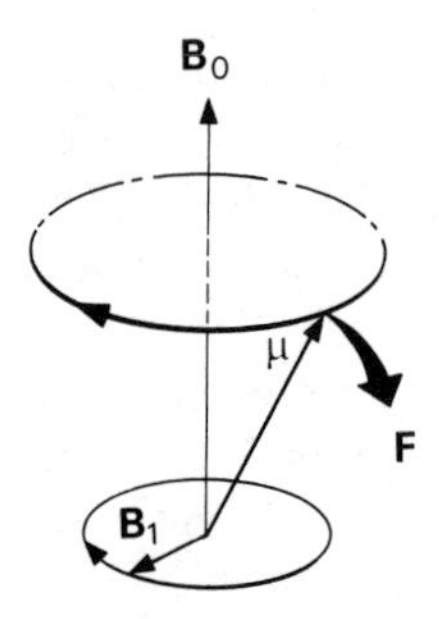

Fig. 1-6 The effect of a rotating magnetic field, $\mathbf{B}_1$, on a precessing magnetic moment, $\boldsymbol{\mu}$. When the $\mathbf{B}_0-\mathbf{B}_1$ plane is perpendicular to the $\mathbf{B}_0-\boldsymbol{\mu}$ plane as in the figure, there is a force $\mathbf{F}$ acting to increase the angle between $\mathbf{B}_0$ and $\boldsymbol{\mu}$. If $\boldsymbol{\mu}$ and $\mathbf{B}_1$ are rotating at the same rate, this force acts always away from $\mathbf{B}_0$ and therefore has a consistent effect.

subscript zero to Δn refers to the equilibrium situation. For all achievable values of the applied field Δn_0 is extremely small (e.g., for ^{1}H in a field of 2·35 T at 300 K, $\Delta n_0/N = 1{\cdot}6 \times 10^{-5}$).

The total magnetic moment or magnetization, **M**, of the sample[4] is the resultant of the individual magnetic moments $\boldsymbol{\mu}$. At equilibrium **M** is along the $+z$ direction and has a magnitude:

$$M_0 = n_\alpha \mu_{z\alpha} + n_\beta \mu_{z\beta} = \Delta n_0 \mu_{z\alpha} = \tfrac{1}{2}\gamma\hbar\,\Delta n_0 \tag{1-21}$$

since $\mu_{z\beta} = -\mu_{z\alpha}$. Substitution for Δn_0 gives expression (1-22):

$$M_0 = \tfrac{1}{4}N\gamma\hbar\,\Delta U/kT = \tfrac{1}{4}N(\gamma\hbar)^2 B_0/kT \tag{1-22}$$

which is analogous to the Curie Equation. As will be discussed in Chapter 3 the magnetization M will be affected by magnetic fields in the same way as individual magnetic moments, i.e. it will precess.

In a steady-state NMR experiment the spin system can absorb energy from the radiofrequency radiation at a rate, R, which depends on three factors, viz.:

(i) the probability, P, per unit time per spin of a transition being induced;
(ii) the population difference between the spin states (since P for transitions $\beta \rightarrow \alpha$ equals that for $\alpha \rightarrow \beta$);
(iii) the energy change appropriate to the transition.

Now the transitions are caused by a magnetic field B_1, conventionally in the x direction, which acts on the component of the spin magnetic moment in that direction. Thus P depends on $\mu_x B_1 \simeq I_x \hbar B_1$. A detailed quantum-mechanical treatment (see Section 2-6) shows that $P \propto \gamma^2 B_1^2 g(\nu)$, where $g(\nu)$ is a signal shape factor (see Section 3-4) to account for the fact that spectral 'lines' are not infinitely sharp, i.e. that there is some transition probability at frequencies differing slightly from exact resonance. Thus

$$R = P\,\Delta U\,\Delta n_0 \propto \gamma^4 B_0^2 N B_1^2 g(\nu)/T \tag{1-23}$$

However, NMR spectrometers do not detect R directly, but rather the rate of induced magnetization change in the direction of the receiver coil, dM_y/dt. This turns out to be R/B_1, so the observed signal height, S, is given by:

$$S \propto \gamma^4 B_0^2 N B_1 g(\nu)/T \tag{1-24}$$

It should be noted that the derivation of Eqn 1-24 has assumed that the experiment has not appreciably changed Δn, which will only be true for low values of B_1 (see Section 3-5). Of course, in practice the noise level is also important, and this depends on the frequency of the experiment. Moreover, optimization of B_1 plus consideration of the form of $g(\nu)$ add further complications which will be discussed in Section 3-6, but at this point it is worth noticing the following:

(i) The strict proportionality of S to N makes the use of NMR for quantitative studies extremely simple. It also means that for isotopes of low natural abundance (such as ^{13}C) improvements in NMR signal intensities can be made by using isotopically enriched material.

(ii) The heavy dependence on γ implies that those nuclei resonating at high frequencies tend to have relatively intense signals. In particular, the popularity of the proton, which has the highest value of γ except for the radioactive tritium nucleus, is readily explained.

(iii) The dependence on B_0 indicates that high-field work is desirable. NMR has, in fact, provided a strong stimulus to the development of stable superconducting magnet systems.

(iv) The inverse proportionality to T (arising from the Boltzmann factor) indicates that low-temperature operation is desirable, but in practice questions of solubility or freezing put a strict limit to the utility of this statement for solution-state studies.

1-8 Electronic shielding

The discussion of Sections 1-5 and 1-6 assumed that the presence of the sample does not perturb the magnetic field. However, this is not strictly true. Indeed, the nuclear Larmor precession itself produces a secondary magnetic field, but this effect is negligible compared to that produced by electrons. In fact, when a substance is placed in a magnetic field it becomes magnetized and modifies the field. The magnetization may be considered in terms of two contributions—a bulk (macroscopic) effect and a local (microscopic) effect. The bulk intensity of magnetization is proportional to the magnetic field, and the constant of proportionality is known as the bulk susceptibility, χ, of the material.[5] For substances with no unpaired electrons, χ is normally negative, and such substances are said to be diamagnetic:[6] the magnetization reduces the magnetic field within the sample. The bulk effect is, however, of little importance for chemical applications of NMR, so in later sections we shall be concerned only with the local effect.

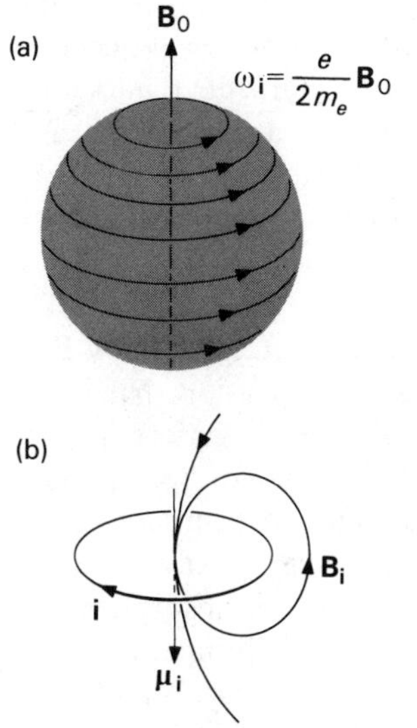

The molecular interpretation of diamagnetism (both bulk and local effects) is straightforward: the behaviour may be traced to motion of electrons induced by the magnetic field. Figure 1-7(a) shows diagrammatically what happens to the electrons in an atom—they are induced to circulate around the nucleus about the direction of the applied field $\mathbf{B}_0$. it can be shown that the angular velocity is given by

$$\boldsymbol{\omega}_i = (e/2m_e)\mathbf{B}_0 \tag{1-25}$$

(a) $\omega_i = \frac{e}{2m_e}\mathbf{B}_0$, $\mathbf{B}_0$

(b) $\mathbf{B}_i$, i, μ_i

Fig. 1-7 Electronic shielding: (a) circulation of the electronic charge cloud under the influence of a magnetic field, (b) the secondary magnetic field produced by the precession. Note that the circulation is equivalent to electric currents in the reverse direction.

This is the electronic orbital counterpart of Eqn 1-19. Note that the sign implies that the circulation is right-handed about $\mathbf{B}_0$ [Fig. 1-7(a)], i.e. in the opposite sense to proton spin precession (Fig. 1-4). Since the circulation involves motion of charge, there will be an associated induced magnetic moment, $\boldsymbol{\mu}_i$, which for a single electron at distance $\mathbf{r}$ from the nucleus is given by Eqn 1-26 (see Problem 1-4):

$$\boldsymbol{\mu}_i = -(e^2/4m_e)r^2\mathbf{B}_0 \sin^2\theta \tag{1-26}$$

where θ is the angle between $\mathbf{r}$ and $\mathbf{B}_0$. In order to find the total magnetic moment, it is of course necessary to sum over the whole electronic distribution. The induced magnetic field produced by $\boldsymbol{\mu}_i$ opposes the primary applied field $\mathbf{B}_0$ [Fig. 1-7(b)]. This is in fact essential if the law of conservation of energy is to hold (the situation is analogous to that in which Lenz's law is applicable). A negative magnetic susceptibility (i.e. diamagnetism) is implied. In principle, if enough information about electron densities and electronic wave functions were available, values of the bulk susceptibility could be calculated, since the intensity of magnetization can be equated to the total magnetic moment induced per unit volume.

It is clear that the electrons will shield (or screen) a nucleus from the influence of the field B_0. This shielding can be taken into account by using an effective field B at the nucleus, given by

$$B = B_0(1-\sigma) \tag{1-27}$$

The dimensionless number σ is a small fraction, usually listed in parts per million, and is known as the shielding constant. In principle it may be calculated for atoms using equations similar to Eqn 1-26. Therefore χ and σ are related quantities, but the relationship is not a simple one. A more detailed treatment of atomic screening will be given in Section 8-2. Clearly, the situation is more complex for molecules, because there is then more than one positively charged centre to influence the circulation of electrons. However, the same type of effect will occur. It is, moreover, possible arbitrarily to divide the total effect into contributions from the electrons associated with the various atoms and bonds in the molecule under consideration. Thus, we may speak of bond induced magnetic moments and bond susceptibilities (see Section 8-8).

1-9 Chemical shifts

If the frequency of NMR transitions were entirely determined by Eqn 1-18, the technique would have little chemical application. Nuclei of different elements or isotopes would give resonances in different parts of the spectrum due to variations in γ, but this fact cannot be exploited easily for elemental analysis, because it is difficult to compare intensities of absorption at the very different frequencies involved. However, in 1949 it was found that the frequency of ^{14}N absorption for solutions or liquids depended on the chemical environment of the nucleus. This feature has since been shown to be general, and such differences in resonance conditions are referred to as chemical shifts. The existence of chemical shifts enables the NMR spectroscopist to distinguish between different chemical environments, and makes NMR spectroscopy a powerful tool for the determination of the structure of molecules.

The phenomenon of the chemical shift arises because of shielding (screening) of the nuclei from the external magnetic field by the electrons, discussed in the previous section. Since the shielding effect is caused by electronic environment, values of σ will vary with the position of the nucleus in the molecule. Factors that affect electron density will also affect σ. Thus the shielding constant for the proton of an aldehyde group is lower than that for the protons of a methyl group. The resonance condition (Eqn 1-18) becomes

$$\nu_j = |\gamma/2\pi|\, B_0(1-\sigma_j) \tag{1-28}$$

The factors ν_j and σ_j refer to the resonance frequency (Larmor frequency) and shielding constant respectively, for nucleus j. Variations in σ thus cause variations in resonance frequencies and are the reason for the occurrence of chemical shifts.

Strictly speaking, the appropriate field B_0 to use in Eqn 1-25 and thereafter is not the applied field, but the applied field *corrected for the effect of bulk magnetization.* The correction depends on the sample shape and bulk susceptibility. Care must be taken in comparing shielding constants for nuclei in different compounds either to eliminate these bulk effects or make allowance for them. This is particularly important because, as discussed in the next section, absolute values of shielding constants are not normally obtained, but chemical shifts are reported relative to that for a standard substance.

For a given compound the appearance of the spectrum is governed by intramolecular chemical shift differences, i.e. differences in resonance frequencies for different nuclei of the same molecule. As an example, Fig. 1-8 shows the 60 MHz proton resonance spectrum of 2,2-dimethoxypropane: the methyl and methoxy protons give absorption at different positions in the spectrum. Figure 1-8 illustrates another of the important features of NMR spectra, namely that the intensity of absorption is strictly proportional to the concentration of the nuclei (in the present case protons) giving rise to the absorption, as shown by Eqn 1-24. This is of great importance for structure determination. For instance, if the compound used in Fig. 1-8 was of unknown structure, the NMR spectrum would immediately show that it contained two types of proton and that there were equal numbers of each type. This use of NMR could be described as providing a relative proton count. Intensities in infrared and ultraviolet spectroscopy cannot be used in this way. Moreover, the proportionality of intensity to concentration for NMR is valid for comparisons between different molecules, thus providing a rapid non-destructive method of obtaining concentrations of impurities, isomer ratios etc. It is also a straightforward matter to obtain absolute concentrations for a given chemical in solution if standard concentrations can be prepared; this technique is

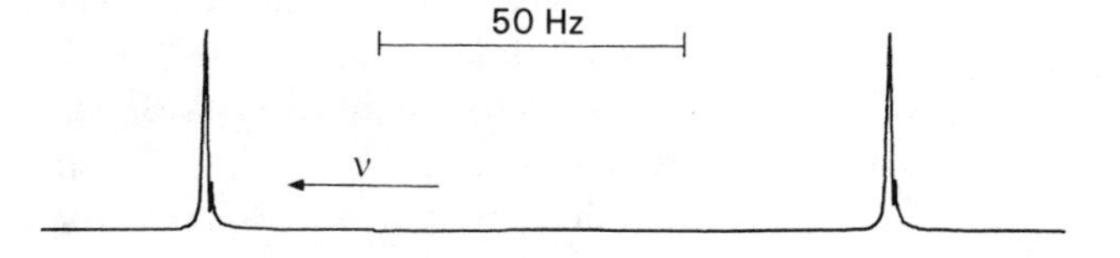

Fig. 1-8 Proton NMR spectrum of 2,2-dimethoxypropane, $(CH_3)_2C(OCH_3)_2$, in solution in CCl_4, at 60 MHz ($\equiv\gamma B_0/2\pi$).

of considerable application in quantitative analytical chemistry, for example for process control in chemical industry.

It may be seen from Eqn 1-28 that resonance frequencies at constant field B_0 are directly proportional to the magnitude of B_0. Thus, all chemical shift differences (if measured in Hz) increase in proportion if B_0 is increased. It is therefore more usual to report chemical shift differences in parts per million (ppm). These are obtained from measurements in Hz by dividing by ν_0, the operating frequency of the spectrometer ($\sim|\gamma B_0/2\pi|$). In such dimensionless units, chemical shift differences are invariant to the spectrometer conditions, and may therefore be used as molecular characteristics. Equation 1-28 may be rewritten as $\nu_j = \nu_0(1-\sigma_j)$, and it becomes clear that a chemical shift difference expressed in ppm is the negative of the corresponding shielding difference, i.e. $\Delta\nu/\nu_0 = -\Delta\sigma$. Thus, in NMR spectra the direction of increasing resonance frequency (customarily to the left in spectral plots) is opposite to that of increased shielding, i.e. a high frequency shift indicates deshielding.[7]

1-10 Reference standards and solvents

Ideally, the chemist would like to know the absolute magnitudes of shielding constants. This would require an accurate knowledge of both B_0 and the resonance frequency. However, it is not normally feasible to measure magnetic fields to a high accuracy; on the other hand, it is usually convenient and accurate to measure small differences in resonance frequency under constant field conditions. Consequently, in solution-state NMR it has become normal practice to measure resonance frequencies relative to that for certain nuclei in a standard chemical. Chemical shift values are thus quoted relative to such a standard. The standard chemical may be either external or internal to the sample being studied. With external referencing the sample is placed in a capillary tube which is surrounded by the reference, contained in a larger-size tube. However, in such cases the magnetic field B_0 for the sample is not strictly comparable with that for the standard, because the liquids will have different magnetic susceptibilities. Consequently bulk-susceptibility corrections need to be made before chemical shift differences can be quoted accurately. This procedure is to be avoided where possible, and therefore we shall discuss internal references only. However, measurements made with an external standard in cases of known chemical shifts may be used in an NMR method of obtaining magnetic susceptibility.

Measurements with internal standards are made by adding a small amount (ca. 5%) of the standard directly to the liquid or solution under study. Susceptibility effects are then the same at both standard and sample molecules (i.e. the value of B_0 is the same), and observed chemical shift differences can be attributed to local influences only. Clearly, a suitable standard should possess as many of the following characteristics as possible:

(a) It should contain only a single type of magnetic nucleus with a single type of environment, so that the NMR spectrum consists of one sharp line.

(b) It should contain many nuclei of that type and environment, so that an intense signal is given, even when the substance is present only in small amount.

(c) The resonance should occur in a region of the spectrum away from the signals given by most common chemicals.

(d) The standard should be chemically inert, so that its resonance position is not affected by the sample or vice versa. For this criterion to apply, there must be no transient complex formation with solvent or solute molecules, and no irreversible chemical reaction. Since most solvent effects are due to electrostatic or steric interactions, the standard chemical should therefore be non-polar and highly symmetrical.

The substance which has become the accepted standard for proton resonance is tetramethylsilane, $Si(CH_3)_4$, which will be referred to as TMS throughout the rest of this book. The most common isotopes of silicon and carbon have zero spin, and the proton resonance of TMS occurs to high frequency of most other proton resonances. Due to rapid internal rotation about the Si–C bonds, all the protons are equivalent. Thus all four criteria are satisfied; TMS has the added advantage that it is a volatile liquid that can be readily handled (though its volatility is for some purposes a little higher than is desirable). Other reference standards are normally necessary for resonances of nuclei other than protons, though TMS is also used for both ^{13}C and ^{29}Si NMR. As with intramolecular chemical shifts, those relative to a standard are proportional to the spectrometer operating frequency if measured in Hz; consequently values are reported in the dimensionless ppm units. Proton resonance chemical shifts in ppm relative to TMS are said to be based on the δ scale. The δ values are positive if the sample absorbs to high frequency of the reference absorption at constant field, as expressed by Eqn 1-29:[8]

$$\delta = 10^6(\nu_{\text{sample}} - \nu_{\text{TMS}})/\nu_{\text{TMS}} \text{ at constant } B_0 \qquad (1\text{-}29)$$

Since shielding constants are small, the value of ν_{TMS} in the denominator may be replaced by the spectrometer operating frequency ν_0 or by $|\gamma B_0/2\pi|$ with little error. Thus δ is related to shielding constants by Eqn 1-30:

$$\delta = 10^6(\sigma_{\text{TMS}} - \sigma_{\text{sample}}) \qquad (1\text{-}30)$$

Figure 1-9 shows the spectrum of a mixture of benzene and acetone, including internal TMS. The δ scale is shown, together with the calibration in Hz at 60 MHz.

Occasionally it becomes necessary to employ secondary standards and to obtain the chemical shifts of the sample indirectly. This is required for work at high temperatures (when TMS vaporizes) or when TMS is insufficiently soluble, as occurs for certain highly polar solvents.

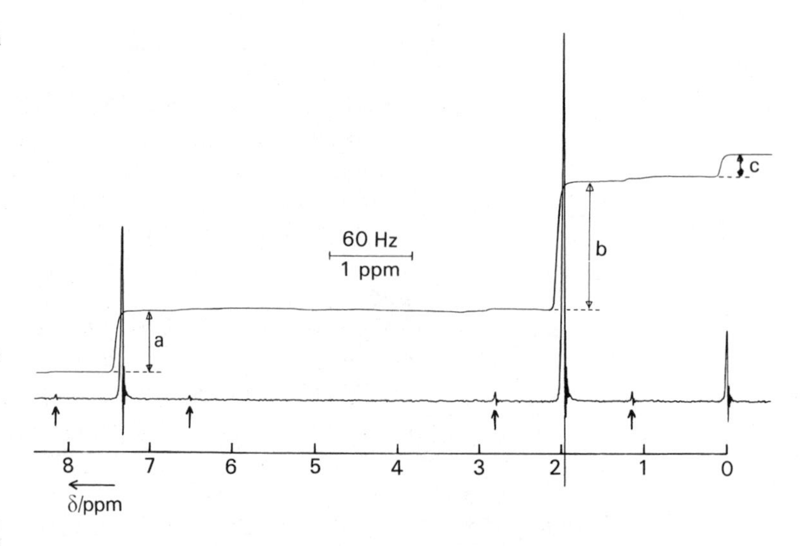

Fig. 1-9 Proton NMR spectrum (at 60 MHz) of a mixture of benzene and acetone with TMS (the three resonances are in that order from left to right). The resonances of benzene and acetone occur at $\delta 7{\cdot}33$ and $\delta 1{\cdot}97$ in this solution. The components of the solution are in the molar ratio 10:22:1, as can be verified from the integral trace superimposed on the spectrum (see Section 1-21)—the steps a, b and c are proportional to concentration in terms of numbers of protons per unit volume. The 'ringing' at the right hand side of each peak is an indication that the scan speed is too fast. The small peaks marked with arrows are spinning sidebands (see Section 1-21); the spinning rate is ca. 50 Hz in this case.

The case most commonly met is aqueous solutions. The salt sodium 2,2-dimethyl-2-silapentane-5-sulphonate, $Me_3SiCH_2CH_2CH_2SO_3Na$, has been suggested for such work (although not all the criteria given above are satisfied) because the resonance of the methyl protons occurs so close to $\delta = 0{\cdot}00$ that the difference can usually be neglected.

Similar scales, also given the symbol δ and using the high-frequency-positive convention, have been established for other nuclei. Each requires the acceptance of a reference standard chemical. However, modern spectrometer systems allow any resonance frequency to be converted to its value in a standard field (in which the protons in TMS resonate at exactly 100 MHz). Such resonance frequencies are given[9] the symbol Ξ (Greek capital xi), and such a procedure in effect makes TMS the primary standard for the NMR of any nucleus (see Table 1-1).

As has been stated earlier, most high-resolution NMR work has been done with samples in the liquid phase. It is therefore desirable to consider suitable solvents, since these will be frequently required for solid samples. An ideal solvent does not contain the isotope whose resonance is being studied. For proton resonance CCl_4 is widely used. However, it is not useful for polar solutes and it is often convenient to use deuterated solvents; the most common ones (to a considerable extent because they are relatively cheap) are D_2O and $CDCl_3$. It is sometimes necessary to use fully deuterated benzene, acetone, or dimethylsulphoxide. Generally, small peaks will be seen in the spectra from residual amounts of undeuterated solvent. The δ values for ^1H and ^{13}C resonance of several suitable solvents are given in Table 1-2. The dielectric constants listed are helpful in deciding the likelihood of a given solute being soluble. A further feature that governs the choice of solvent is the temperature at which the spectrum is to be taken. Thus the melting and boiling points of the solvent must be taken into

Table 1-2 Chemical shifts, melting and boiling points, and dielectric constants of some useful NMR solvents

	δ_H [a]	δ_C	*Melting point* /°C	*Boiling point* /°C	κ [b]
Benzene	7·37	128·5	6	80	2·3
Trichloromethane	7·27	77·2	−64	62	4·8
Trichloroethene	6·45	117·6 & 125·1	−73	87	3·4 (ca. 16 °C)
1,1,2,2-Tetrachloroethane	5·95	75·5	−36	146	8·2
Dichloromethane	5·30	54·0	−95	40	9·1
Water	4·76	—	0	100	80·4
p-Dioxane	3·70	67·4	12	101	2·2 (25 °C)
Dimethylsulphoxide	2·62	40·5	18	189 [c]	46·0 (ca. 35 °C)
Acetone	2·17	30·4 & 204·1	−95	56	20·7 (25 °C)
Acetonitrile	2·00	1·7 & 117·7	−46	82	37·5
Cyclohexane	1·43	27·5	7	81	2·0
Carbon disulphide	—	192·8	−112	46	2·6
Tetrachloromethane	—	96·0	−23	77	2·2
Tetrachloroethene	—	121·3	−19	121	2·3 (25 °C)
Trichlorofluoromethane	—	117·6	−111	24	2·3 (−30 °C)

[a] All δ_H values except for water and trifluoroacetic acid are for ca. 7% solutions in $CDCl_3$ (plus TMS).

[b] Dielectric constants at ca. 20 °C except where stated.

[c] Unstable at high temperatures.

consideration (see Table 1-2). For low-temperature work, CS_2 can take the place of CCl_4; acetone and toluene (preferably perdeuterated) are also useful low-temperature solvents. Solvent mixtures, which have lower melting points than the pure components, can be used. For example, mixtures of $CHCl_2F$ and $CHClF_2$ can be obtained which are still liquid at the boiling point of nitrogen (−196 °C). Highly chlorinated compounds such as tetrachloroethene or *o*-dichlorobenzene (though this is not suitable for ^1H NMR studies of aromatic systems since it is difficult to obtain in perdeuterated form) are frequently employed for high-temperature work.

1-11 Spin–spin coupling

In most spectra additional features are present. In the simplest cases (referred to as first-order spectra) these occur as a splitting of the bands due to each nucleus. If there are only two nuclei with non-zero spin in the molecule under consideration, having, say, spins I_1 and I_2, then it is found that the resonance of spin I_1 is split into $2I_2+1$ lines of equal intensity and that of spin I_2 is similarly split into $2I_1+1$ lines. The line separations are all equal. The nature of the splitting suggests, correctly, that the different lines of I_1 arise from molecules with differing values of m_I for the second spin. In fact the energies of I_1 are modified by a small interaction between the spins. This interaction is known as spin–spin coupling, sometimes further categorized as indirect or scalar, to distinguish it from the direct dipolar interaction discussed

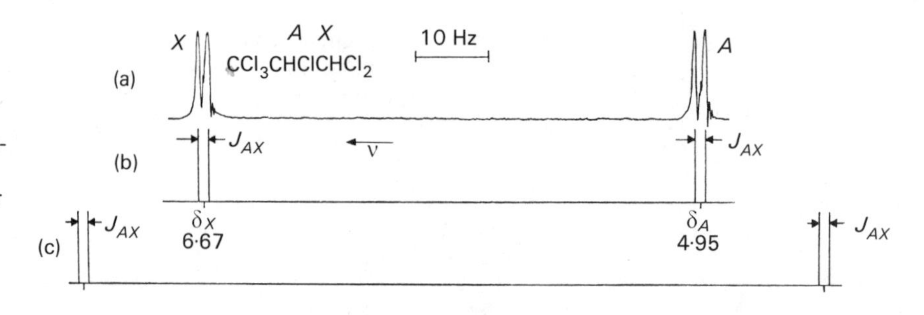

Fig. 1-10 A first-order spectrum of a two-spin system: the proton resonances of 1,1,1,2,3,3-hexachloropropane (a) observed at 40 MHz, (b) first-order lines corresponding to (a), (c) first-order lines calculated for 60 MHz (with the average chemical shift held at the same position in the figure). The coupling constant is 1·4 Hz and the chemical shift difference is 1·72 ppm. [A. B. Dempster and N. Sheppard are thanked for spectrum (a)].

is Section 4-1. Its magnitude is determined by a spin–spin coupling constant (usually simply referred to as a coupling constant, and written J_{jk} for interaction between spins j and k). The transition energies are affected by coupling, and indeed the splittings in the spectra are equal in magnitude to J. Coupling involving nuclei of spin$>\frac{1}{2}$ is relatively rarely seen, so the following discussion relates only to spin-$\frac{1}{2}$ nuclei. For a pair of nuclei with $I=\frac{1}{2}$, the spectrum consists of two doublets, i.e. one doublet assigned to transitions of each spin. The doublet splittings are equal to J. The example in Fig. 1-10(a) is the proton spectrum at 40 MHz of $CCl_3CHClCHCl_2$.

Coupling constants, like chemical shifts, depend on chemical environment and are therefore of great use in structure determination. However, variation of the spectrometer operating frequency does *not* affect the magnitude of the spectral splittings due to coupling, expressed in Hz (contrast the case of chemical shift differences). Therefore coupling constants are always expressed in Hz and never in ppm. Because chemical shifts and coupling constants are affected differently by changes in spectrometer conditions, the spectra themselves change in appearance. Figures 1-10(b) and (c) show theoretical proton spectra at 40 MHz and at 60 MHz to illustrate this point. Individual absorption lines should never be assigned δ values, since their positions are unlikely to be independent of spectrometer operating frequency; the δ scale should be reserved for true chemical shift positions, or simply for a general indication of a region of the spectrum.

Note that coupling is only effective between nuclei in the same molecule. When there are more than two magnetic nuclei in the molecule, coupling may occur between each pair of nuclei, but the coupling constants will normally all differ in magnitude unless some nuclei are related by symmetry. Each coupling constant J_{jk} causes splittings in the resonance of nuclei j and k (but not that of other nuclei). The pattern of lines for a given nucleus may be explained by the *method of successive splittings.* Consider a system consisting of three spin-$\frac{1}{2}$ nuclei, which will be labelled A, M, and X. The A resonance is split by coupling to both M and X, but to different extents. Assume $J_{AM}>J_{AX}$. The pattern of A lines may be predicted by considering first of all a splitting by M, and then a further splitting of each of the resulting pair of lines by X. The final pattern is

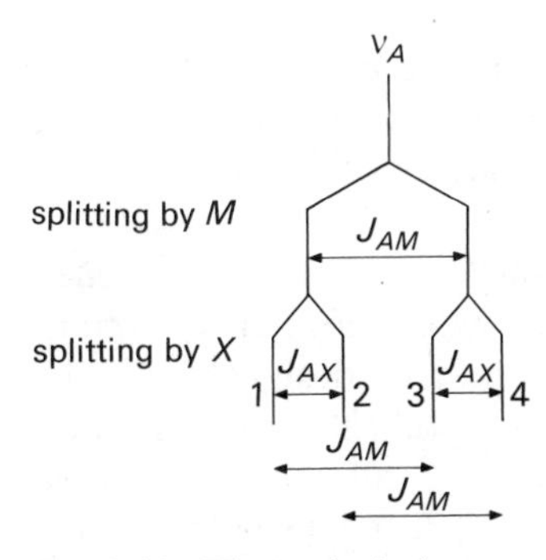

Fig. 1-11 The method of successive splittings for nucleus A of a three-spin system, showing the effect in first order of coupling to nuclei M and X.

of four lines. This procedure is illustrated in Fig. 1-11. Note that J_{AM} is given by the separation of lines 1 and 3 or 2 and 4 while J_{AX} may be measured from the splitting between lines 1 and 2 or 3 and 4. The M resonance will be split similarly but not identically (since this time J_{AM} and J_{MX} are involved). Altogether the spectrum consists of 12 lines (4A, 4M, and 4X). It is important to realize that each part of the spectrum is symmetrical about its mid-point, and that the mid-point corresponds to the Larmor frequency (or unperturbed resonance frequency) of the nucleus concerned. The full spectrum of a three-spin proton system is shown in Fig. 1-12.

The above discussion may be extended to show that for a system consisting of n spin-$\frac{1}{2}$ nuclei (none of which are equivalent), each nucleus gives rise in principle to 2^{n-1} lines. The total number of lines is therefore $n2^{n-1}$. However, the possibilities that some lines may overlap, or that some coupling constants may be negligibly small, should be borne in mind. In fact, as a general rule the magnitude of coupling constants tends to decrease as the number of chemical bonds separating the coupled nuclei increases. This is because the most common mechanism for coupling involves the transmission of the effect through the bonding electrons; the magnitude of the affect is attenuated (when the electrons are localized into two-centre bonds) by the requirement that the spin information must be passed through successive bonds. For saturated hydrocarbons the coupling constant between protons separated by more than three bonds is often negligibly small. Thus, spectral splittings may often be attributed to nearest neighbours only. However, such a generalization must be treated with great reserve, and frequently does not hold when there is unsaturation in the molecule.

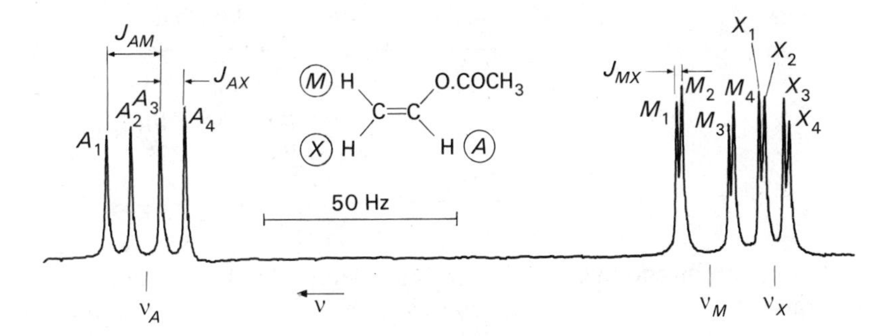

Fig. 1-12 60 MHz proton resonance spectrum for the vinyl protons in vinyl acetate (solution in CCl_4). The methyl protons are sufficiently distant in the molecule not to couple appreciably to the vinyl protons, which therefore form an isolated three-spin system. The spectrum is nearly first-order; the slight variations in intensity are due to second-order effects. The lines are labelled as in the energy level diagram of Fig. 1-14 under the assumption that all the coupling constants are positive, though for this compound the *geminal* coupling constant is known to be negative [see Problem 1-8(b)].

1-12 Basic product functions

As mentioned earlier, the wave functions for spin-$\frac{1}{2}$ nuclei are frequently labelled α for $m_I = +\frac{1}{2}$ and β for $m_I = -\frac{1}{2}$. This notation is very convenient in use and has the merit that it emphasizes our inability to specify the wave function more mathematically (the values of m_I are simply the eigenvalues of the particular operator $\hat{I}_z$—see Chapter 2).

In spectra such as those discussed above, where all the theoretical splittings due to coupling are resolved, each transition may be labelled with the spin states (α or β values) of the nuclei other than the one

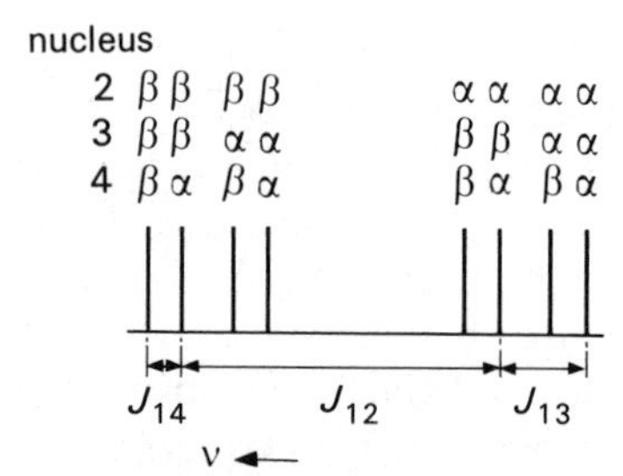

Fig. 1-13 The labelling of the transitions of nucleus 1 of a four-spin first-order system, according to the spin wave functions of the other three nuclei.

undergoing the transition. Thus the eight lines due to one of the nuclei of a four-spin system may be labelled as in Fig. 1-13 (the assignment actually assumes all the coupling constants are positive, which is not necessarily the case).

The spin states for a system of spin-$\frac{1}{2}$ nuclei may also be labelled using the α, β notation. In accordance with normal quantum-mechanical considerations the overall state may be designated by the product of the functions for the individual nuclei. Thus one state of a three-spin system (nuclei A, M, and X, say) may be designated $\alpha_A\beta_M\beta_X$. If the product functions are always listed using the same order for the individual nuclei, the subscripts can be omitted, and the above state referred to simply as $\alpha\beta\beta$. Such functions are termed 'basic product functions'. Since there are two possibilities for the wave function of each nucleus, there are 2^n basic product functions for a spin system of n spin-$\frac{1}{2}$ nuclei. The word 'basic' refers to the fact that the more exact calculations discussed in Chapter 2 frequently use the product functions as a *basis set.*

It is useful to classify basic product functions according to the value of the total component of spin angular momentum, i.e. according to $m_T = \sum_j m_j$, where the summation is over all the nuclei of the system. Table 1-3 lists the basic product functions with their m_T values for a three-spin system.

Table 1-3 The basic product functions of a three-spin system, with classification according to the total component of spin

m_T	*Basic product functions*
$-\frac{3}{2}$	$\beta\beta\beta$
$-\frac{1}{2}$	$\beta\beta\alpha$, $\beta\alpha\beta$, $\alpha\beta\beta$
$+\frac{1}{2}$	$\beta\alpha\alpha$, $\alpha\beta\alpha$, $\alpha\alpha\beta$
$+\frac{3}{2}$	$\alpha\alpha\alpha$

1-13 First-order energies and spectra

The simple spectra described in the preceding sections only arise when all chemical shift differences $|\nu_j - \nu_k|$ are much greater than the corresponding coupling constants $|J_{jk}|$. When nuclei are of different isotopic species this is automatically the case, but if two or more are of the same species, it is not necessarily so. However, when the condition is obeyed the basic production functions are eigenfunctions of the spin Hamiltonian, and the spectra are said to be first order. Each spin-$\frac{1}{2}$ nucleus then gives rise to a set of resonances well-separated from those of the other nuclei. The group of lines due to a given nucleus is symmetrical about its chemical shift position and the splittings within the group are equal to the appropriate coupling constants. In this section some of the basic features of first-order systems will be discussed. Situations

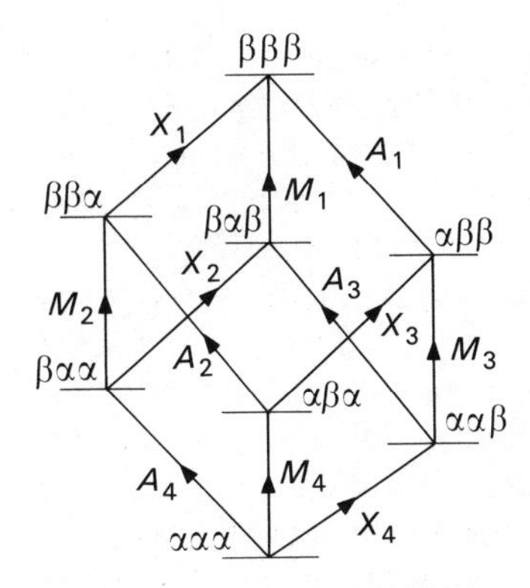

Fig. 1-14 Energy level diagram for a system of three spin-$\frac{1}{2}$ nuclei, with the assignment of the allowed NMR transitions. The relative positions of the levels of the same value of m_T are only diagrammatic. The transition labels are the same as designated in Fig. 1-12.

where second-order effects are important will be treated in Chapter 2.

In Section 1-5 it was stated that the selection rule for NMR transitions of a single nucleus of spin I is $\Delta m_I = \pm 1$. For a spin-$\frac{1}{2}$ nucleus of positive γ this means that the transition $\alpha \rightarrow \beta$ is allowed in absorption. For a spin system consisting of several spin-$\frac{1}{2}$ nuclei the selection rule $\Delta m_T = \pm 1$ is normally rigidly obeyed. Moreover, an allowed transition involves a change in the spin component of only one nucleus. Thus the selection rule for a transition assigned to nucleus j is that of Eqn 1-31:

$$\Delta m_j = \pm 1; \qquad \Delta m_k (k \neq j) = 0 \tag{1-31}$$

The simultaneous change of three spins (e.g. $\alpha\alpha\beta \rightarrow \beta\beta\alpha$ for a three-spin system) is therefore forbidden. With these considerations in mind the energy level diagram (showing allowed transitions) for a system of three spin-$\frac{1}{2}$ nuclei is as shown in Fig. 1-14. This energy level diagram satisfactorily indicates the number and type of transitions discussed in Section 1-10.

For the detailed evaluation of energy levels, however, it is necessary to consider the energy equation. The expression appropriate to the first-order case is given as

$$h^{-1}U = -\sum_j \nu_j m_j + \sum_{j<k} J_{jk} m_j m_k \tag{1-32}$$

where $\nu_j = |\gamma_j/2\pi|\, B_0(1-\sigma_j)$, as in Eqn 1-28, and is the Larmor frequency of nucleus j. The first term is the chemical shift term (often referred to as the nuclear Zeeman energy) and takes account of the interactions of the nuclear magnetic moments with the magnetic field. The second term is the coupling contribution and arises from electron-coupled interactions between the nuclear magnetic moments. The condition $j<k$ is written into the summation to ensure that a given pair of nuclei is included only once.

The way in which the energy level diagram is based on Eqn 1-32 may be best discussed using Fig. 1-15, which shows the one- and two-spin cases. For the two-spin case there are four basic product functions, $\alpha\alpha$, $\alpha\beta$, $\beta\alpha$ and $\beta\beta$; on labelling the nuclei A and X Eqn 1-32 becomes

$$h^{-1}U = -\nu_A m_A - \nu_X m_X + J_{AX} m_A m_X \tag{1-33}$$

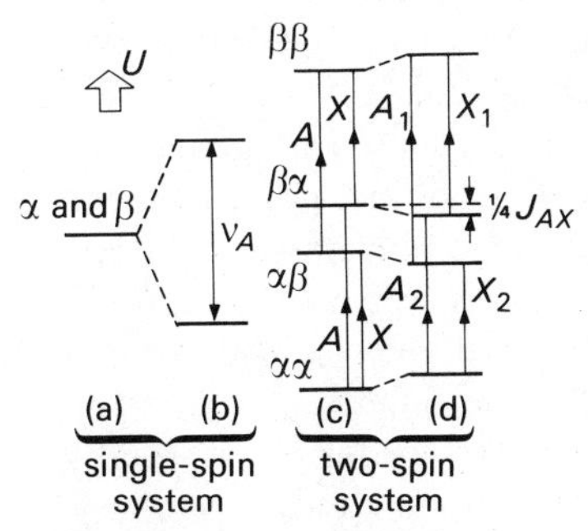

Fig. 1-15 Construction of the energy level diagram for one and two-spin systems on the basis of Eqn. 1-32. For further explanation see the text.

Figure 1-15 shows the energy (a) for nucleus A alone, in the absence of the field **B**, (b) for nucleus A alone, in the presence of **B**, (c) for both nuclei A and X in the field, but with $J_{AX} = 0$ (i.e. using the chemical shift term only), and (d) for both A and X in the field with $J_{AX} \neq 0$. With selection rule 1-31 the allowed transitions are:

A transitions	$\Delta m_A = \pm 1,$	$\Delta m_X = 0$
X transitions	$\Delta m_A = 0,$	$\Delta m_X = \pm 1$

These are indicated on Fig. 1-15, which shows how coupling splits the

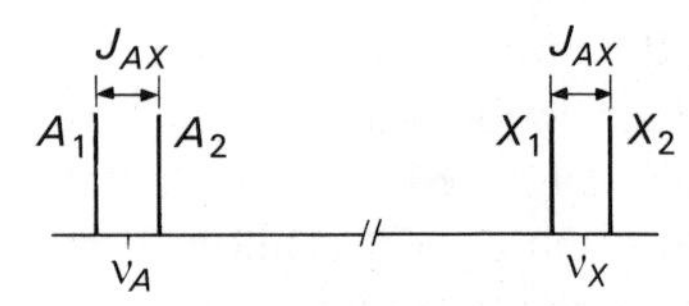

Fig. 1-16 The first-order NMR spectrum of a two-spin system (diagrammatic), derived from the energy level diagram. For the transition numbering see Fig. 1-15. For comparison with an observed spectrum see Fig. 1-10. Note agreement with the method of successive splittings.

two A transitions, which were degenerate for $J_{AX}=0$, such that the splitting is $|J_{AX}|$. The spectrum is indicated in Fig. 1-16 (see also Fig. 1-10). It can be seen, both from Fig. 1-15 and from Eqns 1-32 and 1-33 that the effect of the coupling term is to lower the energy by $\frac{1}{4}J$ if the coupled spins are 'anti-parallel' (i.e. $\alpha\beta$ or $\beta\alpha$) but to increase the energy by $\frac{1}{4}J$ if they are 'parallel' ($\alpha\alpha$ or $\beta\beta$). Actually, this is so only if J is positive (see Chapter 8). In fact Eqn 1-32 defines the sign of a coupling constant as positive if coupling stabilizes anti-parallel spins, and as negative if coupling destabilizes anti-parallel spins.

The above discussion is readily expanded to larger spin systems. The spectra predicted are those described in preceding sections. Actually, Eqn 1-32 is applicable to nuclei of any spin, although it is only applied here to the spin-$\frac{1}{2}$ case. One feature of Eqn 1-32 and the resulting spectra that has not yet been noted is that although the energy levels depend on the signs of the coupling constants, the transition energies do not. Thus first-order spectra are invariant to a change of sign of any or all of the coupling constants. Neither relative nor absolute signs of coupling constants can be obtained from such spectra unless additional techniques, such as double resonance (see Chapter 4), are employed. Thus, strictly speaking, first-order splittings are equal to the *moduli* of the appropriate coupling constants.

1-14 Equivalence

So far only nuclei which differ in their electronic environment have been discussed. Frequently, however, due to molecular symmetry, two or more nuclei may have identical environments. In such cases the nuclei (or atoms) are said to be equivalent. All physical and chemical properties will then be the same for the equivalent nuclei; for example, the rates of any chemical reaction will be the same at equivalent sites. Consequently it can be seen that symmetrically equivalent nuclei will have the same chemical shift (since this depends solely on environment). Nuclei with the same chemical shift are said to be isochronous. However, there are several ways in which nuclei can be equivalent, and it is important for NMR to distinguish two such cases. Consider the way in which two isochronous nuclei are coupled to a third magnetic nucleus which is not equivalent to them. The strength of the coupling may be equal for the two isochronous nuclei or it may be unequal. In the former case (if this holds for coupling to every individual nucleus that is non-equivalent to the two considered) the ischronous nuclei are *magnetically* equivalent, in the latter case they are *chemically* equivalent. Some examples will make this clear. In 1-fluoro-3,4,5-trichlorobenzene, 1-[I], the only spin-$\frac{1}{2}$ nuclei are fluorine and hydrogen. The two protons are isochronous, by symmetry. They are also equally coupled to the fluorine nucleus, again due to symmetry (the relationship is an *ortho* one in the case of each (H, F) pair). Thus the protons are magnetically equivalent. In the case of 1,4-difluoro-2,3-dichlorobenzene, 1-[II], the two protons are again isochronous (as are the two fluorine nuclei). However, if F_d is chosen as the 'third' nucleus, it can

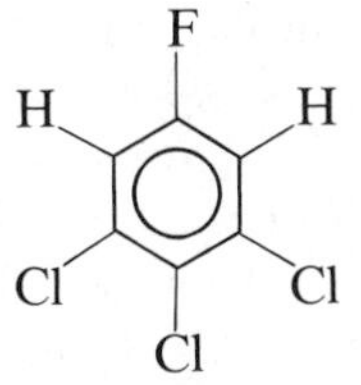

1-[I] 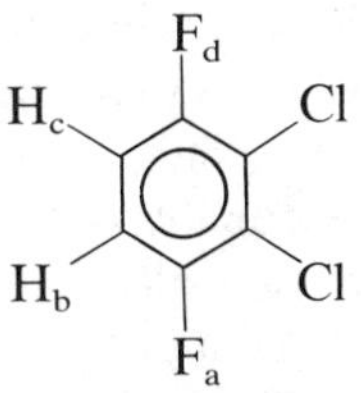 1-[II] 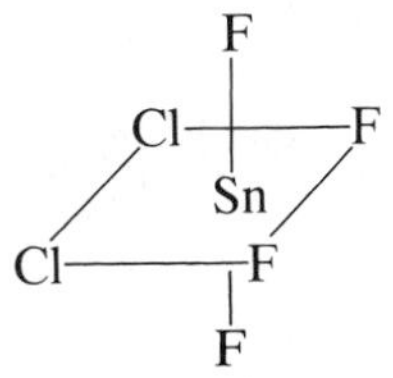 1-[III]

be seen that the (H_b, F_d) relationship is *not* the same as the (H_c, F_d) one—they are *meta* and *ortho* respectively. The corresponding coupling constants are therefore unequal, and the protons are magnetically *non*-equivalent; they are merely *chemically* equivalent. It is perfectly possible to have a situation with two isochronous pairs of nuclei which do have magnetic equivalence. An example is the complex 1-[III], in which the two *trans* fluorine nuclei are magnetically equivalent because they are equally coupled to any one of the remaining fluorines; similarly the two remaining fluorine nuclei form a magnetically equivalent pair.

There are two types of situation which require further explanation. In the first place, if *all* the magnetic nuclei in the molecule are isochronous, they are necessarily magnetically equivalent, since there is no other nucleus to allow distinction between chemical and magnetic equivalence. Benzene, methyl iodide and TMS (provided rotation about the C–Si bonds is assumed to occur rapidly) are examples of this situation. Secondly, the more complicated cases 1-[IV] and 1-[V] will be used to indicate an extension of the above principles. It is clear that

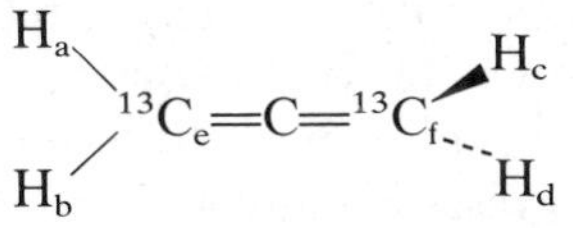

1-[IV] 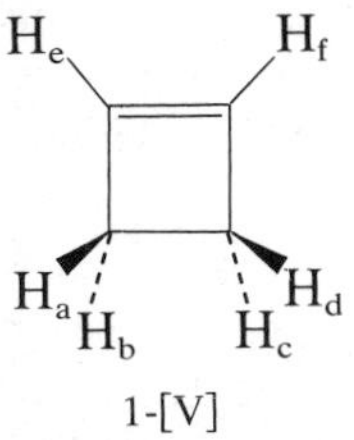 1-[V] 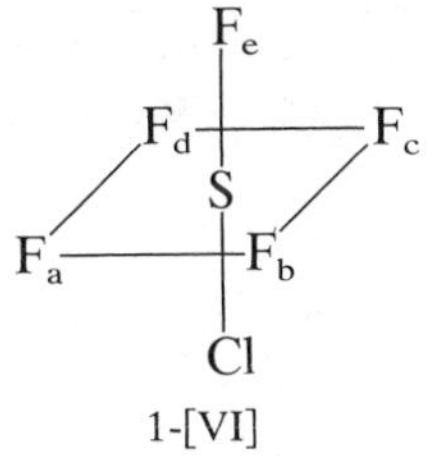 1-[VI]

in both cases the four protons a, b, c, and d are all isochronous (symmetrically equivalent); on the other hand, the pairs a, b are not magnetically equivalent to the pairs c, d, since the two pairs couple differently to nucleus e. It is, however, more difficult to decide whether protons a and b are magnetically equivalent or not. The situation is resolved by considering how a and b couple to c, which is *not* part of the set a, b, since it is not magnetically equivalent to them. In 1-[IV], $J_{ac} = J_{bc}$ by symmetry (and $J_{ad} = J_{bd}$), so that a and b are magnetically equivalent (as are c and d). For 1-[V], on the other hand, $J_{ac} \neq J_{bc}$ and $J_{ad} \neq J_{bd}$; therefore a and b are not magnetically equivalent, but are only chemically equivalent. However, it is not necessary for all the nuclei of a group to be equally coupled together for them to be magnetically equivalent. For instance, pentafluorosulphur chloride, 1-[VI], has four isochronous fluorine nuclei in equatorial positions, a, b, c, and d, plus a lone axial fluorine, e. All the equatorial fluorine nuclei are equally coupled to the axial nucleus; a, b, c, and d are therefore magnetically equivalent. However, these nuclei are not equally coupled to one another, since $J_{ac} = J_{bd}$ (*trans* coupling) $\neq J_{ab} = J_{bc} = J_{cd} = J_{da}$ (*cis* coupling).

In general, both chemical and magnetic equivalence give rise to simplifications in the spectra; however, the simplifications are much greater for magnetic than for chemical equivalence. Indeed, except when there are certain fortuitous relationships between coupling constants, spectra from molecules containing chemical equivalence can never be first order (See Chapter 2). In certain situations, molecules with magnetically equivalent nuclei *can* give rise to first-order spectra. Magnetic equivalence in such circumstances is further considered in Section 1-16 when the appropriate notation has been developed; the effects of magnetic equivalence in second-order spectra will be discussed in Sections 2-10 and 2-11.

1-15 Notation for spin systems

It is important to classify molecules or molecular fragments according to their spin systems, which define the number and type of magnetic nuclei and the relationship between them. The notation which will be described is normally used for the spin-$\frac{1}{2}$ nuclei of a molecule only. Each nucleus is assigned a capital letter of the Roman alphabet. If the chemical shift difference between a pair of nuclei is much greater than the coupling constant between them, they are assigned letters well apart in the alphabet. This always applies if the nuclei are of different isotopes or elements. Thus *cis*-1,2-dichlorofluoroethylene, 1-[VII], is said to possess an AX spin system (it is irrelevant whether A refers to H and X to F or vice versa; the chlorines may be ignored). If all chemical shift differences, $|\nu_j - \nu_k|$, in a spin system are much greater than the corresponding coupling constants, $|J_{jk}|$, then the spectrum is first order, as noted in preceding sections. The spin-system notation can therefore indicate if the spectrum is expected to be first order. This is clearly so for 1-[VII], and also for 1-[VIII], which has an AMX spin system. Conversely, if a chemical shift difference is of the order of, or less than, the corresponding coupling constant, adjacent letters of the alphabet are used for the two nuclei involved. Thus the two protons of 1-[IX] constitute an AB spin system. Note that this does not necessarily imply that the nuclei are close in space. If two nuclei, j and k, in a

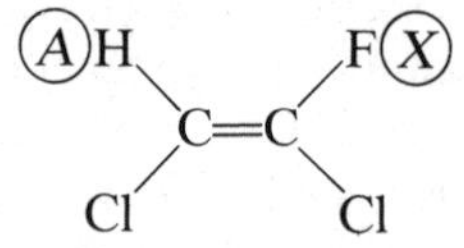

1-[VII] 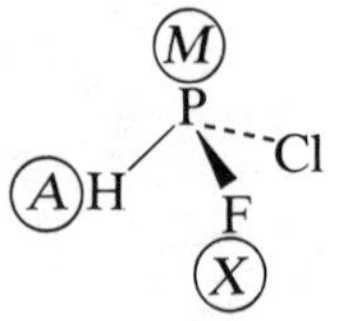   1-[VIII] 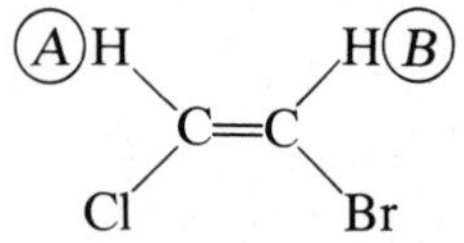   1-[IX]

molecule are of the same isotope and it is not known if $|\nu_j - \nu_k| \gg J_{jk}$, it is usually safest to label the spin system as if $|\nu_j - \nu_k| \approx J_{jk}$, as has been done for 1-[IX] above. This implies that the spectrum may not be first order.

There are several notations for indicating the type of symmetry of the spin system. It is usual to indicate magnetic equivalence by means of a numerical subscript; thus an A_2X system has two magnetically equivalent A nuclei (as for molecule 1-[I]). Similarly the spin system of 1-[III] is A_2X_2 and that of 1-[VI] is AB_4 or AX_4. In the notation of Haigh[10] square brackets are used to indicate repetition through symmetry of a group of nuclei. Thus molecule 1-[II] has an $[AX]_2$ spin system. On occasion it is necessary to use more than one pair of square brackets for a given spin system. Thus molecule 1-[V] possesses the system $[A[B]_2]_2$.

Ideally, any notation for spin systems should also specify the relationships between the coupling constants, since differences in these will affect the form of the spectrum. Thus there are two types of $[AX]_3$ system shown by the hypothetical molecular structures 1-[X] and 1-[XI], which have D_{3h} and C_{3h} symmetry respectively. There are only

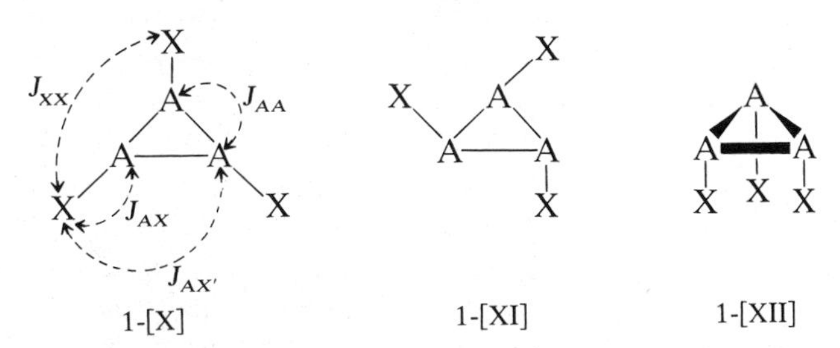

1-[X] 1-[XI] 1-[XII]

two types of (A, X) coupling constant for 1-[X] whereas there are three for 1-[XI]. It should be emphasized that it is the relationships between spin parameters that matter (since they are scalar quantities), however, not necessarily the symmetry point group. Thus several related point groups may give the same spin system—for example, the C_{3v} atomic structure 1-[XII] has the same type of spectrum as that of the D_{3h} structure 1-[X] because once more there are only two types of (A, X) coupling. When it is desired to discuss spin systems in detail it is necessary to distinguish on a molecular diagram (or a spin system diagram such as 1-[X]) the various coupling constants and their notation. Slight difficulties are sometimes encountered if molecular diagrams are used, since it may not be possible to distinguish between certain couplings by means of the spectrum alone. For instance there is an ambiguity between J_{ae} and J_{af} for 1-[V]; the spectrum provides two different (A, B) coupling constants but does not indicate which is which. This arises because, as noted above, it is the spin system which matters for the spectrum, not the molecule itself. The notation $[A[B]_2]_2$ for 1-[V] could indicate protons a, b, and e or protons a, b, and f as the repeated group $A[B]_2$. Normally such ambiguities may be

LSON LIBRARY
ORTHERN MICHIGAN UNIVERSITY
IARQUETTE, MICHIGAN 49855

resolved by consideration of values of coupling constants for related compounds.

1-16 First-order spectra involving magnetic equivalence

As stated in Section 1.14, spectra of spin systems with magnetic equivalence can be very simple. The circumstances under which this happens are analogous to those discussed earlier for first-order spectra, namely when all chemical shift differences between non-equivalent nuclei are much greater than the corresponding coupling constants between them. For this purpose, a set of magnetically equivalent nuclei is treated as a single group. The splitting patterns may be obtained as before, provided it is recognized that the equivalence of some nuclei implies that some coupling constants are equal, and therefore that some lines overlap. This is shown for the simplest case, that of the AX_2 spin system, in Fig. 1-17. This method of successive splittings can however be greatly simplified by recognizing that the overlapping of lines leads to the following rules:

(a) Splitting of a resonance due to coupling of the nucleus concerned with p magnetically equivalent nuclei leads to the appearance of $p+1$ distinct lines.

(b) The intensity distribution within the $p+1$ lines is that of the binomial coefficients, i.e. those of the expansion $(1+x)^p$.

The appropriate relative intensities for various values of p are given by Pascal's triangle, as listed in Table 1-4. Three features need to be pointed out. The first is that if calculated intensities of lines in different parts of the spectrum are to be comparable, then account must be taken of the number of nuclei of each type. Thus in Fig. 1-17 the relative intensities are given in such a fashion that the total intensity in the X part is twice as great as the total intensity in the A part. Secondly, no account need be taken of coupling between nuclei that are magnetically equivalent. Such coupling will normally exist but it does not affect the spectrum, even in second-order cases, as can be rigorously proved. Lastly, the splitting patterns need not be confused with those given by coupling to a single nucleus of spin $>\frac{1}{2}$, since the intensity distributions are not the same even when the number of

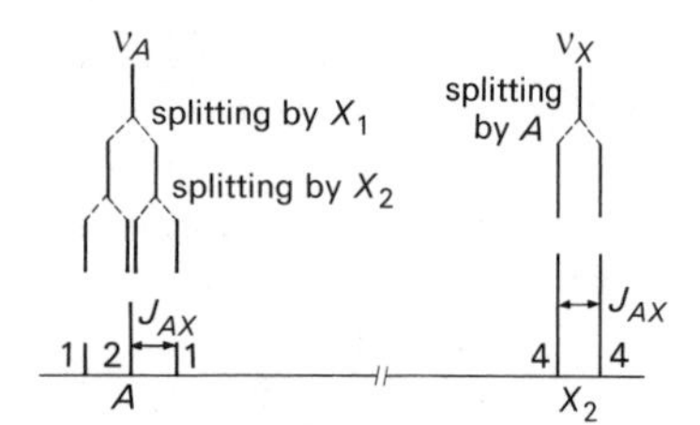

Fig. 1-17 The splitting pattern for an AX_2 spin system. The numbers adjacent to the lines give the intensities relative to that of the weakest A line.

Table 1-4 Relative intensities within NMR multiplets due to coupling to p magnetically equivalent nuclei in first order

p	*Relative intensities (binomial coefficients)*
1	1:1
2	1:2:1
3	1:3:3:1
4	1:4:6:4:1
5	1:5:10:10:5:1
6	1:6:15:20:15:6:1

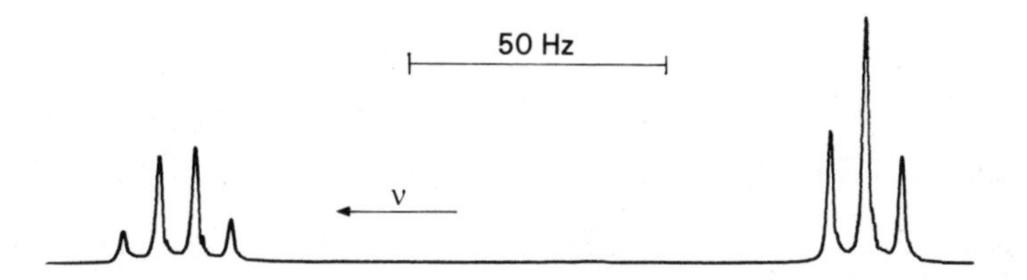

Fig. 1-18 60 MHz spectrum of diethyl ether. Coupling between protons on different ethyl groups may be ignored, so this is an example of an A_3X_2 spectrum.

observed lines is identical. For instance, the A region of Fig. 1-17 is a 1:2:1 triplet because of coupling to two magnetically equivalent spin-$\frac{1}{2}$ nuclei; coupling to a single spin-1 nucleus would have given a 1:1:1 triplet arising from the three equally probable m_I values for the spin-1 nucleus.

Figure 1-18 shows a proton spectrum of diethylether at 60 MHz, which is very nearly of the first-order type. The CH_2 protons give rise to a 1:3:3:1 quartet because of coupling to the three methyl protons which are rendered magnetically equivalent by rapid internal rotation; likewise, the CH_3 protons give a 1:2:1 triplet (or 3:6:3 if considered relative to the 1:3:3:1 CH_2 quartet). The slight deviations from binomial intensity ratios are due to departure from first-order conditions.

1-17 Abundant spins and dilute spins

Sections 1-11 to 1-16 have assumed that all the spins under discussion are of isotopes that are present in ca. 100% natural abundance, such as ^1H, ^{19}F, and ^{31}P. However, NMR studies are often concerned with isotopes of low natural abundance, such as ^{13}C (1·1%). Such species may be referred to as dilute spins. They have the important property that interactions between them are not of significance. In particular, spin–spin coupling between ^{13}C nuclei can generally be ignored since the probability of having more than one ^{13}C in the same molecule is low. Therefore, splittings in the spectra of dilute spins are caused only by spin–spin coupling to abundant spins. Typically, then, ^{13}C spectra are complicated only by (^{13}C, ^1H) coupling.

By the same argument, spectra of abundant spins are primarily affected by coupling only to other abundant spins. However, it is possible to detect weak signals arising from molecules containing dilute spins in addition to the abundant spins. For example, 1·1% of chloroform molecules are $^{13}CHCl_3$. The proton spectrum of such molecules consists of a doublet with splitting $|J_{CH}|$. The lines will be very weak compared to the proton spectrum of $^{12}CHCl_3$ and they are often referred to as '^{13}C satellites'. The $^{12}CHCl_3$ signal is not precisely in the middle of the ^{13}C satellites because there is a small isotopic effect of the ^{13}C on the ^1H chemical shift. For symmetrical molecules, such as $(CHBr_2)_2$ (see Fig. 1-19), the presence of ^{13}C can lift the equivalence of the protons and allow normally inaccessible (H, H) coupling constants to be obtained from the ^{13}C satellites (see Section 2-9(c)).

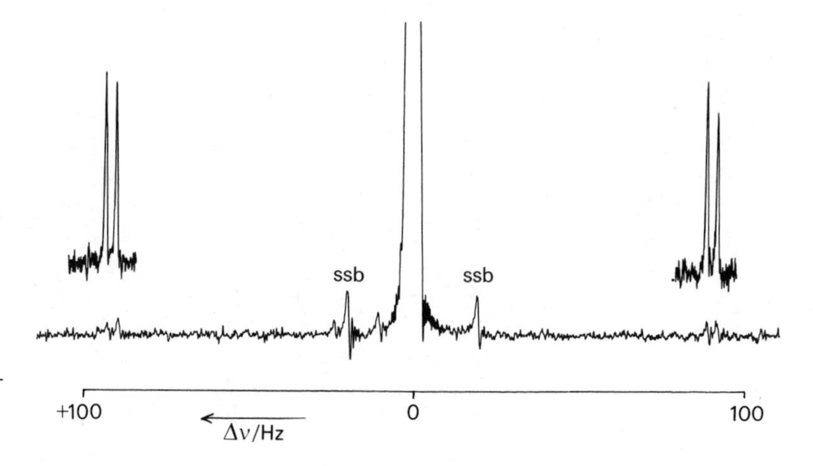

Fig. 1-19 'Carbon-13 satellites' in the ^1H NMR spectrum of 1,1,2,2-tetrabromoethane. The peaks marked ssb are spinning sidebands. The value of $^1J_{CH}$ for this compound is 182 Hz and is given by the separation of the centres of the satellite doublets. The doublet splitting is $^3J_{HH}$, as explained in Section 2-9(c). The central peak, due to the per-^{12}C-isotopomer, has not been taken to its full height.

Some isotopes are, of course, present in natural abundance to extents which are intermediate between the abundant and the dilute categories. One common example is ^{195}Pt (abundance 33·8% spin $I=\frac{1}{2}$). The other stable isotopes of platinum have spin $I=0$. Consequently a ^{13}C bonded to platinum will show a splitting into a doublet of spacing J(Pt, C) for 33·8% of its intensity whereas the remaining 66·2% of intensity (due to ^{13}C bonded to $I=0$ platinum isotopes) will be unsplit.

1-18 Decoupling

Although coupling constants are a useful source of chemical information the splittings caused by their existence complicate the appearance of spectra and reduce the available signal-to-noise (S/N) ratio. It is therefore often advantageous to be able to remove the splittings. This process of decoupling is carried out by a double resonance technique, i.e. two radiofrequencies are needed (see Chapter 4). One r.f. is used to observe signals due to a given nucleus or nuclei. The other irradiates strongly the resonance of the nucleus to be decoupled. If the observed and irradiated nuclei are of the same isotope the experiment is said to be homonuclear; if they are different isotopic species the experiment is heteronuclear. An example of homonuclear decoupling is shown in Fig. 1-20. Such an experiment shows unequivocally which resonances are coupled.

The commonest type of heteronuclear decoupling is observation of a ^{13}C spectrum while protons are decoupled. In fact it is generally desired to decouple *all* protons of the sample when studying ^{13}C spectra. This is done by using a *spread* of frequencies in the ^{1}H region, sufficient to cover all the ^{1}H signals of interest. The range of frequencies is obtained by modulating an appropriate radiofrequency by audiofrequency noise. The procedure is therefore known as noise (or broad-band) decoupling. An example is given in Fig. 1-21. Identical conditions were used for the spectra illustrated, apart from the decoupling. It can be seen that the gain in sensitivity, arising principally from

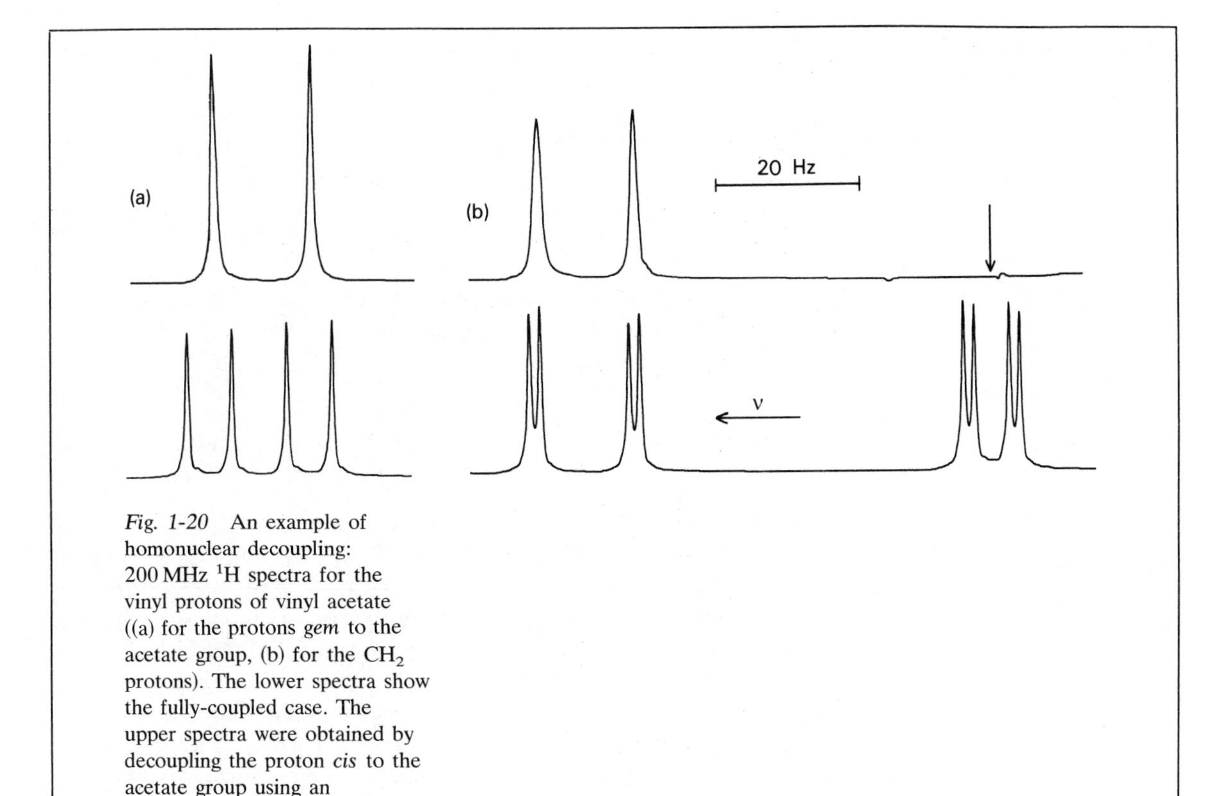

Fig. 1-20 An example of homonuclear decoupling: 200 MHz ^{1}H spectra for the vinyl protons of vinyl acetate ((a) for the protons *gem* to the acetate group, (b) for the CH_2 protons). The lower spectra show the fully-coupled case. The upper spectra were obtained by decoupling the proton *cis* to the acetate group using an irradiating r.f. at the position marked by the vertical arrow. The intensities of the upper and lower spectra are not on comparable scales.

the collapse of multiplets into single peaks (but see Section 4-10), is so substantial that no peaks can be distinguished in the coupled spectrum.

As was pointed out in the preceding section, since ^{13}C is a dilute spin species, only (^{13}C, ^{1}H) coupling generally affects ^{13}C spectra (in the absence of nuclei such as ^{19}F and ^{31}P). Consequently the result of noise decoupling is that the ^{13}C spectrum consists of a single line for each type of carbon in the molecule (see Fig. 1-21(b)). Thus spectral interpretation is greatly simplified, though coupling information is lost. The complexity of the coupled spectrum is not, of course, apparent for Fig. 1-21(a) because of the high noise level.

1-19 The multiscan principle

Since NMR is intrinsically an insensitive technique, it is desirable to use all possible means to improve the signal-to-noise ratio. Several techniques (e.g. increasing B_0, isotopic-enrichment, and decoupling) have already been mentioned. With the ready availability of minicomputers an additional method becomes feasible, namely multiscan operation. The spectrum of interest is obtained repeatedly; each

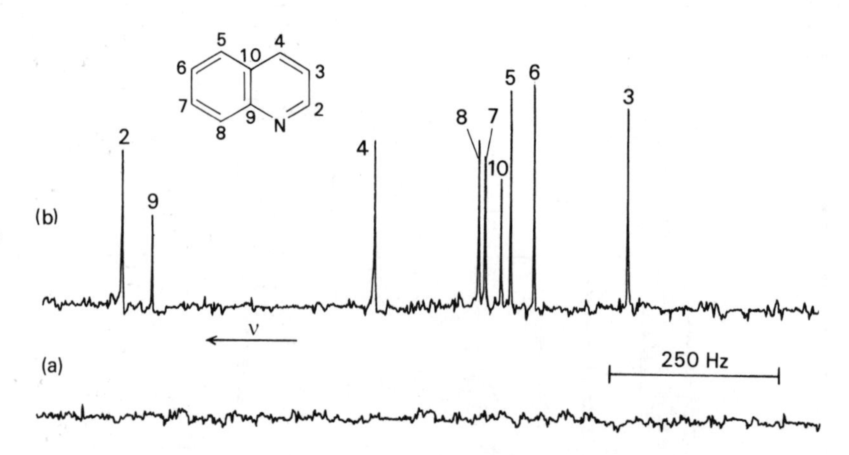

Fig. 1-21 Carbon-13 NMR spectra of quinoline (50% by volume in $CDCl_3$) at 25 MHz: (a) coupled to protons; (b) decoupled from protons. The intensities of (a) and (b) *are* on comparable scales. The line assignments are as given by R. J. Pugmire, D. M. Grant, M. J. Robins & R. K. Robins, *J. Amer. Chem. Soc.*, **91,** 6381 (1969).

time it is digitized and stored in the same channels of the computer in additive mode. Since true signals are always present, their intensity increases proportionately to the number of scans of the spectrum, n. However, random noise increases more slowly, in fact as $\sqrt{n}$. Thus the signal-to-noise ratio improves as $\sqrt{n}$. The only limitation lies in the total spectrometer time available. A typical proton continuous wave (CW) spectrum may take 10 min to record. To gain an order of magnitude in S/N would take 100 scans, i.e. ca. 16 h, which is feasible. A further similar gain in S/N would take longer than is reasonable. Figure 1-22 shows the effectiveness of a modest application of the multiscan principle.

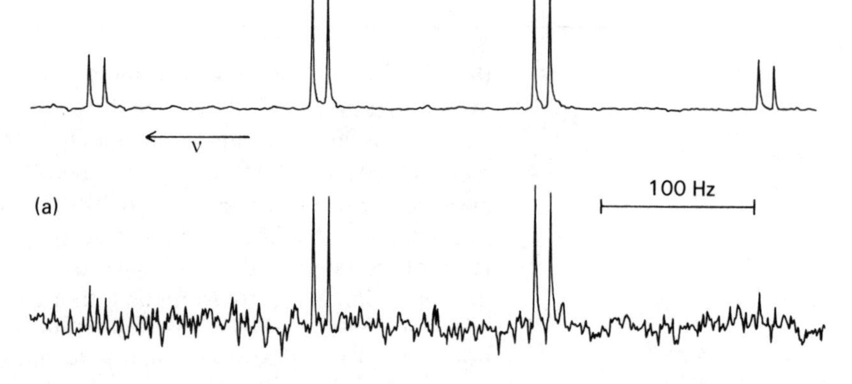

Fig. 1-22 Carbon-13 NMR spectra of trimethylphosphite, $(MeO)_3P$, at 25 MHz: (a) single spectrum; (b) computer sum of 50 spectra. The vertical scales were adjusted to give similar peak heights for the two spectra. Noise levels in (b) should therefore appear lower by a factor of $\sqrt{50} \sim 7$ than for (a). The splitting pattern (1:3:3:1 quartet of doublets) arises from coupling of the ^{13}C to the directly-bonded protons and to the phosphorus nucleus.

1-20 Continuous-wave NMR: the spectrometer

The simplest (and cheapest) way of obtaining high-resolution NMR spectra is to use a monochromatic radiofrequency and to vary its frequency slowly so as to scan all the resonances of interest. This is known as continuous-wave (CW) NMR and is usually only used for the most receptive nuclei, ^1H and ^{19}F. This section describes the elements of a CW spectrometer. Like any other spectroscopic instrument this requires a source of radiation, a sample cell, a detector and a recording device, together with a method of scanning the spectrum. In addition, a magnet giving an extremely stable and homogeneous magnetic field is necessary. The range of magnetic fields usually used is ca. 1·5 to 12 T, giving proton resonance frequencies from 60 to 500 MHz. The resolution normally required is better than 0·5 Hz, so at a proton frequency of 100 MHz the magnet homogeneity (which may be a limiting factor for the resolution) must be at least 1 part in 2×10^8 over the sample volume, and the stability should be equally as good. The magnets themselves are products of precision engineering, but further devices are necessary to reach the required standard. One such device for obtaining good homogeneity is the use of shim (or Golay) coils. These are conducting loops, situated at the pole faces, which carry small variable currents, producing magnetic fields that can be adjusted to compensate for field gradients in the principal field. The magnets themselves may be of three types, viz. permanent magnets, ordinary electromagnets and superconducting electromagnets; the last-mentioned are the only type suitable for use above 2·5 T. The electromagnets have the advantage that the fields produced are far less sensitive to temperature changes than those from permanent magnets; however, the inherent field stability apart from the temperature effect is lower for electromagnets, they are costly to operate, and (except for the superconducting type) they require efficient water-cooling.

The source of electromagnetic radiation at the radiofrequencies necessary is a crystal oscillator, which again must be of high stability. Usually its basic frequency is lower than that actually used in the spectrometer, and electronic circuits are used to provide appropriate frequency multiplication. In some spectrometers the frequency employed to obtain resonance is not that produced directly by multiplication from the crystal frequency (this is known as the centre-band frequency), but is a sideband produced by modulation of the centre band using an audio frequency oscillator. Typical audio frequencies used are in the region of 2000 Hz for proton resonance; the value may be varied linearly, thus providing a method of producing frequency-sweep spectra, while maintaining a constant magnetic field. Added spectrometer stability is obtained by detecting a particular signal using a separate frequency and maintaining it on resonance by electronically compensating for any changes in the magnetic field; such a technique is known as *field-frequency locking*. The magnetic field strength may be brought into the correct region for resonance by employing auxiliary electric currents in conducting loops known as the sweep coils, situated at the pole faces.

Physically, an NMR spectrometer usually consists of two parts: (a) the magnet with associated power supply and cooling unit (if it is an electromagnet) and (b) the operating console, which contains the radiofrequency source and associated electronic circuits, most of the controls, and the chart recorder. However, the sample must be placed

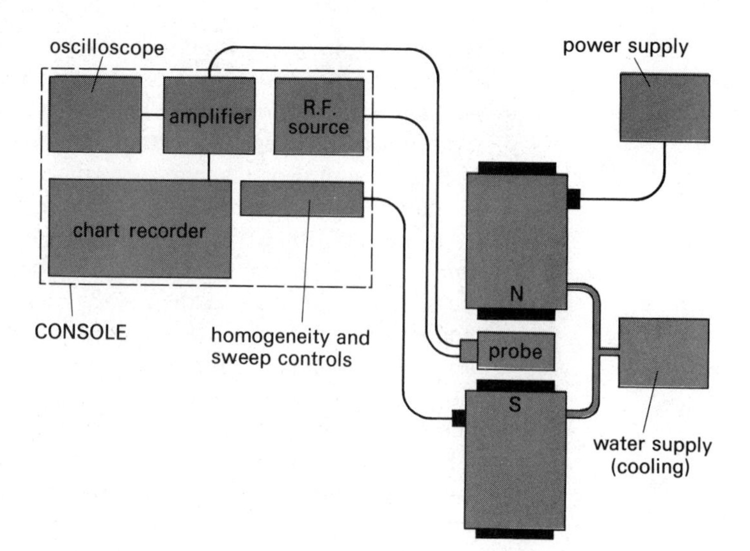

Fig. 1-23 Block diagram of a high-resolution NMR spectrometer using an electromagnet.

in the magnetic field, together with the actual irradiating and detecting devices. These features are located in a unit known as the probe, which is situated in the pole gap. A commonly-used system has separate coils for irradiation and detection and is hence known as the double-coil method. As shown in Section 1-6, the irradiating frequency must correspond to a magnetic field B_1, which is perpendicular to the primary field B_0, but which oscillates with time (see also Chapter 3). The axis of the transmitter coil is therefore perpendicular to B_0; the axis of the receiver coil should be perpendicular to both B_0 and B_1, since there must be no direct induction of current in the receiver from the transmitter, only indirect induction via the magnetic resonance. The receiver coil is normally smaller in diameter than the transmitter coil and it is wound on a vertical tube of glass or other non-magnetic

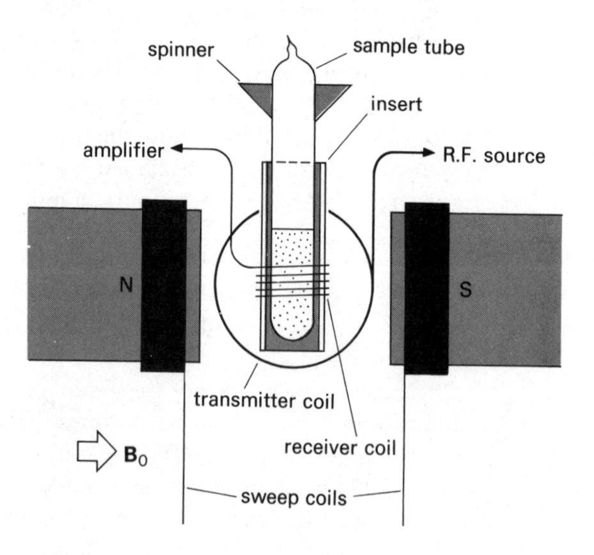

Fig. 1-24 Diagrammatic representation of the arrangements in the region of the pole gap of a crossed-coil NMR spectrometer based on an electromagnet. The shim coils are not shown.

material, forming what is known as the insert. The small current induced in the receiver coil at resonance is amplified many times before being recorded. Figure 1-23 shows a block diagram of an NMR spectrometer using an electromagnet, while Fig. 1-24 illustrates diagrammatically the arrangements in the vicinity of the pole gap.

1-21 Experimental aspects: the sample and the spectrum

As explained previously, high-resolution NMR studies are usually made on liquids or solutions. Fortunately, glass does not give absorptions detectable under normal operation (except for weak broad ^{29}Si signals), so that sample cells are Pyrex-glass tubes placed vertically in the insert. The tubes may either be sealed at the top or left unsealed. The normal size of tube for proton resonance is 5 mm in outside diameter (o.d.) or 12 mm o.d. for ^{13}C studies. About 3 cm depth of sample is used, so that the volume required is ca. 0·5 cm^3 for ^{1}H studies and ca. 3 cm^3 for ^{13}C work; however, it is possible to use less sample (say ca. 0·05 cm^3) if a special microtube is employed. Unfortunately, NMR is not a very sensitive technique, and with the normal tubes solution concentrations of at least 1% are necessary. Except when solvent effects on coupling constants or chemical shifts are being studied, it is normal to use saturated solutions (care being taken to avoid having any solid particles) or neat[11] liquids. To some extent larger tubes (with correspondingly larger receiver coils) can be used to increase the sensitivity, but since homogeneity of B_0 over the entire volume must be maintained, there are limitations to this procedure. In any case larger tubes are only useful if large amounts of sample are available (e.g. for solubility-limited cases). When the quantity of sample is limited it is better to use small tubes and high concentrations. As explained in Section 1-10, a reference compound is normally added to the sample, either directly (internal standard) or in an annulus outside the sample tube (external standard). Special sets of concentric tubes are available for the latter case, the outer tube being of the appropriate outside diameter.

The tubes are normally spun about the long axis at rates of ca. 15 Hz in order to average inhomogeneities of the magnetic field; this procedure results in an improvement of resolution. In order to do this the top of the tube is gripped in a plastic spinner; the spinning is affected by directing a current of air tangentially at the spinner. The spinning process may introduce unwanted spinning sidebands into the spectrum. These are symmetrically placed on either side of a genuine resonance (see Fig. 1-9) and separated from it by the spinning rate (in Hz). In principle, there are further sidebands at multiples of that separation from a genuine band. The spectrometer conditions are adjusted so as to minimize the intensity of spinning sidebands.

For many experiments it is desirable to vary the temperature of the sample (while leaving that of the magnet fixed). This is accomplished by passing a stream of nitrogen gas into the insert past the sample tube; the gas is either pre-heated by passing it over an electrically heated wire coil, or pre-cooled, using a heat exchanger dipped in liquid nitrogen. The temperature is measured using a thermocouple, or, alternatively, making use of a calibrated temperature-dependent chemical shift difference. The range −100 °C to +200 °C is commonly available.

The spectrum can be observed in two ways: either on an oscilloscope or on a paper chart. The oscilloscope is usually used only when adjusting the spectrometer conditions initially; the frequency or magnetic field is repetitively scanned so that its variation coincides with the travel of the light spot across the oscilloscope screen. An absorption will therefore occur at the same position on the screen for every scan until conditions are altered. The spectrum may be permanently obtained using the chart recorder; conventionally the frequency increases from right to left on the charts. With modern stable spectrometers the charts are usually pre-calibrated in Hz (or δ values). By suitable adjustment of the spectrometer, the frequency scale of the chart may be expanded (so as to observe fine structure of bands clearly), and different regions of the spectrum may be brought within the range of the chart (by means of a field or frequency shift).

The intensity scale shows total absorption and not percentage absorption (as is the case for infrared spectroscopy), since there is never an appreciable proportion of the transmitted signal absorbed. Signal areas are therefore a direct measure of the amount of absorption, provided the spectrometer controls are not altered; this is the basis of the use of NMR to 'count' the relative numbers of different nuclei. When linewidths are all the same, signal heights may be used as a measure of intensity. However, because of the effects of spin–spin coupling it is frequently not feasible to measure the total intensities attributed to given nuclei in a simple way. It may be noted that signal *areas* are really required; this is equivalent to an *integration* of the signal above the level of the base-line. It is normal practice to use automatic electronic integration, which may be done so as to trace on the spectrum a line whose height is proportional to the area of the signals scanned (see Fig. 1-9). Even with such a device, relative intensities cannot be obtained very accurately (only to ca. 5% usually); relative frequencies of absorption may be measured to a far greater accuracy (±0·1 Hz). Absolute values of these two parameters are not normally obtained. When dealing with fine-structure details, particularly when lines overlap, electronic integration is not of much use.

In order to obtain accurate values of relative absorption frequencies, it is necessary to have an expanded-scale spectrum and to scan the frequency very slowly—being careful, however, to avoid saturation (see Chapter 3). When a peak is scanned rapidly, three occurrences distort its shape. In the first place, the maximum of the peak is shifted to later in the scan, owing to the delay in response of the system (which does not have time to maintain equilibrium with the radiation as the scan advances). Secondly, the maximum is lower, and thirdly the side of the peak that is scanned last exhibits the phenomenon of 'ringing' instead of a steady decay from maximum absorption (see Fig. 1-9). To observe a lot of ringing during a rapid scan is a good sign, since this indicates that the magnetic field homogeneity is high. In fact, the ringing of a peak (on the oscilloscope) is often used as a monitor of homogeneity while adjusting the shim coils. Moreover, when a single peak is scanned the decay of the ringing should be exponential; departure from such a character in the decay is also an indication of a maladjustment of the shim coils.

Notes and references

[1] E. M. Purcell, H. C. Torrey & R. V. Pound, *Phys. Rev.*, **69,** 37 (1946).
F. Bloch, W. W. Hansen & M. E. Packard, *Phys. Rev.*, **69,** 127 (1946).

[2] W. G. Proctor & F. C. Yu, *Phys. Rev.*, **77,** 717 (1950).
W. C. Dickinson, *Phys. Rev.*, **77,** 736 (1950).

[3] Textbooks sometimes write this equation as $\mu = \gamma\hbar I$, implying discussion of the *maximum component* of $\boldsymbol{\mu}$. This appears to the present author to be unhelpful, though in quantum mechanics, with μ and I treated as operators it is completely correct to write $\hat{\boldsymbol{\mu}} = \gamma\hbar\hat{\mathbf{I}}$.

[4] The total magnetization should be distinguished from the magnetization per unit volume or per mole.

[5] Susceptibilities as normally listed are defined in terms of the magnetic field intensity H and not the magnetic induction field B.

[6] Substances with unpaired electrons have permanent magnetic moments which cause χ to be positive (giving paramagnetism). Such compounds are vital for electron spin resonance spectroscopy but are of relatively minor importance in NMR (see Section 8-14).

[7] The terms 'high field' and 'low field' (equivalent to low frequency and high frequency respectively) should be avoided, since they refer to outmoded methods of recording NMR spectra.

[8] An outdated alternative to the δ scale is that known as the τ scale. This lists chemical shifts in ppm relative to TMS but with increasing values to low frequency such that the resonance of TMS occurs at $\tau = 10{\cdot}00$. The relationship between the τ and δ scales is given by $\tau = 10{\cdot}00 - \delta$. Similar outmoded scales, with a low-frequency-positive convention, have been used for other nuclei, so caution is necessary when reading the literature.

[9] W. McFarlane, *Proc. Roy. Soc. London A*, **306,** 185 (1968).
R. K. Harris & B. J. Kimber, *J. Magn. Reson.*, **17,** 174 (1975).

[10] C. W. Haigh, *J. Chem. Soc. A*, 1682 (1970). Strictly speaking, Haigh's notation involves square brackets in cases 1-[I], 1-[III], and 1-[VI], which would be denoted $[A_2X]$, $[A_2X_2]$, and $[AB_4]$ spin systems respectively.

[11] The term 'neat' is used in order to avoid the connotations of 'pure', particularly as a reference compound is normally added.

Further reading (elementary texts)

N.B. These texts also include an elementary account of Fourier Transform NMR.

NMR Spectroscopy: An Introduction, H. Günther, Wiley & Sons (1980).
Proton and Carbon-13 NMR Spectroscopy: An Integrated Approach, R. J. Abraham & P. Loftus, Heyden & Son Ltd (1978).
NMR and Chemistry: An Introduction to the Multinuclear Fourier Transform Era, J. W. Akitt, Chapman & Hall (1983).

Problems

[N.B. Make use of Tables 1-1 and 1-2 in answering these questions.]

1-1 The proton NMR signal of nitromethane occurs at 259·8 Hz to high frequency of that of TMS for a spectrometer operating at 60 MHz. Give the δ value for nitromethane, and calculate the shift from TMS in Hz for a 100 MHz instrument.

1-2 At what frequency do the following nuclei give NMR absorption for a magnetic field of 1·5 T: (a) ^{29}Si, (b) ^{199}Hg, (c) ^{35}Cl?

1-3 In a certain spectrometer, protons absorb at 42·577 MHz. At what magnetic field does the spectrometer operate? Suppose it is desired to achieve resonance for ^{31}P nuclei using the same electronics. To what value must the field be raised?

1-4 Derive Eqn 1-26 from considerations similar to those used in Section 1-4.

1-5 Describe the splitting patterns of the proton NMR spectra (which may be assumed to be first order) of the following:
(a) CH_3CHO, (b) oxete, 1-[XIII], (c) $CH_3CHClCH_3$

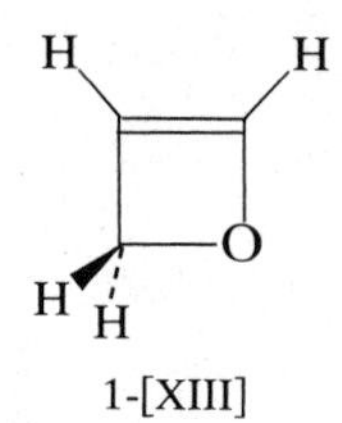

1-[XIII]

Magnetic effects of the chlorine nucleus in (c) may be ignored.

1-6 An *AX* spin system gives rise to proton NMR lines at 29·6, 34·3, 391·1, and 395·8 Hz to high frequency of the resonance of TMS for an instrument operating at 40 MHz. What is the magnitude of the coupling constant? Give the δ values for the two types of proton. At what separations (in Hz) from TMS would resonance occur for an instrument operating at 60 MHz?

1-7 What is the difference between the proton resonance frequencies of benzene and acetone (7% in $CDCl_3$) at a constant magnetic field of $B_0 = 2{\cdot}35$ T? What is the difference in the shielding constant between the two molecules? Hence, what is the difference in the magnetic fields actually experienced by the protons of the two molecules?

1-8 (a) Re-label the transitions (α, β etc.) of Fig. 1-13 for the case of negative J_{13} (all the remaining coupling constants being positive).
(b) Re-label the transitions (A_1, A_2 etc.) of Fig. 1-12 for the case of negative J_{MX} so that the labels remain consistent with Fig. 1-14.

1-9 Furan-2-aldehyde 1-[XIV] has the following spin parameters:

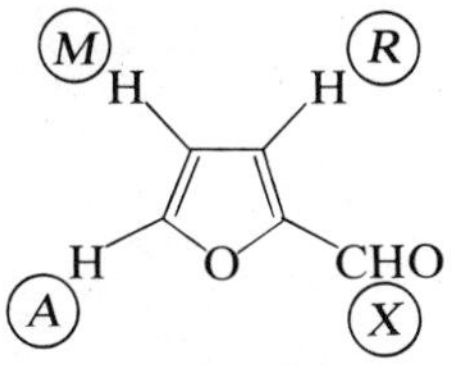

1-[XIV]

$\delta_A = 7{\cdot}81$ $\quad J_{AM} = 1{\cdot}65$ Hz
$\delta_M = 6{\cdot}67$ $\quad J_{MR} = 3{\cdot}62$ Hz
$\delta_R = 7{\cdot}23$ $\quad J_{AR} = 0{\cdot}80$ Hz
$\delta_X = 9{\cdot}66$ $\quad J_{AX} = 0{\cdot}81$ Hz
$J_{MX} \sim J_{RX} \sim 0$

Sketch the appearance of each band of the spectrum in first order.

1-10 Draw an energy-level diagram for a first-order three-spin system with all coupling constants zero. Clearly label the levels with the appropriate basic product spin wave functions. Indicate the effect on the energy levels of introducing (a) $J_{AM} = +10$ Hz, $J_{AX} = +6$ Hz, $J_{MX} = +2$ Hz and (b) $J_{AM} = -10$ Hz, $J_{AX} = -6$ Hz, $J_{MX} = -2$ Hz. Show that the *spectrum* is the same for (a) and (b), except for the labelling of the lines.

1-11 Suggest the notation to be used for the spin systems (for the nuclei with $I = \frac{1}{2}$ only) of the following molecules:
(a) *p*-bromochlorobenzene, (b) cyclopropyl chloride,
(c) cyclohexane (as if fixed in a single chair form), (d) SF_5CF_3,
(e) *o*-chlorofluorobenzene, (f) PF_3Cl_2

1-12 Derive the first-order pattern of lines, with the intensities relative to the weakest line in the whole spectrum, for an $A_3M_4X_2$ spin system. Assume that $J_{AM} > J_{AX} > J_{MX} > 0$.

1-13 If each first-order transition has unit intensity, what is the total intensity in the A region of an A_pX_q spectrum?

1-14 The ^{19}F NMR spectrum of ClF_3 at low temperatures consists of a doublet considerably to high frequency of a 1:2:1 triplet. The total intensities in the two bands are in the ratio 2:1, and the splittings are all the same. Explain the spectrum and state what information may be derived about the molecular structure of ClF_3.

1-15 Show that, in the high temperature limit ($kT \gg \gamma\hbar B_0$) the population difference between the α and β spin states of a spin-$\frac{1}{2}$ nucleus is given by Eqn 1-20.

1-16 Generalize Eqn 1-21 for a nucleus of spin I. Hence show that at sufficiently high temperatures the equilibrium magnetization in a magnetic field B_0 is given by $M_0 = \frac{1}{3}N\gamma^2\hbar^2 I(I+1)B_0/kT$.

1-17 Show that a nucleus of spin-$\frac{1}{2}$ at thermal equilibrium at temperature T in a field B has a probability $[1 + \exp(\gamma\hbar B/kT)]^{-1}$ of being in the state β. Calculate the difference in the population (as nuclei per million) of the two states of ^{13}C in a field of 3 T at 27 °C.

2 Analysis of NMR spectra for isotropic solutions

2-1 Introduction

Values for coupling constants and chemical shifts can be obtained from first-order spectra by direct measurement of line positions, which are discussed in terms of the method of successive splittings in Section 1-3. However, first-order spectra only occur if the relative chemical shift of every pair of nuclei is large compared with their mutual coupling (except for magnetically equivalent spins). Such spectra may be recognized by the fact that the lines either have equal intensities, or show the binomial intensity patterns described in the previous chapter.

This chapter will be concerned with the more complex spectra that are obtained when coupling constants and chemical shift differences are of comparable sizes or when there is chemical equivalence. The practical problem of finding values of the spin parameters from an observed spectrum is usually most readily approached by solving the converse problem—'given a set of J's and σ's what is the appearance of the spectrum?' As spectral lines arise from transitions between nuclear spin states, quantum-mechanical methods must be used to find the answer to this question.

The allowed energies and wave functions for molecules and atoms are found by solving the Schrödinger equation which may be written

$$\hat{\mathscr{H}}\psi = U\psi \tag{2-1}$$

The equation which determines the nuclear spin states is formally the same but $\hat{\mathscr{H}}$ and ψ are very different. Instead of involving the differential operators ∇^2 and functions of position, the Hamiltonian $\hat{\mathscr{H}}$ contains spin angular momentum operators. Consequently, the method of solution of the equation is quite different from the techniques used for the hydrogen atom electronic energies. In fact we must go one step back towards the basic fundamentals of quantum mechanics and use the operator methods introduced by Dirac.

As quantum mechanics is concerned with predicting the results of experiments, observable quantities are particularly important. Every observable property corresponds to an operator, $\hat{A}$, and the results of the observation are deduced from the result of operating with $\hat{A}$ on the wave function. The latter contains all the information necessary for predicting the results of any observation. However, these often cannot be predicted exactly, so that only the relative probabilities of different results can be forecast. The replacement of certainties by probabilities is one of the differences between classical and quantum systems.

Unfortunately, most chemists only meet quantum mechanics in the form of Schrödinger's equation. The operator techniques that must be used to understand spin systems are inherently simpler and involve less mathematics than the solution of the differential equation for the hydrogen atom. The only problem is unfamiliarity. The first sections of this chapter summarize the necessary background and definitions. The form of the spin Hamiltonian is given in Section 2-5, followed by a description of the calculation of the energy levels and spectra of one-, two- and many-spin systems (Sections 2-6 to 2-8). Finally, a few special techniques applicable to certain types of spectra are introduced.

2-2 Operators

The idea of an operator is frequently used in mathematics. An operator acts on a function to give another function.

$$\hat{A}\mathrm{f}(x) = \mathrm{g}(x) \tag{2-2}$$

We shall distinguish operators with a circumflex. There are many different types of operators; a few examples are:

(a) multiply by $x : \hat{x}$

$$\hat{x}\mathrm{f}(x) = x\mathrm{f}(x) \tag{2-3}$$

(b) differentiate with respect to $x : \dfrac{\hat{\mathrm{d}}}{\mathrm{d}x}$

$$\frac{\hat{\mathrm{d}}}{\mathrm{d}x}\mathrm{f}(x) = \mathrm{f}'(x) \tag{2-4}$$

(c) displace the origin of the coordinate system by a units: $\hat{T}$

$$\hat{T}\mathrm{f}(x) = \mathrm{f}(x-a) \tag{2-5}$$

Usually the new function g is different from f, but if it is only a multiple of f, so that

$$\hat{A}\mathrm{f}(x) = a\mathrm{f}(x) \tag{2-6}$$

then $\mathrm{f}(x)$ is said to be an *eigenfunction* of $\hat{A}$ with *eigenvalue a*. For example $\exp kx$ is an eigenfunction of the operator $\hat{\mathrm{d}}/\mathrm{d}x$ with eigenvalue k.

$$\frac{\hat{\mathrm{d}}}{\mathrm{d}x}\exp kx = k \exp kx \tag{2-7}$$

As any value of k satisfies this equation, this operator has an infinite number of eigenvalues. Other operators, such as the spin operators discussed in Section 2-4, may only have a finite number of eigenfunctions. As the stationary states of a quantum mechanical system are eigenfunctions of the energy (Hamiltonian) operator $\hat{\mathscr{H}}$, calculations of NMR spectra start with the problem of finding the eigenfunctions and eigenvalues of $\hat{\mathscr{H}}$.

The operators which correspond to observables in quantum mechanics are linear and Hermitian, that is, they satisfy the equations

$$\hat{A}[c_1 f_1(x)+c_2 f_2(x)]=c_1\hat{A}f_1(x)+c_2\hat{A}f_2(x) \tag{2-8}$$

and

$$\int g^*(x)\hat{A}f(x)\,dx=\int f^*(x)\hat{A}g(x)\,dx \tag{2-9}$$

where $f(x)$, $g(x)$ are any suitable functions, and $f^*(x)$ is the complex conjugate of $f(x)$. Two important consequences of Eqn (2-9) which we shall use are, first, that the eigenvalues of Hermitian operators are real, and secondly that their eigenfunctions corresponding to different eigenvalues are orthogonal to each other.

The effect of a sum of operators operating on a function $f(x)$ is the sum of the functions produced by the individual operators. Thus, if

$$\hat{A}_1 f(x)=g_1(x) \qquad \text{and} \qquad \hat{A}_2 f(x)=g_2(x) \tag{2-10}$$

then

$$(\hat{A}_1+\hat{A}_2)f(x)=g_1(x)+g_2(x) \tag{2-11}$$

A product of operators operating on $f(x)$ is defined as successive operation by the operators starting with the one nearest the function.

$$\hat{A}_1\hat{A}_2 f(x)=\hat{A}_1 g_2(x)=g_{12}(x) \tag{2-12}$$

The order in which the operators act is important, as the result of operating first with $\hat{A}_1$ and then with $\hat{A}_2$ is not necessarily the same as operating in the reverse order. For example

$$\frac{\hat{d}}{dx}\hat{x}f(x)\neq\hat{x}\frac{\hat{d}}{dx}f(x) \tag{2-13}$$

Two operators are said to commute if the order of operation makes no difference. The commutator $[\hat{A}_1, \hat{A}_2]$ of $\hat{A}_1$ and $\hat{A}_2$ is an operator defined to be

$$[\hat{A}_1, \hat{A}_2]=\hat{A}_1\hat{A}_2-\hat{A}_2\hat{A}_1 \tag{2-14}$$

For example

$$\left[\hat{x}, \frac{\hat{d}}{dx}\right]=\hat{1} \tag{2-15}$$

where $\hat{1}$ is the unit operator which leaves any function unchanged.

2-3 Wave functions

The functions on which quantum-mechanical operators operate are known as wave functions and are functions of the variables of the particular system considered. For example, a particle in a box has a wave function which is a function of position (x, y, z). Dirac's notation for wave functions will be used in this book because, although it may be less familiar, it is particularly suitable for spin functions. In this notation $|r\rangle$ is a wave function or ket and $\langle r|$ is its complex conjugate or bra. Whenever a bracket is completed integration over all variables is implied. Thus the normalization condition is

$$\langle r \mid r\rangle = 1 \tag{2-16}$$

and the expressions $\langle s| \hat{\mathcal{H}} |r\rangle$ and $\int \psi_s^* \hat{\mathcal{H}} \psi_r \, d\tau$ are equivalent.

The wave function itself cannot be observed, but the results of any observation can be predicted from it. If $|r\rangle$ is an eigenfunction of an operator $\hat{A}$

$$\hat{A} \, |r\rangle = a \, |r\rangle \tag{2-17}$$

then the result of the observation corresponding to $\hat{A}$ is always a. The wave function of the system is unaltered by this observation. If $|r\rangle$ is not an eigenfunction of $\hat{A}$ a further hypothesis must be introduced. Now $|r\rangle$ can always be written as a linear combination of eigenfunctions of $\hat{A}$. Suppose

$$|r\rangle = c_1 \, |a\rangle + c_2 \, |b\rangle \tag{2-18}$$

where $|a\rangle$ and $|b\rangle$ are normalized eigenfunctions of $\hat{A}$, with differing eigenvalues a and b. Then the hypothesis is that the probabilities of observing a and b are $c_1 c_1^*$ and $c_2 c_2^*$ respectively.[1] The mean or *expectation* value $\langle A \rangle$ for the observation corresponding to $\hat{A}$ is

$$\langle A \rangle = c_1 c_1^* a + c_2 c_2^* b \tag{2-19}$$

which can be written

$$\langle A \rangle = \langle r| \, \hat{A} \, |r\rangle \tag{2-20}$$

because eigenfunctions such as $|a\rangle$ and $|b\rangle$, which have different eigenvalues, are always orthogonal, that is $\langle a \mid b\rangle = \langle b \mid a\rangle = 0$. Equation 2-20 is universally true provided $|r\rangle$ is normalized.

A wave function may in certain circumstances be an eigenfunction of two (or more) different operators. If two operators commute, there is a complete set of functions which are simultaneously eigenfunctions of both operators. On the other hand if they do not commute and

$$[\hat{A}, \hat{B}] = \hat{C} \tag{2-21}$$

then no functions are simultaneously eigenfunctions of $\hat{A}$ and $\hat{B}$ (unless they have eigenvalue zero for each operator and for the commutator $\hat{C}$) (Problem 2-5).

Expansion of a function in terms of a complete set In Eqn 2-18 the function $|r\rangle$ was expanded in terms of a number of eigenfunctions of an operator $\hat{A}$ in order to predict the possible results of the observation corresponding to $\hat{A}$. This technique of expansion in terms of a set of normalized and orthogonal functions is very useful, and we shall use it repeatedly in discussing spin functions. The set of functions chosen must satisfy certain conditions: they must be possible wave functions for the system (not necessarily stationary states) and there must be enough of them to make up the desired function. Often one uses a complete set which contains enough functions (sometimes an infinite number) to describe any possible wave function of the system as some linear combination of this basis set.

The coefficients in the expansion

$$|r\rangle = \sum_j |j\rangle c_j \tag{2-22}$$

can be found by completing the bracket with each $\langle j|$ in turn. As the set of functions $|j\rangle$ are orthogonalized and normalized the coefficient c_j is the overlap integral of $\langle j|$ and $|r\rangle$:

$$\langle j \mid r\rangle = c_j \tag{2-23}$$

2-4 Spin operators and spin functions

The spin of an electron or a nucleus gives rise to an internal angular momentum of the particle. There are three independent components of spin angular momentum with corresponding operators[2] $\hbar\hat{I}_x$, $\hbar\hat{I}_y$, and $\hbar\hat{I}_z$. These act on the spin wave functions of the particles, which depend on the orientation of the particles and not on their positions.

All the properties of the spin operators can be deduced from the commutation relations between the spin operators. These are as follows ($i^2 = -1$):

$$\begin{aligned} [\hat{I}_x, \hat{I}_y] &= i\hat{I}_z \\ [\hat{I}_y, \hat{I}_z] &= i\hat{I}_x \\ [\hat{I}_z, \hat{I}_x] &= i\hat{I}_y \end{aligned} \tag{2-24}$$

An indication of how to derive these is given in text books on quantum mechanics.

The first deduction that can be made from these equations is that there are no states which are simultaneous eigenstates of $\hat{I}_x$, $\hat{I}_y$, and $\hat{I}_z$ (except states with zero spin angular momentum). This means that if a spin function is an eigenstate of $\hat{I}_z$ with eigenvalue m, the result of an observation of the z component of angular momentum can be predicted exactly, while the x and y components are uncertain (Problem 2-5). If two operators do commute it is possible to choose states which are simultaneously eigenstates of both operators. The result of each operation is its eigenvalue and the order of the operations is immaterial. Thus

$$\hat{A}\hat{B}\,|a, b\rangle = \hat{A}b\,|a, b\rangle = ab\,|a, b\rangle$$

and

$$\hat{B}\hat{A}\,|a, b\rangle = \hat{B}a\,|a, b\rangle = ba\,|a, b\rangle \tag{2-25}$$

The square of the magnitude of the angular momentum corresponds to the operator $\hbar^2\mathbf{I}^2$ where

$$\hat{\mathbf{I}}^2 = \hat{I}_x^2 + \hat{I}_y^2 + \hat{I}_z^2 \tag{2-26}$$

As this operator commutes with $\hat{I}_x$, $\hat{I}_y$, and $\hat{I}_z$ (Problem 2-6), it is possible to define both the magnitude of the spin angular momentum and its component in one direction simultaneously. Thus we may choose a set of normalized functions $|I, m\rangle$ which are eigenfunctions of both $\hat{\mathbf{I}}^2$ and $\hat{I}_z$. These are the $2I+1$ states with total angular momentum $\hbar[I(I+1)]^{1/2}$ and with z component of angular momentum $\hbar m$ which were discussed in Chapter 1.

It is possible to find the allowed values of I and m from the commutation relations 2-24. This is done by introducing the raising and lowering operators $\hat{I}_+$ and $\hat{I}_-$ which convert the state $|I, m\rangle$ into $|I, m+1\rangle$ and $|I, m-1\rangle$ respectively.[3] These operators are

$$\begin{aligned} \hat{I}_+ &= \hat{I}_x + i\hat{I}_y \\ \hat{I}_- &= \hat{I}_x - i\hat{I}_y \end{aligned} \tag{2-27}$$

which obey the commutation relations

$$\begin{aligned} [\hat{I}_+, \hat{I}_z] &= -\hat{I}_+ \\ [\hat{I}_-, \hat{I}_z] &= \hat{I}_- \\ [\hat{I}_+, \hat{I}_-] &= 2\hat{I}_z \end{aligned} \tag{2-28}$$

To show that $\hat{I}_+$ increases the m value by one unit, we use the first commutation relation acting on $|I, m\rangle$

$$(\hat{I}_+\hat{I}_z - \hat{I}_z\hat{I}_+)\,|I, m\rangle = -\hat{I}_+\,|I, m\rangle \tag{2-29}$$

but $\hat{I}_z\,|I; m\rangle = m\,|I, m\rangle$, so Eqn (2-29) becomes

$$m\hat{I}_+\,|I, m\rangle - \hat{I}_z\hat{I}_+\,|I, m\rangle = -\hat{I}_+\,|I, m\rangle \tag{2-30}$$

or

$$\hat{I}_z(\hat{I}_+\,|I, m\rangle) = (m+1)(\hat{I}_+\,|I, m\rangle) \tag{2-31}$$

This shows that the state $(\hat{I}_+\,|I, m\rangle)$ is an eigenfunction of $\hat{I}_z$ with eigenvalue $(m+1)$. It can also be shown that it is an eigenfunction of $\hat{\mathbf{I}}^2$ with the same eigenvalue as $|I, m\rangle$, as $\hat{I}_+$ commutes with $\hat{\mathbf{I}}^2$. However, as this new state is not necessarily normalized, we must write

$$\hat{I}_+\,|I, m\rangle = c_{Im}\,|I, (m+1)\rangle \tag{2-32}$$

where c_{Im} is a normalization factor which has the value $[I(I+1)-m(m+1)]^{1/2}$ (Problem 2-8). Similarly

$$\hat{I}_-|I, m\rangle = [I(I+1) - m(m-1)]^{1/2}\,|I, m-1\rangle \tag{2-33}$$

As the magnitude of a component of angular momentum cannot be greater than the total angular momentum, there must be a maximum value of m, m_{max}, and $\hat{I}_+$ acting on this state must give zero.

$$\hat{I}_+ |I, m_{max}\rangle = 0 \tag{2-34}$$

By considering the effect of operating with $\hat{\mathbf{I}}^2$ on $|I, m_{max}\rangle$ and $|I, m_{min}\rangle$ it can be shown (Problem 2-7) that

$$\hat{\mathbf{I}}^2 |I, m\rangle = m_{max}(m_{max}+1) |I, m\rangle = m_{min}(m_{min}-1) |I, m\rangle \tag{2-35}$$

As it must be possible to go from $|I, m_{max}\rangle$ to $|I, m_{min}\rangle$ by successive use of the lowering operator, $(m_{max}-m_{min})$ must be zero or a positive integer. The only possible solution is that

$$m_{max} = -m_{min} = I \tag{2-36}$$

where I has possible values $0, \frac{1}{2}, 1, \frac{3}{2}, \ldots$, and

$$\hat{\mathbf{I}}^2 |I, m\rangle = I(I+1) |I, m\rangle \tag{2-37}$$

All types of angular momentum are quantized in this way in quantum mechanics. Normally, different letters are used for different types of angular momentum, e.g. I for nuclear spin, S for electron spin. Orbital angular momenta have only integral quantum numbers (for example, s orbitals have zero angular momentum, p orbitals have $\ell = 1$, and d orbitals have $\ell = 2$).

The most important class of nuclei for nuclear resonance is that comprising those with quantum number $I = \frac{1}{2}$. The properties of the various spin-$\frac{1}{2}$ operators may be summarized using the Pauli spin matrices. With the states $|\alpha\rangle$ and $|\beta\rangle$ ($|\frac{1}{2}, \frac{1}{2}\rangle$ and $|\frac{1}{2}, -\frac{1}{2}\rangle$ respectively) as a basis, these are

$$\hat{\mathbf{I}}^2 \equiv \frac{3}{4}\begin{pmatrix} 1 & 0 \\ 0 & 1 \end{pmatrix} \qquad \hat{I}_+ \equiv \begin{pmatrix} 0 & 0 \\ 1 & 0 \end{pmatrix} \qquad \hat{I}_- \equiv \begin{pmatrix} 0 & 1 \\ 0 & 0 \end{pmatrix}$$

$$\hat{I}_x \equiv \frac{1}{2}\begin{pmatrix} 0 & 1 \\ 1 & 0 \end{pmatrix} \qquad \hat{I}_y \equiv \frac{1}{2i}\begin{pmatrix} 0 & -1 \\ 1 & 0 \end{pmatrix} \qquad \hat{I}_z \equiv \frac{1}{2}\begin{pmatrix} -1 & 0 \\ 0 & 1 \end{pmatrix}$$

The four elements of these matrices are, for a given operator $\hat{0}$,

$$\begin{pmatrix} \langle\beta|\hat{0}|\beta\rangle & \langle\beta|\hat{0}|\alpha\rangle \\ \langle\alpha|\hat{0}|\beta\rangle & \langle\alpha|\hat{0}|\alpha\rangle \end{pmatrix}$$

In the preceding discussion of angular momentum operators and their eigenvalues we did not have to specify the form of the operator or the coordinates of the wave function. All the information about eigenvalues and eigenfunctions was derived from the commutation relations. It is not necessary to know anything about the internal structure of nuclei in order to interpret nuclear resonance spectra. Nuclear physicists, however, relate the observed values of I and γ to the nuclear structure.

2-5 The nuclear spin Hamiltonian operator

The energy of a system is observable, and the corresponding operator $\mathscr{H}$ is known as the Hamiltonian operator. It plays an important role in quantum mechanics, as its eigenfunctions describe the stationary states of the system. These are the solutions of the time-independent form of Eqn 2-1.

The complete Hamiltonian for a molecule contains terms describing the kinetic energy of each particle, the potential energy of interactions between particles, and interactions of the particles with any external electric or magnetic fields. It can be divided approximately into terms representing electronic, vibrational, rotational, and nuclear spin contributions to the energy, and the wave function can be written as a product of electronic, vibrational, rotational, and nuclear spin functions, each of which is an eigenfunction of its own *effective* Hamiltonian. The nuclear spin Hamiltonian contains nuclear spin operators, together with constants (the coupling constants and shielding constants) whose values all depend on the electronic, vibrational and rotational states of the molecule. The nuclear spin states are eigenstates of this Hamiltonian.

The spin Hamiltonian used to interpret nuclear resonance spectra obtained from isotropic fluid samples is

$$h^{-1}\hat{\mathcal{H}} = -\sum_{j} (2\pi)^{-1}\gamma_j B_0(1-\sigma_j)\hat{I}_{jz} + \sum_{j<k} J_{jk}\hat{\mathbf{I}}_j \cdot \hat{\mathbf{I}}_k \tag{2-38}$$

This equation is usually written in a form in which the terms (as above) have the dimensions of frequency and not energy. The differences in eigenvalues therefore give transition frequencies directly. This Hamiltonian is the quantum-mechanical equivalent of the first-order equation 1-32. The first term represents the interaction of the nuclear magnetic moments with the applied field $\mathbf{B}_0$. The energy of a nucleus in a magnetic field $\mathbf{B}$ is given classically by Eqn 1-13. As the magnetic dipole of a nucleus is related to its angular momentum by the magnetogyric ratio γ, the operator corresponding to the magnetic moment of nucleus j is

$$\hat{\boldsymbol{\mu}}_j = \gamma_j \hbar \hat{\mathbf{I}}_j \tag{2-39}$$

and the quantum-mechanical equivalent of Eqn 1-13 is

$$\hat{\mathcal{H}} = -\gamma_j \hbar \mathbf{B} \cdot \hat{\mathbf{I}}_j \tag{2-40}$$

The first term in Eqn 2-38 is identical to this with coordinate axes chosen[4] with the z direction along $\mathbf{B}_0$ and B replaced by $B_0(1-\sigma)$ to represent the shielding discussed in Section 1-8. The second term is the spin–spin coupling between different nuclei in the same molecule, discussed in Section 1-11.

The Hamiltonian 2-38 is appropriate for molecules which make frequent random collisions. Isolated molecules in definite rotational and vibrational states (e.g. in a molecular beam), and molecules in solids, have more complex spin Hamiltonians. In a gas at a pressure of about 1 atmosphere or more, molecules frequently collide, changing their vibrational and rotational states and their relative positions in a random way; the observed spin Hamiltonian is the average over this motion. The values of J and σ are the mean values and show a small temperature variation due to changes of population of individual states. For normal (i.e. isotropic) fluids, all terms in the spin Hamiltonian which depend on the relative orientations of the molecule and the applied field average to zero. In a liquid crystal solvent or in an electric field, however, some orientations are more probable than others and additional terms appear. The spin Hamiltonian for a solid also contains orientation-dependent terms, and, in addition, includes interactions between molecules (see Chapter 6).

2-6 The nuclear resonance spectrum of a single nucleus

The simplest type of nuclear resonance spectrum is that of a molecule containing a single magnetic nucleus A. The spin Hamiltonian is

$$h^{-1}\hat{\mathscr{H}} = -\nu_A \hat{I}_{Az} \tag{2-41}$$

where from the general spin Hamiltonian 2-38 the constant ν_A is

$$\nu_A = (2\pi)^{-1}\gamma_A B_0(1-\sigma_A) \tag{2-42}$$

The eigenstates of this Hamiltonian are the functions $|I, m\rangle$ with eigenvalues $-m\nu_A$. The energy-level diagram consists of a series of $2I+1$ equally-spaced levels, as already described in Section 1-5 and shown in Fig. 1-2.

To predict the spectrum, the frequencies and relative probabilities of all possible transitions between states with different m values must be calculated. In the NMR spectrometer a radiofrequency field $2B_1 \cos \omega t$ is applied perpendicular to the static field B_0, say in the x direction. This adds a small time-dependent term to the Hamiltonian

$$h^{-1}\hat{\mathscr{H}}(t) = (2\pi)^{-1}\gamma B_1 \hat{I}_x 2 \cos \omega t = h^{-1}\hat{\mathscr{H}}_1 2 \cos \omega t \tag{2-43}$$

which induces transitions between the states. The resulting probability per unit time of a spin in state $|s\rangle$ changing to state $|r\rangle$ can be found from time-dependent perturbation theory (see Appendix 3). It is

$$(2\pi/\hbar)\, |\langle r|\, \hat{\mathscr{H}}_1\, |s\rangle|^2\, \delta(U_r - U_s - h\nu) \tag{2-44}$$

where $\delta(x)$ is one if x is zero, and zero otherwise. As discussed in Section 1-7, the rate of absorption of energy is proportional to the above transition probability. The factor $\delta(x)$ is, however, to be replaced by a lineshape $g(\nu)$ as discussed in Sections 1-7 and 3-4. Consideration of Eqns 2-43 and 2-44 shows that in a spectrum the intensity for transition $|r\rangle \rightarrow |s\rangle$ is proportional to

$$|\langle r|\, \hat{I}_x\, |s\rangle|^2 \tag{2-45}$$

The selection rules and relative transition probabilities can be most readily derived by rewriting $\hat{I}_x$ in terms of the raising and lowering operators.

$$\langle r|\, \hat{I}_x\, |s\rangle = \tfrac{1}{2}\langle r|\, \hat{I}_+ + \hat{I}_-\, |s\rangle \tag{2-46}$$

As $\hat{I}_+$ and $\hat{I}_-$ raise and lower m by one unit, the selection rule is

$$\Delta m = \pm 1 \tag{2-47}$$

The energy levels are in the order of their m values, with maximum m at minimum energy (if γ is positive). Net absorption of energy occurs when m decreases, so the selection rule is effectively $\Delta m = -1$. For a single nucleus the spectrum consists of a single line at ν_A.

2-7 Spectra of molecules with two coupled nuclei

For the remainder of the chapter attention will be confined to nuclei of spin-$\frac{1}{2}$. Using the notation discussed in Section 1-15 a molecule with two spin-$\frac{1}{2}$ nuclei contains an AB nuclear spin system. The Hamiltonian for this system is

$$h^{-1}\hat{\mathscr{H}}_{AB} = -\nu_A \hat{I}_{Az} - \nu_B \hat{I}_{Bz} + J\hat{\mathbf{I}}_A \cdot \hat{\mathbf{I}}_B \tag{2-48}$$

where ν_A and ν_B are defined by Eqn 2-42. The scalar product of the spin operators can be expanded to give

$$h^{-1}\hat{\mathscr{H}}_{AB} = -\nu_A \hat{I}_{Az} - \nu_B \hat{I}_{Bz} + J\hat{I}_{Az}\hat{I}_{Bz} + \tfrac{1}{2}J(\hat{I}_{A+}\hat{I}_{B-} + \hat{I}_{A-}\hat{I}_{B+}) \tag{2-49}$$

The products $\hat{I}_{A+}\hat{I}_{B-}$ and $\hat{I}_{A-}\hat{I}_{B+}$ are often referred to as flip-flop operators, since they only link states related by raising one spin and lowering the other.

The spin states of the AB system can be written as linear combinations of the simple product states $|\alpha\alpha\rangle$, $|\alpha\beta\rangle$, $|\beta\alpha\rangle$, and $|\beta\beta\rangle$ introduced in Section 1-12. The effect of an A spin operator on one of these states depends on the first symbol, while the B spin operator acts on the second symbol, leaving the first unchanged. For example

$$\hat{I}_{Az}\,|\alpha\beta\rangle = \tfrac{1}{2}\,|\alpha\beta\rangle \tag{2-50}$$

$$\hat{I}_{B+}\,|\alpha\beta\rangle = |\alpha\alpha\rangle \tag{2-51}$$

The problem of obtaining the linear combinations of these basis states which are stationary states is simplified by finding operators which commute with the Hamiltonian, as the stationary states are also eigenstates of any operators which do so (Section 2-3). One such operator is $\hat{F}_z$, the z component of the total spin angular momentum:

$$\hat{F}_z = \hat{I}_{Az} + \hat{I}_{Bz} \tag{2-52}$$

Both $\hat{I}_{Az}$ and $\hat{I}_{Bz}$ commute with the first three terms of Eqn 2-49 as these contain $\hat{I}_z$ operators only. The commutators of $\hat{I}_{Az}$ and $\hat{I}_{Bz}$ with the fourth term are

$$[\hat{I}_{Az}, \hat{I}_{A+}\hat{I}_{B-}] = \hat{I}_{A+}\hat{I}_{B-} \quad \text{and} \quad [\hat{I}_{Bz}, \hat{I}_{A+}\hat{I}_{B-}] = -\hat{I}_{A+}\hat{I}_{B-} \tag{2-53}$$

As these are equal and opposite, the sum, $\hat{F}_z$, commutes with the complete Hamiltonian, and the stationary states are also eigenstates of $\hat{F}_z$ with eigenvalues m_T. The m_T values can be obtained by inspection of the basic product functions (see Table 2-1).

As $\alpha\alpha$ and $\beta\beta$ are the only functions with m_T values of 1 and -1 respectively, they must be stationary states. Their energies are found

Table 2-1 m_T values of product functions for two nuclear spins

Function	$\beta\beta$	$\alpha\beta$	$\beta\alpha$	$\alpha\alpha$
m_T	-1	0	0	1

from the first three terms of the Hamiltonian

$$h^{-1}\hat{\mathcal{H}}\,|\alpha\alpha\rangle=[-\tfrac{1}{2}(\nu_A+\nu_B)+\tfrac{1}{4}J]\,|\alpha\alpha\rangle \tag{2-54}$$

$$h^{-1}\hat{\mathcal{H}}\,|\beta\beta\rangle=[+\tfrac{1}{2}(\nu_A+\nu_B)+\tfrac{1}{4}J]\,|\beta\beta\rangle \tag{2-55}$$

The functions $\alpha\beta$ and $\beta\alpha$ have the same value of m_T. They are not stationary states, as they are mixed by the flip-flop terms of the Hamiltonian 2-49 as follows:

$$h^{-1}\hat{\mathcal{H}}\,|\alpha\beta\rangle=[-\tfrac{1}{2}(\nu_A-\nu_B)-\tfrac{1}{4}J]\,|\alpha\beta\rangle+\tfrac{1}{2}J\,|\beta\alpha\rangle \tag{2-56}$$

$$h^{-1}\hat{\mathcal{H}}\,|\beta\alpha\rangle=[\tfrac{1}{2}(\nu_A-\nu_B)-\tfrac{1}{4}J]\,|\beta\alpha\rangle+\tfrac{1}{2}J\,|\alpha\beta\rangle \tag{2-57}$$

The stationary states are combinations $c_1\,|\alpha\beta\rangle+c_2\,|\beta\alpha\rangle$ which satisfy the equation

$$\hat{\mathcal{H}}(c_1\,|\alpha\beta\rangle+c_2\,|\beta\alpha\rangle)=U(c_1\,|\alpha\beta\rangle+c_2\,|\beta\alpha\rangle) \tag{2-58}$$

If we write

$$\hat{\mathcal{H}}\,|\alpha\beta\rangle=\mathcal{H}_{11}\,|\alpha\beta\rangle+\mathcal{H}_{21}\,|\beta\alpha\rangle, \tag{2-59}$$

then integration with $\langle\alpha\beta|$ and $\langle\beta\alpha|$ shows that H_{11} and H_{21} are the integrals

$$H_{11}=\langle\alpha\beta|\,\hat{\mathcal{H}}\,|\alpha\beta\rangle,\qquad H_{21}=\langle\beta\alpha|\,\hat{\mathcal{H}}\,|\alpha\beta\rangle \tag{2-60}$$

which are often known as diagonal and off-diagonal matrix elements of $\hat{\mathcal{H}}$. Equation 2-58 becomes

$$[H_{11}c_1+H_{12}c_2]\,|\alpha\beta\rangle+[H_{21}c_1+H_{22}c_2]\,|\beta\alpha\rangle - U(c_1\,|\alpha\beta\rangle+c_2\,|\beta\alpha\rangle)=0 \tag{2-61}$$

This equation is satisfied if the coefficients of the functions $|\alpha\beta\rangle$ and $|\beta\alpha\rangle$ are separately zero. This gives simultaneous equations for c_1 and c_2 (the secular equations)

$$\begin{aligned}(H_{11}-U)c_1+H_{12}c_2&=0\\ H_{21}c_1+(H_{22}-U)c_2&=0\end{aligned} \tag{2-62}$$

which can be written in matrix form

$$\begin{bmatrix}H_{11}-U & H_{12}\\ H_{21} & H_{22}-U\end{bmatrix}\begin{bmatrix}c_1\\ c_2\end{bmatrix}=0 \tag{2-63}$$

where the coefficients form what is known as an eigenvector. These equations are only consistent (two equations with one unknown, i.e.

c_1/c_2) if the determinant

$$\begin{vmatrix} H_{11}-U & H_{12} \\ H_{21} & H_{22}-U \end{vmatrix} = 0 \tag{2-64}$$

i.e.

$$(H_{11}-U)(H_{22}-U)-H_{12}H_{21}=0 \tag{2-65}$$

The two values of U for which this secular determinant is zero are the two allowed energy levels, with energies (using $H_{12}=H_{21}$) given by

$$U_1=\tfrac{1}{2}(H_{11}+H_{22})+\tfrac{1}{2}[(H_{11}-H_{22})^2+4H_{12}^2]^{1/2} \tag{2-66}$$

$$U_2=\tfrac{1}{2}(H_{11}+H_{22})-\tfrac{1}{2}[(H_{11}-H_{22})^2+4H_{12}^2]^{1/2} \tag{2-67}$$

The spin functions corresponding to these two values of U are now found by substituting in the secular equations 2-63, and using the subsidiary condition that the function is normalized

$$c_1^2+c_2^2=1 \tag{2-68}$$

The two spin functions are distinguished by an additional subscript. The eigenvector $[c_{11}\ c_{21}]$ satisfies Eqn (2-58) when $U=U_1$, while $[c_{12}\ c_{22}]$ is the solution when $U=U_2$. The normalization condition may be incorporated by defining an angle of θ such that $c_{11}=\cos\theta$, $c_{21}=\sin\theta$. As $[c_{12}c_{22}]$ must be orthogonal to $[c_{11}c_{21}]$, $c_{21}=-c_{12}$; $c_{11}=c_{22}$. The value of θ for the AB system is given by

$$\sin 2\theta = J/D \tag{2-69}$$

$$\cos 2\theta = (\nu_A-\nu_B)/D \tag{2-70}$$

$$D=[(\nu_B-\nu_A)^2+J^2]^{1/2} \tag{2-71}$$

Table 2-2 shows the spin functions and energies of the stationary states of the AB system.

One extreme of the AB system is the AX system where $|\nu_A-\nu_B|$ is very large compared to $|J|$. It can be seen from Eqn 2-69 to 2-71 that this corresponds to $\theta=0$. In this case the functions $\alpha\beta$ and $\beta\alpha$ are not mixed, as the last term of the Hamiltonian is negligible compared to the difference $H_{11}-H_{22}$ of the diagonal elements. The flip-flop terms can therefore be omitted from the calculation of an AX spectrum and of all first-order spectra (see Section 1-13).

Table 2-2 Stationary states of the AB spin system

Level number[a]	*Spin function*	*Energy*[b]
4	$\beta\beta$	$\frac{1}{2}(\nu_A+\nu_B)+\frac{1}{4}J$
3	$\cos\theta\,\alpha\beta+\sin\theta\,\beta\alpha$	$\frac{1}{2}D-\frac{1}{4}J$
2	$-\sin\theta\,\alpha\beta+\cos\theta\,\beta\alpha$	$-\frac{1}{2}D-\frac{1}{4}J$
1	$\alpha\alpha$	$-\frac{1}{2}(\nu_A+\nu_B)+\frac{1}{4}J$

[a]Arbitrary.
[b]In frequency units.

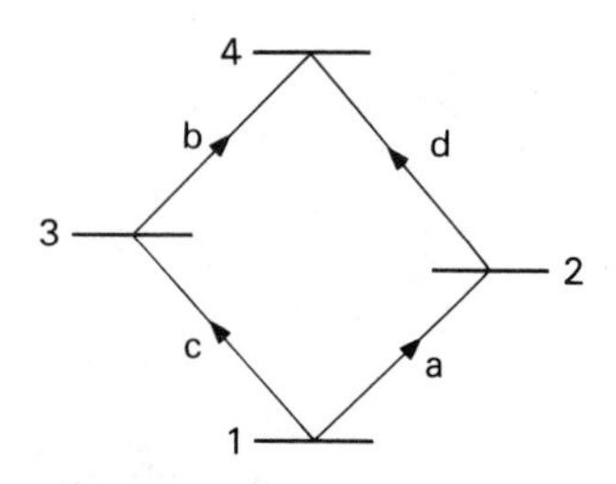

Fig. 2-1 Energy level diagram of an AB spin system. See Tables 2-2 and 2-3 for the labelling. It is assumed that $\nu_B > \nu_A$.

As $|J|$ increases, θ increases and the functions are mixed. The energy levels are separated by more than if there were no mixing. The other extreme of the AB system is reached when $\nu_A = \nu_B$, the A_2 system. Now θ is 45° and there is maximum mixing to give one state symmetric with respect to exchange of nuclei, $2^{-1/2}(\alpha\beta + \beta\alpha)$, with energy $\frac{1}{4}J$ and an anti-symmetric state $2^{-1/2}(\alpha\beta - \beta\alpha)$ with energy $-\frac{3}{4}J$ (Problem 2-11).

For like nuclides (i.e. those with the same magnetogyric ratio) the relative intensities of the transitions in the spectrum are given by (cf. Eqn 2-47)

$$|\langle r| \hat{F}_{-} |s\rangle|^2 \tag{2-72}$$

where $\hat{F}_{-}(=\hat{I}_{A-} + \hat{I}_{B-})$ lowers the value of m_T by one unit, giving a selection rule

$$\Delta m_T = -1 \tag{2-73}$$

The AB system therefore has four allowed transitions which are shown in the energy-level diagram (Fig. 2-1). Their relative intensities are found by operating with $\hat{F}_{-}$ on the lower state. For example

$$\hat{F}_{-} |3\rangle = \cos\theta\, \hat{F}_{-} |\alpha\beta\rangle + \sin\theta\, \hat{F}_{-} |\beta\alpha\rangle \tag{2-74}$$

but

$$\begin{aligned} \hat{F}_{-} |\alpha\beta\rangle &= \hat{I}_{A-} |\alpha\beta\rangle + \hat{I}_{B-} |\beta\alpha\rangle \\ &= |\beta\beta\rangle + 0 \end{aligned} \tag{2-75}$$

so

$$\langle 4| \hat{F}_{-} |3\rangle = (\cos\theta + \sin\theta) \tag{2-76}$$

and the intensity of this transition is proportional to $(1 + \sin 2\theta)$.

Table 2-3 shows the transition frequencies and intensities of the AB system. Figure 2-2 gives an example, and Fig. 2-3 shows the way in which the AB spectrum varies as θ increases. When $J/(\nu_B - \nu_A)$ is zero (the AX limit), the spectrum consists of four lines of equal intensity. Lines a and b arise from transitions in which the spin of the A nucleus changes and that of the X nucleus is unaltered. These can be described accurately as A lines, and are centred on ν_A and split by $|J|$. This is the first-order situation described in Chapter 1.

Table 2-3 NMR spectrum of the AB system

Transition	*Frequency*	*Relative intensity*
d $4 \leftarrow 2$	$\frac{1}{2}(\nu_A + \nu_B) + \frac{1}{2}J + \frac{1}{2}D$	$1 - \sin 2\theta$
b $4 \leftarrow 3$	$\frac{1}{2}(\nu_A + \nu_B) + \frac{1}{2}J - \frac{1}{2}D$	$1 + \sin 2\theta$
c $3 \leftarrow 1$	$\frac{1}{2}(\nu_A + \nu_B) - \frac{1}{2}J + \frac{1}{2}D$	$1 + \sin 2\theta$
a $2 \leftarrow 1$	$\frac{1}{2}(\nu_A + \nu_B) - \frac{1}{2}J - \frac{1}{2}D$	$1 - \sin 2\theta$

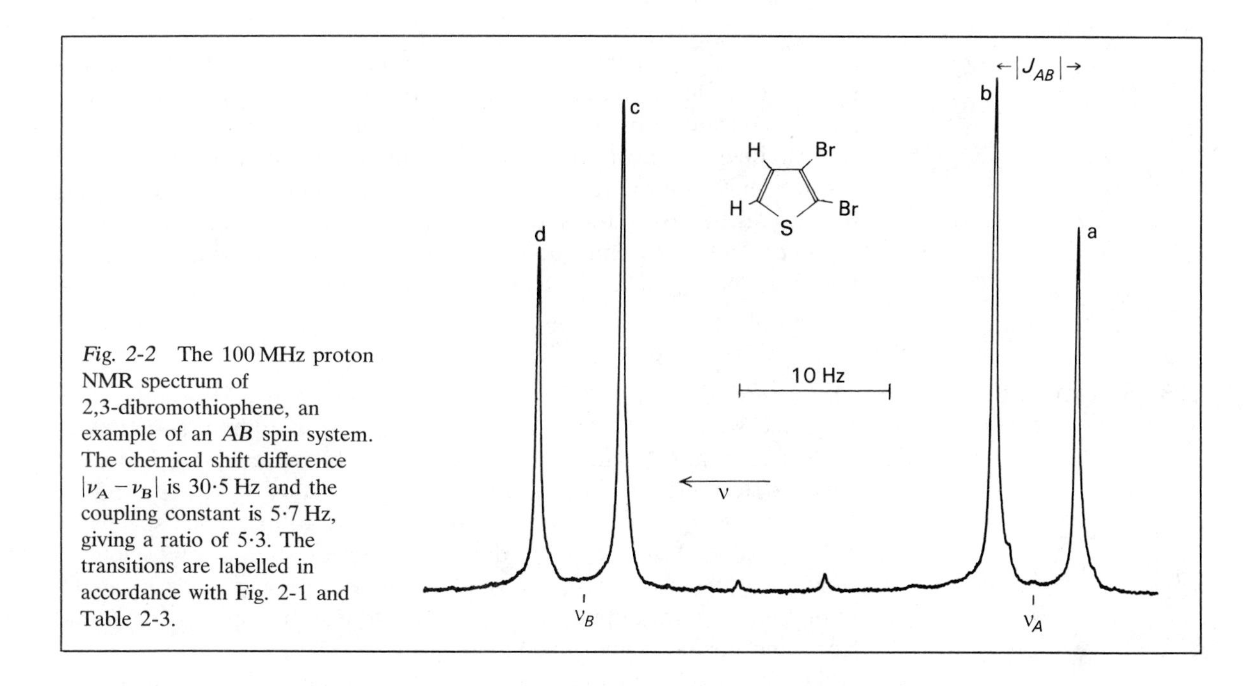

Fig. 2-2 The 100 MHz proton NMR spectrum of 2,3-dibromothiophene, an example of an *AB* spin system. The chemical shift difference $|\nu_A - \nu_B|$ is 30·5 Hz and the coupling constant is 5·7 Hz, giving a ratio of 5·3. The transitions are labelled in accordance with Fig. 2-1 and Table 2-3.

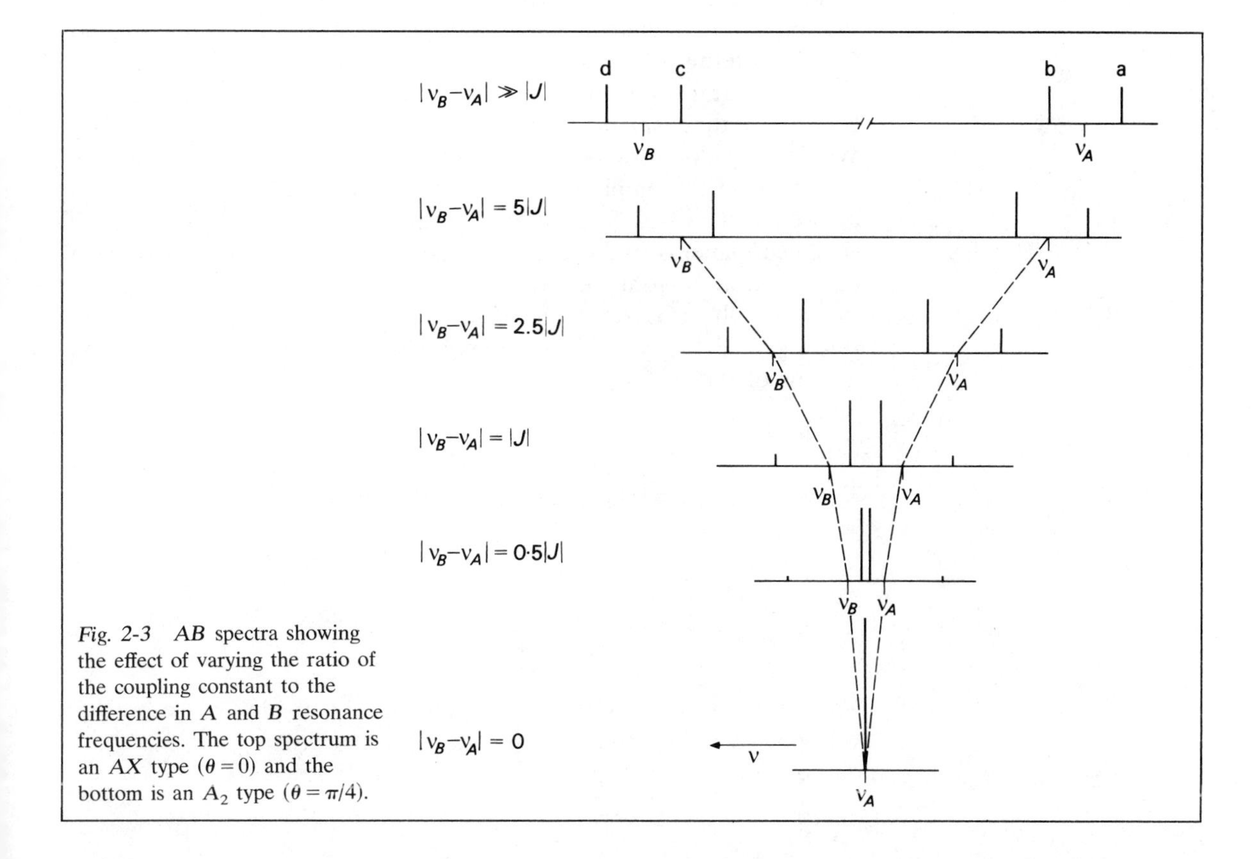

Fig. 2-3 *AB* spectra showing the effect of varying the ratio of the coupling constant to the difference in *A* and *B* resonance frequencies. The top spectrum is an *AX* type ($\theta = 0$) and the bottom is an A_2 type ($\theta = \pi/4$).

As $|J/(\nu_B - \nu_A)|$ increases, the states are mixed and the line intensities are no longer equal. Furthermore, lines c and d are no longer strictly A transitions. In this system, J is given exactly by the separation of lines c and d (or a and b), but this is not generally true in systems with more nuclei. The most convenient way of obtaining $|\nu_B - \nu_A|$ is by taking the square root of the product of line separations $(ad)(bc)$. In the A_2 limit the inner lines coalesce at ν_A, while the outer lines (a and d) have zero intensity. Thus in this limit J cannot be measured.

2-8 Systems with more than two nuclei

The spectra of systems with more than two nuclei become increasingly complex, with more transitions involving more parameters—a system of N nuclei has N chemical shifts and $\frac{1}{2}N(N-1)$ coupling constants which determine the line positions and intensities of the transitions. The relative signs of coupling constants may also affect the spectrum, although the 'absolute' signs do not. It is, for example, possible to show that the *cis* and *trans* couplings in vinyl compounds have the same sign, but whether this is positive or negative cannot be found from line positions and intensities.

Although the algebraic and numerical manipulation becomes more tedious as the number of nuclei increases, the method of calculation of the spectrum remains the same. Eigenfunctions of the Hamiltonian are found in terms of a basis set of functions—usually simple product functions. Transition frequencies are then given by the differences in energies of these spin states and intensities determined from Eqn 2-45. We now develop formal expressions for these processes for N nuclei.

The set of 2^N simple product functions such as $|\alpha_1\alpha_2\beta_3 \ldots \alpha_N\rangle$ can be used as the basis. These are eigenfunctions of $\hat{F}_z\,(\hat{\mathbf{F}} = \sum_j \hat{\mathbf{I}}_j)$ which, by similar arguments to those used in the previous section for two nuclei, can be shown to commute with $\hat{\mathcal{H}}$. If $|\phi_1\rangle \ldots |\phi_n\rangle$ are the set of product functions which have the eigenvalue m_T for $\hat{F}_z$, then the n stationary states $|\psi_1\rangle \ldots |\psi_n\rangle$ which correspond to the same m_T are linear combinations of these $|\phi_j\rangle$.

$$|\psi_k\rangle = \sum_j |\phi_j\rangle c_{jk} \tag{2-77}$$

The stationary states $|\psi_k\rangle$ are eigenstates of $\hat{\mathcal{H}}$

$$\hat{\mathcal{H}}\,|\psi_k\rangle = U_k\,|\psi_k\rangle \tag{2.78}$$

which becomes

$$\sum_j \hat{\mathcal{H}}\,|\phi_j\rangle\, c_{jk} = \sum_j U_k\,|\phi_j\rangle c_{jk} \tag{2-79}$$

when written in terms of the basis functions. Completing the brackets with each $\langle\phi_\ell|$ in turn we obtain equations such as

$$\sum_j \langle\phi_\ell|\,\mathcal{H}\,|\phi_j\rangle c_{jk} = U_k c_{\ell k} \tag{2-80}$$

These are the secular equations which can be written in matrix form

$$\begin{bmatrix} H_{11}-U & H_{12} & \ldots \\ H_{21} & H_{22}-U & \\ \cdot & & \\ \cdot & & \\ \cdot & & \end{bmatrix} \begin{bmatrix} c_1 \\ c_2 \\ \cdot \\ \cdot \\ \cdot \end{bmatrix} = 0 \tag{2-81}$$

where we have dropped the subscript k. They are consistent only if the secular determinant

$$\begin{vmatrix} H_{11}-U & H_{12} & \ldots \\ H_{21} & H_{22}-U & \\ \cdot & & \\ \cdot & & \\ \cdot & & \end{vmatrix} = 0; \tag{2-82}$$

solving this nth order equation for U gives n possible values of U and n spin eigenfunctions. The process is repeated for every value of m_T.

This process of solving a set of secular equations is often referred to as 'diagonalizing the Hamiltonian'. If a complete set of functions $(|\phi_1\rangle \ldots |\phi_n\rangle)$ for a problem is given, then all the information about the Hamiltonian is contained in the matrix

$$\begin{bmatrix} H_{11} & H_{12} & \ldots \\ H_{21} & H_{22} & \\ \cdot & & \\ \cdot & & \\ \cdot & & \end{bmatrix} \tag{2-83}$$

which is said to represent the Hamiltonian in this basis. If another basis is chosen, then the form of the matrix changes; if eigenstates $(|\psi_1\rangle \ldots |\psi_n\rangle)$ are chosen as basis there are no off-diagonal elements

$$\begin{bmatrix} H_{11} & 0 & 0 & \ldots \\ 0 & H_{22} & 0 & \\ 0 & 0 & H_{33} & \ldots \\ \cdot & & & \\ \cdot & & & \\ \cdot & & & \end{bmatrix} \tag{2-84}$$

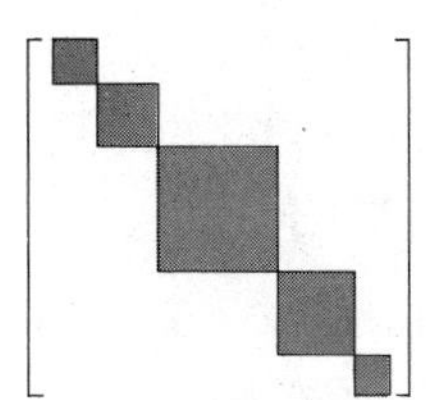

Fig. 2-4 Form of Hamiltonian matrix when factorized into sub-matrices corresponding to different values of m_T.

and the Hamiltonian is diagonal. Diagonalizing the Hamiltonian is not always easy. If the basis is the set of simple product functions, the Hamiltonian matrix consists of a number of blocks or sub-matrices, each of which corresponds to a different value of m_T (Fig. 2-4). If any of these blocks is bigger than 2×2, it cannot normally be diagonalized explicitly in terms of the chemical shifts and coupling constants. Instead, numerical values of the constants are inserted and the eigenvalues found by an iterative process. This is tedious, but can conveniently be performed by a computer (see Section 2-12).

In systems where some nuclei are equivalent it is possible to reduce the size of the sub-matrices further. If a group of nuclei are magnetically equivalent, they can be replaced by a single 'composite particle', as described in Section 2-10. As chemically equivalent nuclei are related by symmetry operations of the point group of the molecule, group theoretical methods can be used to construct a basis of symmetry functions in a similar way to that used in constructing vibrational normal modes of molecules or molecular orbitals. Further details are given in the books listed at the end of this chapter as 'Further Reading'.

Once the Hamiltonian has been diagonalized and the energies and spin functions of the stationary states are known, the intensities of the transitions must be calculated. If the nuclei all have the same magnetogyric ratios, the relative intensities are given by the values of expression 2-72. Nuclei with different magnetogyric ratios are discussed in the next section.

2-9 The *X* approximation

(a) General principles The resonance frequencies of nuclei with different magnetogyric ratios, such as ^{13}C and ^{31}P, always differ by a large amount compared to the coupling between them. It will be shown in this section that transitions in such systems can correctly be described as ^{13}C or ^{31}P transitions. In such cases, the X approximation discussed in this section aids the diagonalization of the Hamiltonian.

The AB Hamiltonian discussed in Section 2-7 was simplified for an AX system [in which $|J| \ll |\nu_X - \nu_A|$], by omitting the last term to give

$$h^{-1}\hat{\mathscr{H}}_{AX} = \nu_A \hat{I}_{Az} - \nu_X \hat{I}_{Xz} + J\hat{I}_{Az}\hat{I}_{Xz}. \tag{2-85}$$

A similar approximation can be made in any system containing spin-$\frac{1}{2}$ nuclei of more than one element or isotope. Coupling between such different types of nuclei only effectively contributes terms like

$$J_{AX}\hat{I}_{Az}\hat{I}_{Xz} \tag{2-86}$$

to the Hamiltonian, while coupling between nuclei of the same type gives the complete term

$$J_{AB}[\hat{I}_{Az}\hat{I}_{Bz} + \tfrac{1}{2}(\hat{I}_{A+}\hat{I}_{B-} + \hat{I}_{A}\hat{I}_{B+})] \tag{2-87}$$

As the simplified Hamiltonian commutes with $\hat{F}_{Az} (= \sum_{A_j} \hat{I}_{jz})$ and $\hat{F}_{Xz}$, the stationary states have definite values of both $m_T(A)$ and $m_T(X)$, the total A and X spins. The simple product functions can be classified according to the values of $m_T(A)$ and $m_T(X)$ and mix only if both $m_T(A)$ and $m_T(X)$ are the same. This leads to additional factorization of the Hamiltonian matrix. The intensity of a transition $r \rightarrow s$ depends on

$$|\langle r| \, \gamma_A \hat{F}_{A-} + \gamma_X \hat{F}_{X-} \, |s\rangle|^2 \tag{2-88}$$

As the states $|m\rangle$ and $|n\rangle$ have definite values of $m_T(A)$ and $m_T(X)$, and as $\hat{F}_{A-}$ reduces $m_T(A)$ by one unit, the selection rules are

$$\begin{array}{lll} \text{either} & \Delta m_T(A) = -1, & \Delta m_T(X) = 0 \quad (A \text{ transition}) \\ \text{or} & \Delta m_T(X) = -1, & \Delta m_T(A) = 0 \quad (X \text{ transition}) \end{array} \tag{2-89}$$

so that all transitions belong to a given type of nucleus. The relative intensities of the A lines depend on

$$|\langle r| \hat{F}_{A-} |s\rangle|^2 \tag{2-90}$$

The X approximation is not confined to different nuclides, but can be applied to any groups of nuclei whose relative chemical shifts much exceed their mutual coupling.

(b) Systems with one X-nucleus The spectra of such systems are of considerable practical importance, because of the existence of such dilute spins as ^{13}C. Organic molecules with ^{13}C in natural abundance automatically fall into the category of spin systems discussed in this sub-section. There are, of course, many other molecules with one X nucleus, such as vinyl fluoride and dimethylphosphine. The spectra of such systems can be analysed in terms of sub-spectra of familiar types. This concept can best be illustrated by the ABX system, the AB portion of which consists of two ab sub-spectra (Fig. 2-5).

The ABX spin Hamiltonian is

$$h^{-1}\hat{\mathcal{H}}_{ABX} = -\nu_A \hat{I}_{Az} - \nu_B \hat{I}_{Bz} + J_{AB} \hat{\mathbf{I}}_A \cdot \hat{\mathbf{I}}_B - \nu_X \hat{I}_{Xz} + J_{AX} \hat{I}_{Az} \hat{I}_{Xz} + J_{BX} \hat{I}_{Bz} \hat{I}_{Xz} \tag{2-91}$$

Fig. 2-5 The 24·29 MHz ^{31}P NMR spectrum of the diphosphite anion. This illustrates the AB part of an ABX spin system. The two ab sub-spectra are indicated.

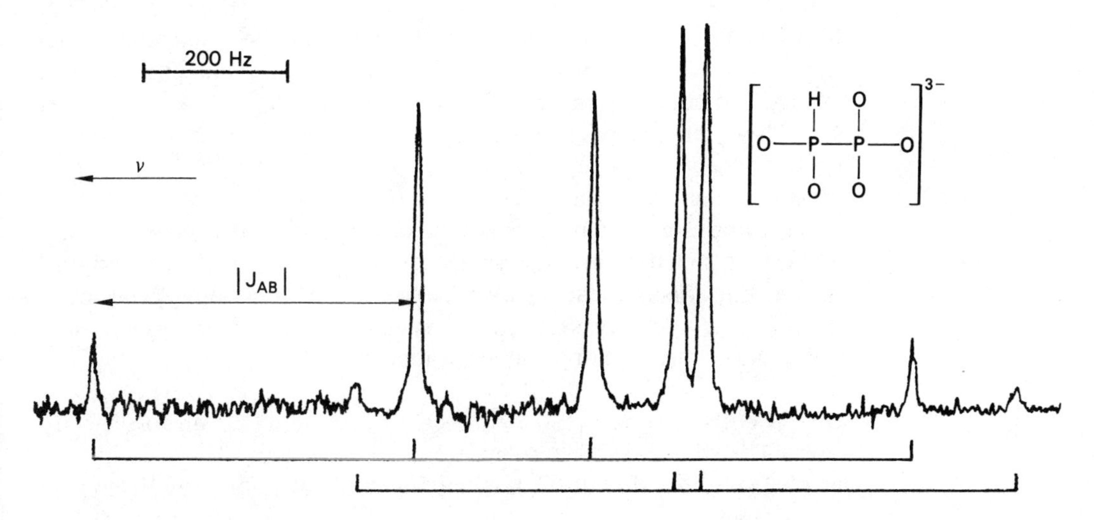

This commutes with $\hat{I}_{Xz}$ and $\hat{F}_{ABz}$ $(=\hat{I}_{Az}+\hat{I}_{Bz})$. The part of the spectrum involving transitions of the A and B nuclei has the selection rule

$$\Delta m_{AB} = -1, \qquad \Delta m_X = 0 \tag{2-92}$$

so the transitions can be divided into two sets, those between states in which $m_X = +\frac{1}{2}$ (α transitions) and those for which $m_X = -\frac{1}{2}$ (β transitions). The molecules in which the X spin is the state $|\alpha\rangle$ will have a Hamiltonian

$$h^{-1}\hat{\mathscr{H}}^{X\alpha}_{AB} = -(\nu_A - \tfrac{1}{2}J_{AX})\mathbf{I}_{Az} - (\nu_B - \tfrac{1}{2}J_{BX})\hat{I}_{Bz} + J_{AB}\mathbf{I}_A \cdot \mathbf{I}_B - \tfrac{1}{2}\nu_X \tag{2-93}$$

As this is like an AB Hamiltonian with ν_A replaced by $(\nu_A - \frac{1}{2}J_{AX})$, ν_B by $(\nu_B - \frac{1}{2}J_{BX})$ and with a constant term added, the α transitions form an *ab* sub-spectrum which is identical to an AB spectrum with *effective* chemical shifts $\nu_A - \frac{1}{2}J_{AX}$ and $\nu_B - \frac{1}{2}J_{BX}$. Similarly, the β transitions form an *ab* sub-spectrum with chemical shifts $\nu_A + \frac{1}{2}J_{AX}$ and $\nu_B + \frac{1}{2}J_{BX}$. Small letters are used to designate sub-spectral types.

The X part of an ABX spectrum is a little more difficult to discuss, but half the intensity lies in two lines, symmetrically placed about ν_X and with separation $|J_{AX} + J_{BX}|$. These arise from X transitions for spin states with the A and B nuclei both α on the one hand and both β on the other. Such states are not mixed with any others because they correspond to unique values of $m_A + m_B$. In addition to the intense doublet, there are four lines symmetrically placed in pairs about ν_X, with separations $|D_+ + D_-|$ and $|D_+ - D_-|$ and intensities $\frac{1}{2}[1 \mp \cos(\phi_+ - \phi_-)]$ respectively, where

$$D_\pm = \tfrac{1}{2}[\{(\nu_A - \nu_B) \pm \tfrac{1}{2}(J_{AX} - J_{BX})\}^2 + J_{AB}^2]^{1/2} \tag{2-94}$$

$$\tan \phi_\pm = J_{AB}/[(\nu_A - \nu_B) \pm (J_{AX} - J_{BX})] \tag{2-95}$$

Proofs of these equations are given in the suggestions for Further Reading at the end of this chapter. It should be commented that assignment of the lines of an ABX pattern can be ambiguous, but if carried out correctly the analysis yields the relative signs of J_{AX} and J_{BX}. However, the relative sign of J_{AB} cannot be obtained by simple spectral analysis.

Any system $ABC\ldots NX$ with one X nucleus has two *abc–n* sub-spectra in the non-X regions corresponding to the two X spin states. These sub-spectra have the same coupling constants as the original system but effective chemical shifts $\nu_j \pm \frac{1}{2}J_{jX}$. This simplifies the interpretation of such spectra. The X spectra of such spin systems are, however, generally complicated in appearance.

(c) *Satellite spectra* As mentioned in Section 1-17, proton spectra contain weak peaks due to molecules containing ^{13}C. These peaks involve splittings due to (C, H) coupling, and inevitably arise from spin systems of the type discussed in this section, so that the X approximation may be used. If a molecule contains n non-equivalent groups of carbons, the ^1H spectrum will contain satellite spectra due to n different but related spin systems. The ^1H chemical shifts will be similar but not identical to those for the per-^{13}C isotopomer, since there are small isotope effects on shielding. The (H, H) coupling constants are invariably within experimental error of those for the per-^{12}C isotopomer, since isotopic effects on coupling constants are usually negligible. Proton spectral analysis can yield (C, H) coupling constants and isotopic effects on chemical shifts, but the procedure is usually hampered by the presence of intense lines due to the per-^{12}C-isotopomer (but see Section 7-4(c)). However, since values of $^1J_{CH}$ are relatively large (usually ca. 125–150 Hz for saturated carbons), peaks due to protons directly bonded to ^{13}C are frequently visible (see Fig. 1-19).

For molecules containing equivalent hydrogens, extra splittings may be visible in the ^{13}C satellites since the presence of the ^{13}C effectively lifts the equivalence. Consider, for example, the molecule 1,1,2,2-tetrabromoethane, $(CHBr_2)_2$, used for Fig. 1-19. For the per-^{12}C isotopomer, the protons are equivalent, the spin system is therefore A_2, and the spectrum consists of a single line, giving no information about J_{HH}. However 2.2% of the molecules are $Br_2H^{13}C.^{12}CHBr_2$, which contains an ABX spin system (with $\nu_A \sim \nu_B$). Since $|^1J_{CH}| \gg |^2J_{CH}|$ and $|^3J_{HH}|$, the satellites are ab sub-spectra with substantial values for $|\nu_b - \nu_a|$. They therefore appear first order, and consist of four lines, with splittings giving $|^3J_{HH}|$ directly, as shown clearly in Fig. 1-19.

On occasion satellite spectra may be second order even when the spectrum of the per-^{12}C isotopomer is first order, since it is the *effective* chemical shifts that determine the former.

(d) Coupled carbon-13 spectra When proton-decoupling is not used, ^{13}C spectra of even quite small organic molecules may be complicated. In principle, when there are n non-equivalent carbons, the spectrum is actually the superposition of the X parts of n $A_aB_bC_c \ldots X$ spin systems, where ABC, etc. refer to the protons. If the proton spectrum of a given isotopomer is first order, then the corresponding ^{13}C spectrum will be also. Second-order characteristics in the ^1H spectrum inevitably lead to second-order features in the ^{13}C spectrum also, so that interpretation of coupled ^{13}C spectra can be difficult. Figure 2-6 shows part of a ^{13}C spectrum exhibiting second-order features.

2-10 Equivalent nuclei

In Section 1-16, it was stated that any coupling within a set of magnetically equivalent nuclei does not affect the spectrum. This is most easily understood by considering a set of *completely* equivalent nuclei, which have the same chemical shift and the same coupling to any nucleus in the molecule. In addition to satisfying the criteria for

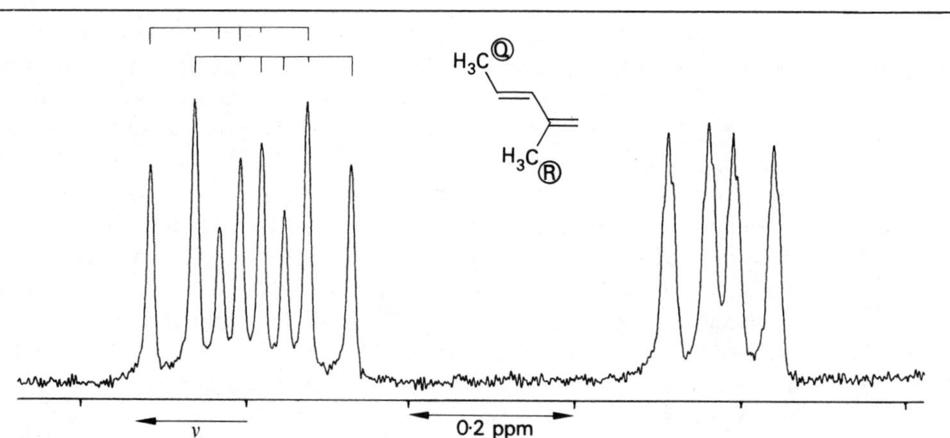

Fig. 2-6 Part of the methyl region of the proton-coupled ^{13}C spectrum of *trans*-2-methyl-1,3-pentadiene at 100 MHz. Only the lowest-frequency bands of the 1:3:3:1 quartets caused by coupling to the directly-bonded hydrogens are shown. Direct coupling between the methyl groups is negligible. The band due to the methyl marked *Q* is effectively first order, split by coupling to H^5 and H^6 with a hint of further coupling. That due to methyl R is definitely second order because of the tight coupling between H^3 and H^4; it can be interpreted (as indicated) as the *x* part of two *abx* sub-spectra separated by $|J_{15}| = 4{\cdot}88$ Hz.

magnetic equivalence, all coupling constants between nuclei of the equivalent set are equal. For example, the three protons of a freely rotating methyl group are *completely* equivalent, as the coupling is the same between any pair of protons, but the six methyl protons of $(CH_3)_2PH$ are only *magnetically* equivalent, as the coupling within the methyl group is different from the coupling between methyl groups.

The Hamiltonian for a molecule containing a set of completely equivalent A nuclei labelled $(1 \ldots j \ldots p)$ and one other nucleus B (an $[A_pB]$ system) is

This can be rewritten in terms of the total spin operator of the A nuclei $\hat{\mathbf{F}}_A$ defined as

$$\hat{\mathbf{F}}_A = \sum_j \hat{\mathbf{I}}_j \tag{2-97}$$

using

$$\hat{\mathbf{F}}_A^2 = \sum_j \hat{\mathbf{I}}_j^2 + 2\sum_{j<k} \hat{\mathbf{I}}_j \cdot \hat{\mathbf{I}}_k \tag{2-98}$$

to give

$$h^{-1}\hat{\mathcal{H}} = -\nu \hat{F}_{Az} - \nu_B \hat{I}_{Bz} + \tfrac{1}{2}J_{AA}(\hat{\mathbf{F}}_A^2 - \sum_j \hat{\mathbf{I}}_j^2) + J_{AB}\mathbf{F}_A \cdot \hat{\mathbf{I}}_B \tag{2-99}$$

As $\hat{\mathbf{F}}_A$ is an angular momentum operator it obeys all the usual commutation relations; $\hat{\mathbf{F}}_A^2$ commutes with $\hat{F}_{Az}$, $\hat{F}_{A+}$, and $\hat{F}_{A-}$; it also commutes with $\hat{\mathbf{I}}_B$ as the latter acts on different coordinates. Hence it commutes with $\hat{\mathcal{H}}$ and the stationary states are eigenfunctions of $\hat{\mathbf{F}}_A^2$ with eigenvalues of the form $[A(A+1)]$. The energy can be written

$$h^{-1}U = h^{-1}U'(\nu_A, \nu_B, J_{AB}) + \tfrac{1}{2}J_{AA}[A(A+1) - \tfrac{3}{4}p] \tag{2-100}$$

The intensity of a transition from state r to state s depends on

$$|\langle r|\, \hat{F}_- |s\rangle|^2 = |\langle r|\, \hat{F}_{A-} + \hat{I}_{B-}\, |s\rangle|^2 \tag{2-101}$$

As $\hat{F}_-$ commutes with $\hat{\mathbf{F}}_A^2$, this integral only connects states with the same eigenvalue of $\hat{\mathbf{F}}_A^2$. The frequency of an allowed transition is therefore

$$\nu_{rs} = h^{-1}(U'_r - U'_s) \tag{2-102}$$

which does not depend on J_{AA}. Thus J_{AA} does not affect the spectrum. This proof can easily be extended to a system with more nuclei (e.g. A_pBC).

The proof for magnetically equivalent nuclei follows the same lines. The complete Hamiltonian is written

$$\hat{\mathcal{H}} = \hat{\mathcal{H}}_1(\nu_A, \nu_B, J_{AB}) + \hat{\mathcal{H}}_2(J_{AA}) \tag{2-103}$$

and the two parts are shown to commute; $\hat{F}_+$ commutes with the second term which therefore cannot affect the spectrum (Problem 2-15).

For first-order spectra of the A_pX type the effective energy equation may be derived from Eqns 2-99 and 2-100 as

$$h^{-1}U' = -\nu_A m_T(A) - \nu_X m_X + J_{AX} m_T(A) m_X \tag{2-104}$$

and it can be seen that this leads to the features discussed in Section 1-16. The relative intensities of the X lines depend on the number of A states with each value of $m_T(A)$, and these are the binomial coefficients ${}^pC_m$.

2-11 The composite particle method for systems containing magnetically equivalent nuclei

As the coupling within a set of magnetically equivalent nuclei does not affect the spectrum, it can be omitted from the calculation. The remaining terms in the spin Hamiltonian involve only the spin operators for the total spin of the group and not the individual spins. The spectrum of the $[A_pB]$ system, for example, can be calculated

Table 2-4 Composite particle states for *p* equivalent nuclei

F	*p* = 1	2	3	4	5	6
2	—	S	—	2S	—	5S
$\frac{1}{2}$	D	—	2D	—	5D	—
1		T	—	3T	—	9T
$\frac{3}{2}$			Q	—	4Q	—
2				Qt	—	5Qt
$\frac{5}{2}$					Sx	—
3						Sp

from the effective Hamiltonian

$$h^{-1}\hat{\mathscr{H}} = -\nu_A \hat{F}_{Az} - \nu_B \hat{I}_{Bz} + J_{AB} \hat{\mathbf{F}}_A \cdot \hat{\mathbf{I}}_B \tag{2-105}$$

As this Hamiltonian does not contain the individual A nuclear spin operators, but only $\hat{\mathbf{F}}_A$, the A_p group can be replaced by a composite particle $\mathscr{A}$ with spin states that are eigenfunctions for $\hat{\mathbf{F}}_A^2$ and $\hat{F}_{Az}$. There is one important difference between an ordinary nucleus and a composite particle, which is that several possible eigenvalues of $\hat{\mathbf{F}}_A^2$ must be considered. Thus the $[A_pB]$ spectrum is a superposition of $\mathscr{A}B$ sub-spectra with different values of the spin of the composite particle $\mathscr{A}$.

The spin states of a composite particle are the states $|F_A, m_A\rangle$. These are used as the basis for setting up the Hamiltonian matrix. Possible values of F are found by coupling the p nuclear spins together in the same way that electronic angular momenta are coupled to find total L and S values in the Russell–Saunders scheme for many-electron atoms (see Problem 2-16). As some values of F can be obtained in several different ways, the corresponding composite particle has a degeneracy.

Table 2-4 shows the F values allowed for different numbers of magnetically equivalent spin-$\frac{1}{2}$ nuclei and their degeneracies. The letters used for the various F values are derived from singlet (S), doublet (D), triplet (T) etc. A singlet state with $F = 0$ has no spectrum and shows no coupling to other nuclei. It should also be noted that the D_AT_B sub-spectra of $[AB_2]$, $[AB_4]$, $[AB_6]$ etc. systems are identical, but they provide a different fraction of the total intensity. The correct calculated intensity of each line in a sub-spectrum is obtained only after multiplication by the degeneracies appropriate to the spin system.

To demonstrate the application of this method the form of the $[A_2B]$ spectrum will be calculated. The $[A_2]$ group can be replaced by composite particles T_A and S_A, both of which are non-degenerate. The S_AD_B sub-spectrum is a b spectrum, as a composite particle of spin-$\frac{1}{2}$ behaves in the same way as a single nucleus of spin-$\frac{1}{2}$. The T_AD_B sub-spectrum is similar to the spectrum of two nuclei with spin-1 and -$\frac{1}{2}$. Table 2-5 shows the matrix elements of the spin Hamiltonian for the

Table 2-5 Hamiltonian matrix elements for the T_AD_B states of the $[A_2B]$ spin system

Basis state				
number[(a)]	*function*[(b)]	m_T	$h^{-1}\mathcal{H}_{jj}$	$h^{-1}\mathcal{H}_{jk}$
1	-1β	$-\frac{3}{2}$	$\nu_A+\frac{1}{2}\nu_B+\frac{1}{2}J_{AB}$	—
2	-1α	$-\frac{1}{2}$	$\nu_A-\frac{1}{2}\nu_B-\frac{1}{2}J_{AB}$	$J_{AB}/\sqrt{2}$
3	0β	$-\frac{1}{2}$	$\frac{1}{2}\nu_B$	
4	0α	$\frac{1}{2}$	$-\frac{1}{2}\nu_B$	$J_{AB}/\sqrt{2}$
5	1β	$\frac{1}{2}$	$-\nu_A+\frac{1}{2}\nu_B-\frac{1}{2}J_{AB}$	
6	1α	$\frac{3}{2}$	$-\nu_A-\frac{1}{2}\nu_B+\frac{1}{2}J_{AB}$	—

[(a)]Arbitrary.
[(b)]The three T states $|1,-1\rangle$, $|1,0\rangle$ and $|1,1\rangle$ are here denoted by their m values.

T_AD_B states. The off-diagonal elements come from the term

$$\tfrac{1}{2}J_{AB}(\hat{F}_{A+}\hat{I}_{B-}+\hat{F}_{A-}\hat{I}_{B+}) \qquad (2\text{-}106)$$

$\hat{F}_{A+}$ raises the value of m_A by one unit, but also gives a factor $[F_A(F_A+1)-m_A(m_A+1)]^{1/2}$. The intensities of the allowed transitions are given by $|\langle r|\,\hat{F}_{A-}+\hat{I}_{B-}\,|s\rangle|^2$.

Table 2-5 shows that the maximum size of Hamiltonian sub-matrix that needs to be diagonalized for the A_2B spin system is 2×2 (as is so for the general A_pB system), so solution of the problem can be carried out algebraically as in Section 2-7. Details can be found in the texts mentioned for further reading at the end of the chapter. There is, of course, only one transition from the S_AD_B state, and that occurs precisely at ν_B. It can be seen from Table 2-5 (bearing in mind the selection rule $\Delta m_T=\pm1$) that there are eight possible transitions for the T_AD_B state, making nine altogether. If these are numbered in sequence in the spectrum, from high to low frequency if $\nu_B>\nu_A$, then line 3 is that arising from the S_AD_B state. The mean position of lines 5 and 7 gives ν_A, and the coupling constant can be obtained from the expression

$$3\,|J_{AB}|=2(\nu_4-\nu_1)-(\nu_7-\nu_5) \qquad (2\text{-}107)$$

Thus all the parameters governing the spin system may be obtained by direct measurement. Line 9 is generally too weak to observe. Figure 2-7 gives a typical example of an A_2B spectrum. In the first order A_2X limit lines 2 and 3 merge to become the centre line of the X triplet, while the pairs of lines 5 and 6 on the one hand and 7 and 8 on the other correlate with the two A lines.

2-12 Analysis of observed spectra

In order to determine chemical shifts and coupling constants from an observed spectrum, the first step is to determine the type of spectrum (*ABC*, *AX* etc.) This is normally known from the molecular structure.

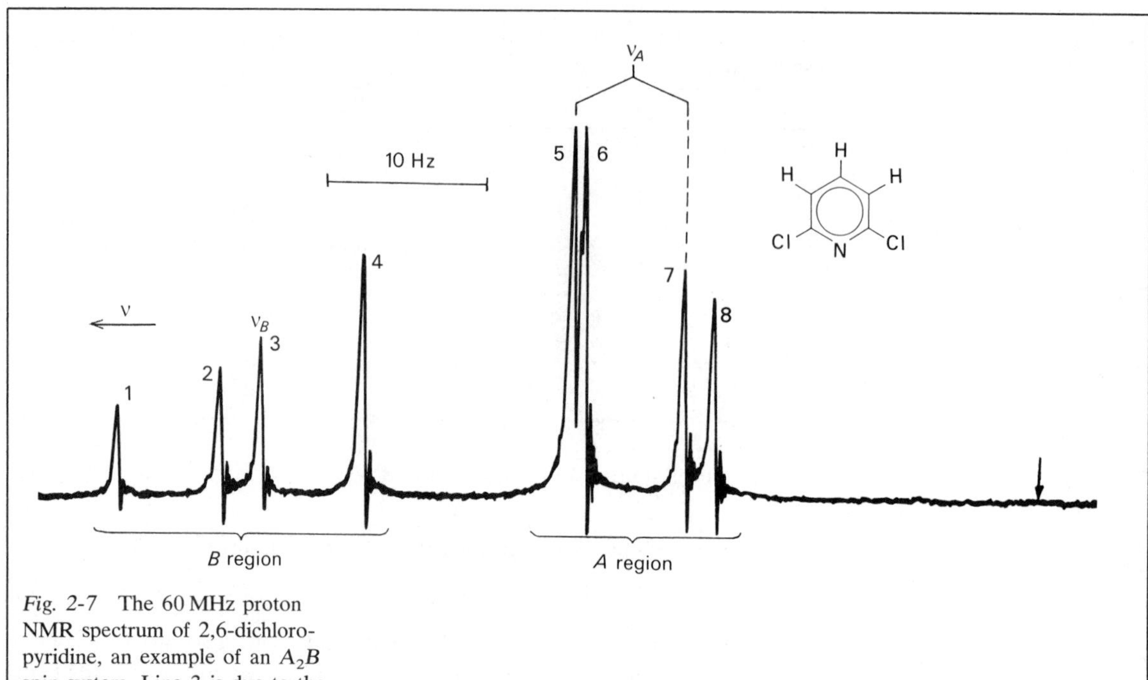

Fig. 2-7 The 60 MHz proton NMR spectrum of 2,6-dichloropyridine, an example of an A_2B spin system. Line 3 is due to the DS spin states. The calculated position of line 9, which has negligible intensity, is indicated by the vertical arrow.

If the system is simple (e.g., AB) then the line positions can be written explicitly in terms of the Larmor frequencies and coupling constants. Once the lines are assigned, the numerical values of these parameters can be found by simple mathematics. If the system is more complex it is less easy to determine the parameters. Some or all of them may be obtainable by a partial analysis, for in most cases some of the line positions can be written explicitly in terms of the Larmor frequencies and coupling constants. These values may then be used to calculate the complete spectrum using a suitable computer program.[5] In other situations the parameters are obtained by a process of iteration. Probable values, chosen by guesswork or from experience, are used to compute a spectrum. These trial parameters are then varied until observed and calculated spectra agree. Figure 2-8 shows an example. The computed spectrum includes provision of a suitable lineshape to enable it to be compared readily with the experimental spectrum. Sophisticated features can be built into the computer program, such as factorization for the appropriate symmetry and calculation of probable errors in the parameters.[6] Sometimes ambiguities are encountered, in that more than one set of parameters fits the line positions of the observed spectrum. Often these have different relative signs for the coupling constants. It is possible in favourable cases to distinguish

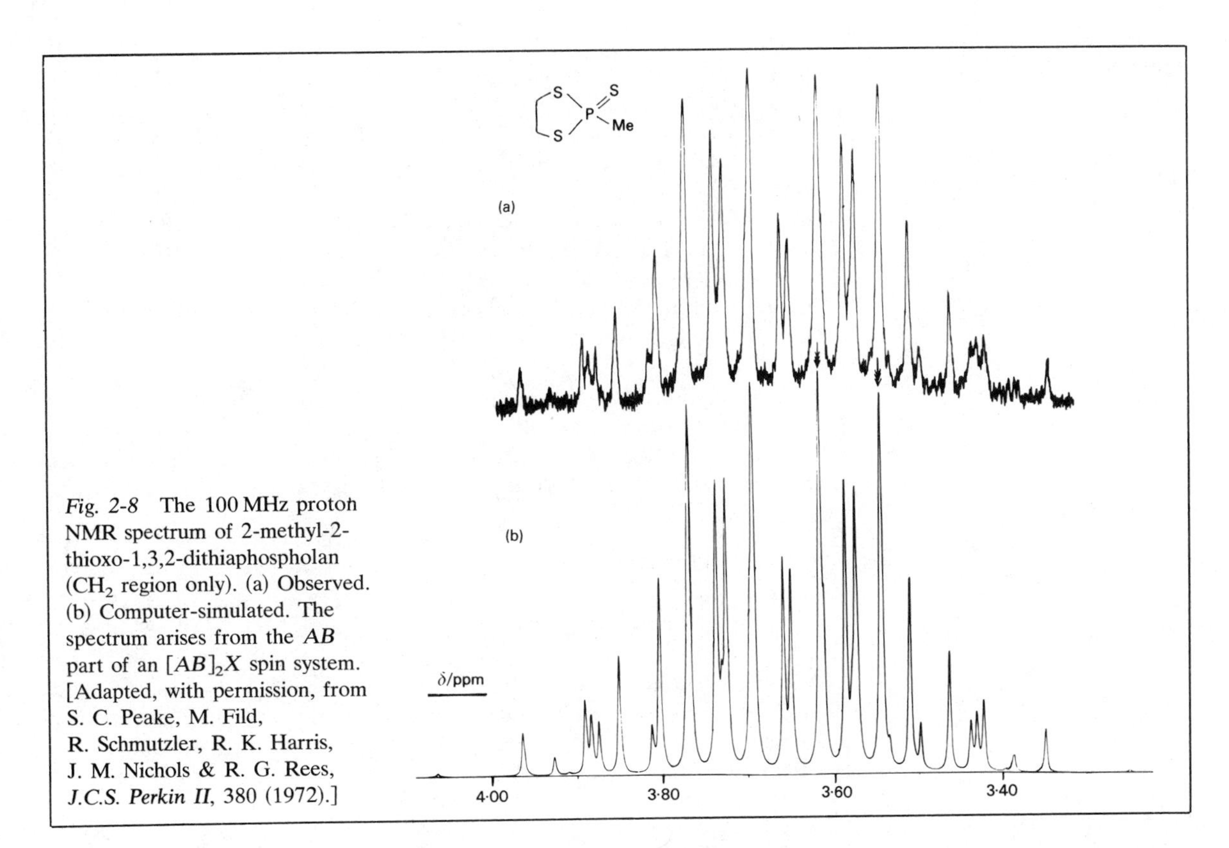

Fig. 2-8 The 100 MHz proton NMR spectrum of 2-methyl-2-thioxo-1,3,2-dithiaphospholan (CH_2 region only). (a) Observed. (b) Computer-simulated. The spectrum arises from the AB part of an $[AB]_2X$ spin system. [Adapted, with permission, from S. C. Peake, M. Fild, R. Schmutzler, R. K. Harris, J. M. Nichols & R. G. Rees, *J.C.S. Perkin II*, 380 (1972).]

between these by comparing the observed and calculated intensity patterns.

Notes and references

[1] This hypothesis leads to the familiar interpretation of $\psi^*(x, y, z)\psi(x, y, z)\,dv$ as the probability of finding the particle in volume dv centred on (x, y, z).

[2] Some texts incorporate $\hbar$ into the definition of the spin operators themselves.

[3] Note that the terms 'raising' and 'lowering' refer to angular momentum rather than energy.

[4] Some books use the opposite convention with the z axis in the $-\mathbf{B}_0$ direction. The first term in the Hamiltonian then has the opposite sign, but the spectrum is unchanged.

[5] A library of suitable computer programs has been compiled and is described in a manual entitled *The NMR Program Library* by D. J. Loomes, R. K. Harris and P. Anstey, published in 1979 by the Science Research Council Daresbury Laboratory, Daresbury, Warrington, Cheshire, WA4 4AD, United Kingdom. There are also later bulletins. Full details may be obtained from Dr W. Smith of the Daresbury Laboratory.

[6] It is even possible to dispense with the stage of finding a suitable trial spectrum since programs have been written to calculate Larmor frequencies and coupling constants directly from spectral information [D. S. Stephenson & G. Binsch, *J. Magn. Reson.*, **37,** 395 (1980); **37,** 409 (1980)].

Further reading

'^{13}CH satellite NMR spectra', J. H. Goldstein, V. S. Watts & L. S. Rattet, *Prog. NMR Spectrosc.*, **8,** 103 (1972).

The Analysis of High-resolution NMR Spectra, R. J. Abraham, Elsevier (1971).

'Subspectral analysis', P. Diehl, R. K. Harris & R. G. Jones, *Prog. NMR Spectrosc.*, **3,** 1 (1967).

Structure of High-resolution NMR Spectra, P. L. Corio, Academic Press (1966).

'Analysis of NMR spectra', R. A. Hoffmann, S. Forsén & B. Gestblom, *NMR Basic Principles & Prog.*, **5,** 1 (1971).

'Computer assistance in the analysis of high-resolution NMR spectra', P. Diehl, H. Kellerhals & E. Lustig, *NMR Basic Principles & Prog.*, **6,** 1 (1972).

Problems

2-1 Show that the functions $\exp m\phi$ are eigenfunctions of the operators $\hat{d}/d\phi$ and $\hat{R}(\phi')$ where the latter rotates the coordinate system by an angle $-\phi'$ about the z axis. For what values of m are these functions single valued? The operator $\hat{I}_z$ is given by $-i(\hat{d}/d\phi)$ in spherical polar coordinates. Comment on the allowed values of m.

2-2 Find the values of

$$\left[\hat{x}, \frac{\hat{d}^2}{dx^2}\right]; \left[\hat{y}, \frac{\hat{d}^2}{dx^2}\right]$$

2-3 If $\hat{A}$ is a Hermitian operator (Section 2-3) show that it has real eigenvalues [substitute $f = g =$ an eigenfunction in Eqn 2-9] and that the eigenfunctions belonging to different eigenvalues are orthogonal [substitute different eigenfunctions for f and g in Eqn 2-9].

2-4 Show
(a) $\langle a \mid b \rangle = \langle b \mid a \rangle^*$
(b) The condition for an operator to be Hermitian can be written

$$\langle a | \hat{A} | b \rangle = \langle b | \hat{A} | a \rangle^*$$

(c) $|r\rangle = \sum_j |j\rangle\langle j \mid r\rangle$ where the functions $|j\rangle$ form a complete set of orthonormal functions.

2-5 If $[\hat{A}, \hat{B}] = \hat{C}$ and $\hat{A}\,|a, b\rangle = a\,|a, b\rangle$, $\hat{B}\,|a, b\rangle = b\,|a, b\rangle$, show that either $\hat{C} = 0$ or $a = b = 0$ and $\hat{C}\,|a, b\rangle = 0$.

2-6 Find the values of

$$[\hat{I}_z^2, \hat{I}_+], \quad [\hat{I}_z^2, \hat{I}_z], \quad [\hat{I}_z, \hat{I}_+\hat{I}_-], \quad [\mathbf{I}^2, \hat{I}_{z+}]$$

2-7 Show that

$$\hat{\mathbf{I}}^2 \equiv \hat{I}_z^2 + \tfrac{1}{2}(\hat{I}_+\hat{I}_- + \hat{I}_-\hat{I}_+) \equiv \hat{I}_z^2 + \hat{I}_z + \hat{I}_-\hat{I}_+$$
$$\equiv \hat{I}_z^2 - \hat{I}_z + \hat{I}_+\hat{I}_-$$

Hence show that

$$\mathbf{I}^2\,|I, m_{\max}\rangle = m_{\max}(m_{\max}+1)\,|I, m_{\max}\rangle$$

$$\mathbf{I}^2\,|I, m_{\min}\rangle = m_{\min}(m_{\min}-1)\,|I, m_{\min}\rangle$$

and derive Eqn 2-37.

2-8 If $\hat{I}_+\,|I, m\rangle = c\,|I, m+1\rangle$ show that $c = \sqrt{[I(I+1) - m(m+1)]}$. (Show $c^*c = \langle I, m|\,\hat{I}_-\hat{I}_+\,|I, m\rangle$ and write this in terms of $\hat{\mathbf{I}}^2$ and $\hat{I}_z$ operators).

2-9 What are the values of the following?

(a) $\langle I, m+2|\, \hat{I}_+ \,|I, m\rangle$ (d) $\langle I, m+1|\, \hat{I}_x \,|I, m\rangle$

(b) $\langle I, m-1|\, \hat{I}_y \,|I, m\rangle$ (e) $\langle \frac{3}{2}, \frac{3}{2}|\, \hat{I}_+ \,|\frac{3}{2}, \frac{1}{2}\rangle$

(c) $\langle \frac{3}{2}, \frac{1}{2}|\, \hat{I}_+ \,|\frac{3}{2}, -\frac{1}{2}\rangle$ (f) $\langle \frac{3}{2}, -\frac{1}{2}|\, \hat{I}_- \,|\frac{3}{2}, \frac{1}{2}\rangle$

2-10 Calculate the values of

$$\langle \alpha\beta\alpha|\, \hat{I}_{2+}\hat{I}_{3-} \,|\alpha\alpha\beta\rangle, \langle \alpha\beta\alpha|\, \hat{I}_{2+}\hat{I}_{3-} \,|\beta\alpha\beta\rangle$$

$$\langle \alpha\beta|\, \hat{I}_{2z} \,|\alpha\beta\rangle, \langle \alpha\alpha|\, \hat{I}_{1z} \,|\alpha\beta\rangle,$$

$$\langle \alpha\alpha\alpha|\, \hat{F}_+ \,|c_1\alpha\alpha\beta + c_2\alpha\beta\alpha + c_3\beta\alpha\alpha\rangle$$

2-11 Show that the states $2^{-1/2}(\alpha\beta+\beta\alpha)$ and $2^{-1/2}(\alpha\beta-\beta\alpha)$ are eigenstates of the A_2 Hamiltonian with eigenvalues $\frac{1}{4}J$ and $-\frac{3}{4}J$. What is the effect of the operator $\hat{\mathbf{F}}^2[=(\hat{\mathbf{I}}_1+\hat{\mathbf{I}}_2)^2]$ on these states and on the state $\alpha\alpha$? These are T and S composite particle states (Section 2-11).

2-12 Describe the energy level diagram for the $[A_2B_2]$ spin system, using the total spin ('composite particle') method of factorization. Thence derive the number of transitions allowed, including a designation of their origin in the first-order limit. Give any further information you can about the transition frequencies.

2-13 The ^{1}H NMR spectrum of an *AB* spin system consists of four lines at frequencies (relative to the signal of TMS) 423·0 Hz, 418·5 Hz, 416·0 Hz and 411·5 Hz for a spectrometer operating at 60 MHz.

(a) What is the coupling constant, J_{AB}, and what are the two chemical shifts, δ_A and δ_B?

(b) Calculate the relative intensities of the transitions.

(c) Calculate the transition frequencies (relative to the signal of TMS) for a spectrometer operating at 100 MHz.

2-14 Construct the Hamiltonian matrix for the *ABX* system.

2-15 If $\hat{\mathbf{F}}_A = \sum_{j=1}^{p} \hat{I}_j$ show that

$$[\hat{F}_{Az}, \hat{F}_{A+}] = \hat{F}_{A+}$$

$$[\hat{F}_A^2, \hat{I}_{j+}] = 2(\hat{I}_{j+}\hat{F}_{Az} - \hat{I}_{jz}\hat{F}_+)$$

$$[\hat{F}_A^2, \hat{F}_{A+}] = 0$$

2-16 Construct a table of product functions of 4 nuclei classified according to the values of m_T. Show that the values of F are 2, 1 and 0 and find the degeneracy of each value. Deduce that four electrons in the configuration $1s^12s^12p^13s^1$ form two ^1P, three ^3P and one ^5P levels.

3 Relaxation and Fourier transform NMR

3-1 Introduction

In the previous chapters we have discussed how observed NMR spectra can be related to nuclear spin energy levels. As in other branches of molecular spectroscopy, the frequencies at which electromagnetic radiation is absorbed depend on the separation of the energy levels. Apart from the difference in the type of energy level, the main contrast with microwave, infrared, or ultraviolet spectroscopy lies in the necessity for an external magnetic field. In addition, we have seen that the transitions are induced by the magnetic part of the electromagnetic field rather than the electric part.

In this chapter we shall deal with effects which, although they are not confined to NMR, are most easily observable in this type of spectroscopy. Because differences in energy between nuclear spin states are small at normal temperatures, these states are nearly equally populated (see Section 1.7). The oscillating magnetic field induces transitions upwards and downwards with equal probability; the net absorption of energy is due to the small difference in population of upper and lower levels. As the quanta are small, it is easy to apply a high intensity of coherent electromagnetic radiation to the system. This leads to multiple quantum and other non-linear effects normally associated with lasers.

For a CW NMR experiment, the act of energy absorption reduces the population difference, and, if the power in the radiofrequency field is too high, the population difference decreases, weakening the spectrum. This phenomenon, which is sometimes a limiting factor in an attempt to observe weak signals, is known as saturation and is discussed in Section 3-5. Analogous phenomena occur in pulsed Fourier transform NMR, as is also discussed later in this chapter.

3-2 The Bloch equations

The pioneers of nuclear resonance were mainly concerned with dynamic processes and lineshapes. Indeed, the first observations of chemical shifts and coupling constants in liquids were unexpected. Bloch's approach was macroscopic and classical and is complementary to the use of a spin Hamiltonian. The Bloch equations, which we shall derive in this section, can be used to discuss the lineshape of a simple spectrum, but they are not applicable to a complex spectrum.

In order to derive the equations, we discuss the behaviour of a macroscopic sample containing many identical molecules, each with one magnetic nucleus. The total magnetic moment or magnetization **M** of the sample is the resultant of the nuclear moments $\boldsymbol{\mu}$, as discussed in Section 1-7.

$$\mathbf{M} = \sum_j \boldsymbol{\mu}_j \tag{3-1}$$

or in terms of the total spin angular momentum **P**

$$\mathbf{M} = \gamma \mathbf{P} \tag{3-2}$$

As the sample is macroscopic, the effect of an applied magnetic field **B** can be predicted from classical mechanics. The interaction of the field and moment gives a torque on the system, changing the angular momentum **P**.

$$\frac{\mathrm{d}}{\mathrm{d}t}\mathbf{P} = -\mathbf{B} \wedge \mathbf{M} \tag{3-3}$$

but as **P** and **M** are related by Eqn 3-2,

$$\frac{\mathrm{d}}{\mathrm{d}t}\mathbf{M} = -\gamma \mathbf{B} \wedge \mathbf{M} \tag{3-4}$$

The classical motion of the nuclear moments $\boldsymbol{\mu}$ discussed in Section 1-6 is described by an equation of motion similar to 3-4. The steady-state solution is a precession of the system about the field direction with angular velocity $\boldsymbol{\omega}_l = -\gamma \mathbf{B}$. This is known as Larmor precession, and the corresponding frequency, $\gamma B/2\pi$, is the Larmor frequency.

Any system with a magnetic moment will show Larmor precession. The magnetic moment of the sample in nuclear resonance may, in addition, change by internal realignment of the individual nuclear spins. Indeed, when the sample is placed in a magnetic field B_0 its moment changes from zero to M_0, the equilibrium value. For spin-$\frac{1}{2}$ nuclei the magnitude of M_0 is given by Eqn 1-22. The equation of motion for the magnetic moment of the sample must include the approach to thermal equilibrium as well as Larmor precession. Bloch assumed that the components of **M** decay to $\mathbf{M}_0$ exponentially, but he allowed the components of **M** parallel and perpendicular to $\mathbf{M}_0$ to decay with different time constants T_1 and T_2, so that with the z axis chosen along $\mathbf{B}_0$

$$\frac{\mathrm{d}}{\mathrm{d}t}M_z = -(M_z - M_0)/T_1 \tag{3-5}$$

$$\frac{\mathrm{d}}{\mathrm{d}t}M_x = -M_x/T_2 \qquad \frac{\mathrm{d}}{\mathrm{d}t}M_y = -M_y/T_2 \tag{3-6}$$

The approach to thermal equilibrium is known as relaxation and T_1 and T_2 are relaxation times. The decay of the longitudinal component

(M_z) may differ from the decay of the transverse components (M_x and M_y) because the energy of the spin system depends on M_z. Any change in M_z is accompanied by an energy flow between the nuclear spin system and the other degrees of freedom of the system known for historical reasons as the 'lattice'. The relaxation time T_1 which describes this flow is usually known as the *spin–lattice relaxation time*. An alternative name is the *longitudinal relaxation time*, emphasizing the relationship of M_z and the applied field. T_2 is known as the *transverse relaxation time* or the *spin–spin relaxation time*. The latter name arises because direct interactions between the spins of different nuclei can cause relaxation of M_x and M_y without energy transfer to the lattice—an important effect for protons in solids. In high-resolution NMR of fluids T_1 and T_2 are both affected by energy exchange between the spin system and the lattice. Section 3-14 contains a further discussion of these processes.

In the nuclear resonance experiment there is a small oscillating magnetic field, $2B_1 \cos \omega t$, in the x direction. This can be resolved into two components rotating in opposite directions with angular velocities $\pm\omega$. These may be considered independently provided B_1 is small, and the component rotating in the opposite direction to the Larmor precession (the one with angular velocity $+\omega$) can be neglected. The rotating field induces a rotating magnetization in the x, y plane which is monitored by the spectrometer. The field $\mathbf{B}$ acting on the sample is

$$\mathbf{B} = B_1 \cos \omega t \, \mathbf{i} - B_1 \sin \omega t \, \mathbf{j} + B_0 \mathbf{k} \qquad (3\text{-}7)$$

where $\mathbf{i}$, $\mathbf{j}$, and $\mathbf{k}$ are unit vectors in the x, y, and z directions.

Combining Eqns 3-4 for the Larmor precession and 3-5 and 3-6 for the relaxation with this value of $\mathbf{B}$, we obtain the Bloch equations:

$$\frac{dM_z}{dt} = -\gamma[B_1 \cos \omega t \; M_y + B_1 \sin \omega t \, M_x] - (M_z - M_0)/T_1 \qquad (3\text{-}8)$$

$$\frac{dM_y}{dt} = -\gamma[B_0 M_x - B_1 \cos \omega t \; M_z] - M_y/T_2 \qquad (3\text{-}9)$$

$$\frac{dM_x}{dt} = +\gamma[B_1 \sin \omega t \; M_z + B_0 M_y] - M_x/T_2 \qquad (3\text{-}10)$$

3-3 The rotating frame of reference

The equations take a simpler form if they are referred to a set of axes (x', y', z) rotating with the applied field, with an angular velocity $-\omega$ about the z axis (Fig. 3-1). The components of magnetization in the plane perpendicular to B_0 will be designated u and v, where u is the component in the direction of B_1, i.e. along x' (the in-phase component), whilst v is the out-of-phase component along the y' axis.

$$u = M_x \cos \omega t - M_y \sin \omega t \qquad (3\text{-}11)$$

$$v = M_x \sin \omega t + M_y \cos \omega t \qquad (3\text{-}12)$$

A signal is induced in the receiver coils of the spectrometer by a

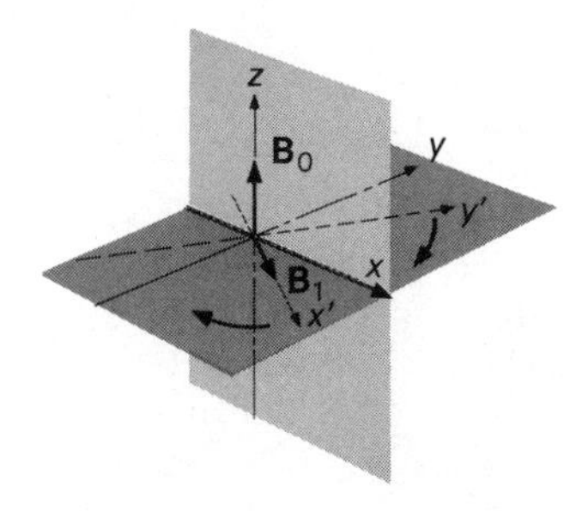

Fig. 3-1 Systems of rotating axes (x', y', z) used for solving the Bloch equations. These axes rotate about the z axis with constant angular velocity $-\omega$; in this system $\mathbf{B}_1$ is constant and is in the x' direction.

change in the magnetization M_y. Phase-sensitive detection is usually used, which means that either u or v may be detected, the latter being normal. From Eqns 3-8 to 3-12

$$\frac{du}{dt}=\frac{dM_z}{dt}\cos\omega t-\frac{dM_y}{dt}\sin\omega t-\omega(M_x\sin\omega t+M_y\cos\omega t)$$

$$=+\gamma B_0 v-\frac{u}{T_2}-\omega v$$

dv/dt can be found in a similar way. The complete set of Bloch equations in rotating axes may be written as:

$$\frac{dM_z}{dt}=-\gamma B_1 v-(M_z-M_0)/T_1 \qquad (3\text{-}13)$$

$$\frac{du}{dt}=+(\omega_i-\omega)v-u/T_2 \qquad (3\text{-}14)$$

$$\frac{dv}{dt}=-(\omega_i-\omega)u+\gamma B_1 M_z-v/T_2 \qquad (3\text{-}15)$$

where $\omega_i=\gamma B_0$, the Larmor angular velocity[1] about B_0.

A simple physical picture of events in the rotating frame of reference can be obtained if they are considered as described by relative motion. Since the reference axes are rotating at the same rate as the effective component of the r.f. magnetic field B_1, this field appears to be static. Moreover, the precessional motion of the magnetization (i.e. $\omega_i=\gamma B_0$) appears to be reduced to a value $(\omega_i-\omega)$. Such motion corresponds to precession in an apparent field $(\omega_i-\omega)/\gamma\equiv B_0(1-\omega/\omega_i)$. Thus, although the total field, B_t, is the vector sum of B_0 and B_1, it is useful to define an effective field, B_{eff}, which is the vector sum of $(\omega_i-\omega)/\gamma$ and B_1, as shown in Fig. 3-2. Since $B_0\gg B_1$, the field B_t is always nearly B_0 and is close to the z direction, whereas B_{eff} varies considerably in direction depending on the offset frequency, $\omega_i-\omega$, between the r.f. and resonance.

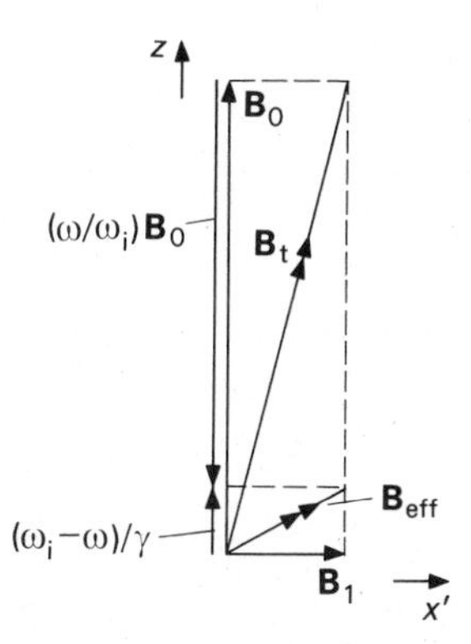

Fig. 3-2 Magnetic fields in the rotating frame of reference. As B_0, the static field, is always much greater than the rotating field (B_1), the total field ($\mathbf{B}_t$) is nearly parallel to $\mathbf{B}_0$. However, the effective field ($\mathbf{B}_{\text{eff}}$) may be in any direction in the $x'z$ plane depending on the ratio ω/ω_i of the applied and Larmor angular frequencies. The magnitude of $\mathbf{B}_1$ is exaggerated with respect to $\mathbf{B}_0$ in the figure.

The motion of the magnetization in the rotating frame in the absence of relaxation is precession about B_{eff}, as is illustrated in Fig. 3-3. In the diagram 3-3(a) ω is less than ω_i, and B_{eff} is in a direction between B_1 and B_0, so that, neglecting relaxation, the magnetization precesses about B_{eff} in a conical path at a rate $-\gamma B_{\text{eff}}$. Diagram 3-3(b) shows the resonance situation where $\omega=\omega_i$ and $B_{\text{eff}}(\equiv B_1)$ is in the x' direction. Now M moves in a circle in the zy' plane at a rate $-\gamma B_1$, to give a large signal. It is only when ω is close to resonance that B_1 has an appreciable effect on the system. When the field is first turned on, the magnetization precesses in the way described here, but after a short period it relaxes to a definite position at rest in the rotating axes. The signals observed in steady state (CW) experiments depend on the magnitudes of u and v when this has occurred. In pulse experiments the r.f. is switched off and the system is observed before the steady state is reached, and the effects of T_1 and T_2 can be seen directly.

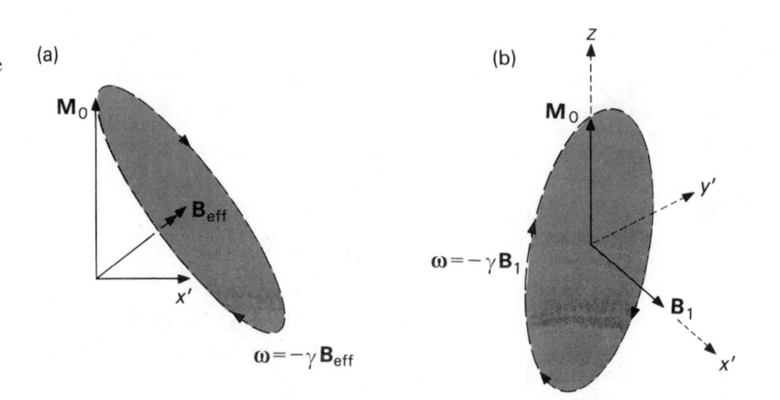

Fig. 3-3 Effective fields and the initial precession of the nuclear magnetization vector $\mathbf{M}_0$ in rotating axes with an applied radiofrequency ω_1 (a) below resonance, (b) at resonance. The axes are considered to be rotating at frequency ω_1 so that $\mathbf{B}_1$ appears to be static. The precession is eventually damped by relaxation, so that the picture is only valid for times $\ll T_1$ and T_2.

3-4 Steady state (CW) experiments

In a normal CW NMR experiment, the spectrometer is tuned to observe the component of magnetization 90° out of phase to the rotating field, B_1, giving the *absorption mode* signal. Occasionally, the in-phase component u is observed, which gives the *dispersion mode* signal. In a typical slow-passage experiment, ω is swept slowly through the resonance value and the signal is recorded continuously. Provided this is done sufficiently slowly, the absorption mode signal at each frequency corresponds to the steady state value of ωv when M has come to rest in rotating axes. The values of u, v, and M_z are given by the solutions of the Bloch equations in rotating axes (Eqns 3-13 to 3-15) with the time derivatives equal to zero. They are

$$M_z = \frac{M_0[1 + T_2^2(\omega_i - \omega)^2]}{T_2^2(\omega_i - \omega)^2 + 1 + T_1 T_2 \gamma^2 B_1^2} \tag{3-16}$$

$$u = \frac{M_0 \gamma B_1 T_2^2(\omega_i - \omega)}{T_2^2(\omega_i - \omega)^2 + 1 + T_1 T_2 \gamma^2 B_1^2} \tag{3-17}$$

$$v = \frac{M_0 \gamma B_1 T_2}{T_2^2(\omega_i - \omega)^2 + 1 + T_1 T_2 \gamma^2 B_1^2} \tag{3-18}$$

Analogous expressions in frequency units may be obtained by the replacements $\omega_i = 2\pi\nu_i$ and $\omega = 2\pi\nu$. Figure 3-4 shows the shape of the u and v signals predicted by the Bloch equations when $\gamma B_1 \ll (T_1 T_2)^{-1/2}$. In such a case the populations of the energy levels involved in the transition do not depart appreciably from their values in the absence of B_1 (i.e. $\Delta n \simeq \Delta n_0$—see Section 1-7) and the absorption mode signal is proportional to

$$g(\nu) = \frac{2T_2}{1 + 4\pi^2 T_2^2(\nu_i - \nu)^2} \tag{3-19}$$

This is the lineshape factor mentioned in Section 1-7. The width of the absorption line at half-height and the peak-to-trough separation of the dispersion line are both equal to $(\pi T_2)^{-1}$ in frequency units (see Problem 3-1).

The v-mode signal is proportional to the power absorbed from the

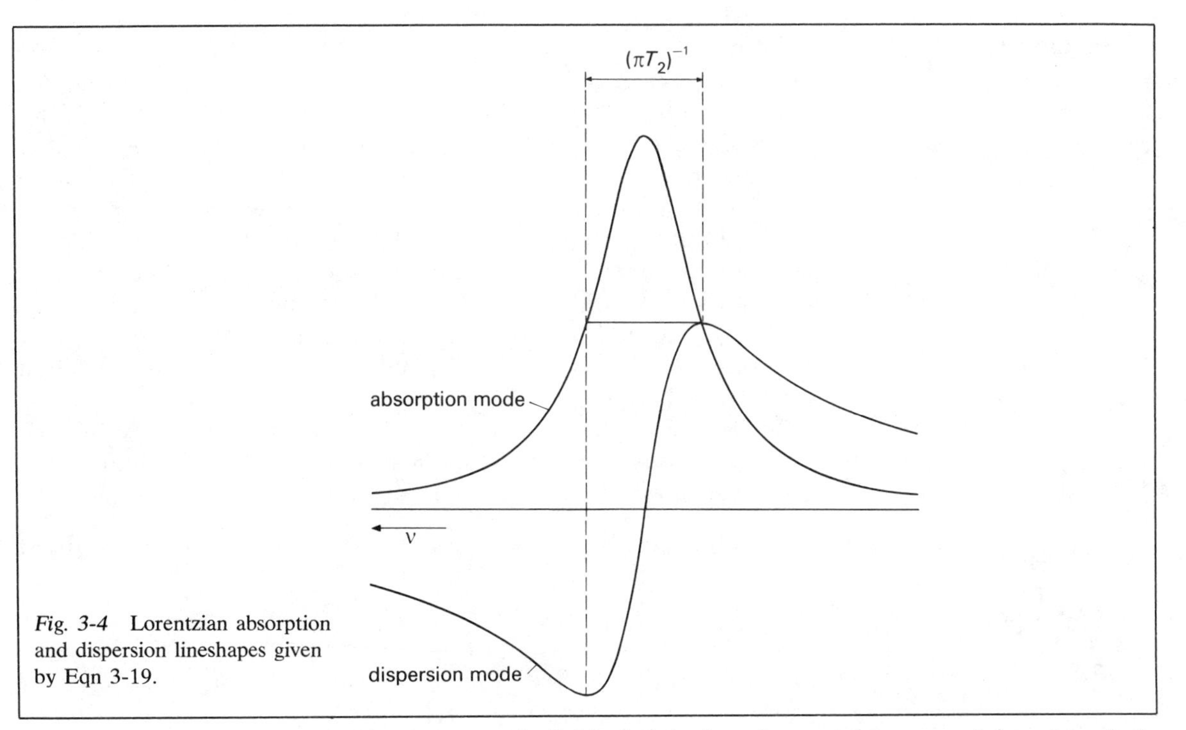

Fig. 3-4 Lorentzian absorption and dispersion lineshapes given by Eqn 3-19.

electromagnetic field, and is therefore analogous to infrared and ultraviolet absorption spectra, except that the NMR signal depends on B_1 and not on B_1^2 (see Section 1-7). The u-mode signal gives the dispersion spectrum, which can be observed in other types of spectroscopy by measuring the refractive index of the sample as a function of frequency. This is only common in optical rotatory dispersion spectra, which depend on the difference of refractive indices of right- and left-handedly polarized light.

The lineshape 3-19 predicted by the Bloch equation is known as *Lorentzian* and is the shape that would be observed in an ideal experiment if the relaxation of M_x and M_y were truly exponential with a single value of T_2. In practice, the lines in NMR experiments are often not Lorentzian and may even by unsymmetrical. This is because the 'natural' width of the lines $(\pi T_2)^{-1}$ is so small that the observed width is due to instrumental effects, such as the variation of B_0 by a few parts per hundred million (0·6 Hz at 60 MHz) over the area of the sample. The lineshape is an indication of the field inhomogeneity over the sample. This is sometimes known as inhomogeneity broadening, as the signal is a composite of lines with slightly different Larmor frequencies. The effects of field inhomogeneities can be minimized by spinning the cylindrical sample about its long axis at a rate of ca. 15 Hz. Frequently the total observed width of a line, $\Delta\nu_{1/2}$, is used to define an effective relaxation time T_2^* by means of the relation

$$\Delta\nu_{1/2} = (\pi T_2^*)^{-1} \tag{3-20}$$

even though the lineshape is not exactly Lorentzian.

3-5 Saturation

If the power in the oscillating field in a CW NMR experiment is too high, the peak height in the absorption spectrum decreases, owing to a reduction in the difference in population between the two levels. This occurs if the rate of energy absorption is comparable to, or greater than, the rate of relaxation between energy levels, T_1^{-1}. The Bloch equation 3-18 for a single nucleus shows this effect when the term $\gamma^2 B_1^2 T_1 T_2$ in the denominator is no longer much less than one. As B_1 increases the lines also get broader. The total integrated intensity increases linearly with B_1 for low B_1, but tends to a constant value. Figure 3-5 shows the variation of peak height and total intensity as a function of B_1. From a practical point of view, a strong signal is best observed with $B_1 \ll (\gamma^2 T_1 T_2)^{-1/2}$, since this gives the best resolution, but if the concentration of spins is too low for this to be possible, the optimum conditions correspond to maximum peak height (although the signal will show saturation broadening) with

$$B_1(\text{opt}) = (\gamma^2 T_1 T_2)^{-1/2} \tag{3-21}$$

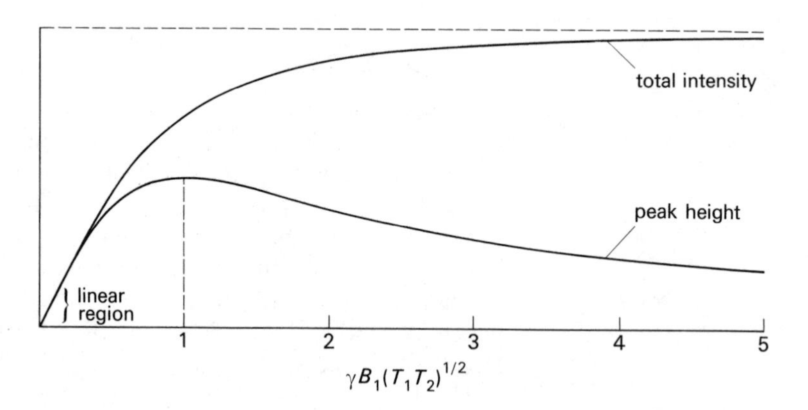

Fig. 3-5 Variation of peak height and total intensity with the amplitude of the observing field B_1 for CW-mode NMR. The vertical scale is in arbitrary units.

In a spectrum of a molecule with several lines with different relative intensities (such as those discussed in Chapter 2), the conditions for saturation vary from line to line. If the lines do not overlap and are not degenerate, each can be described by equations similar to the Bloch equations, but with different values of T_1 and T_2, and with B_1, u, and v replaced by fB_1, u/f, and v/f, where f^2 is the intensity of the line, $|\langle r|F_+|s\rangle|^2$. The absorption line shape is proportional to ωv, i.e. to

$$\frac{\gamma^2 f^2 B_1 T_2 M_0 B_0}{T_2^2(\omega_0 - \omega)^2 + 1 + f^2 T_1 T_2 \gamma^2 B_1^2} \tag{3-22}$$

From an experimental point of view it should be noted that the size of the observing field B_1 can be increased to observe lines which have low transition probabilities, even if more intense lines in the same spectrum are saturated. However, lines which are weak because of concentration effects (such as ^{13}C satellites) cannot be enhanced in this way.

3-6 Nuclear receptivity

In order to compare the suitability of different spin-$\frac{1}{2}$ nuclei for NMR, it is sensible to take the maximum peak height as given under the condition 3-21 and to substitute into Eqn 1-24. In these circumstances $g(\nu) \propto T_2$, and Eqn 3-23 is obtained:[2]

$$S^{\max}(\nu = \nu_i) \propto |\gamma^3 B_0^2 N \sqrt{T_2}/T\sqrt{T_1}| \tag{3-23}$$

If $T_1 = T_2$, as is usually the case for non-viscous solutions, and inhomogeneities in B_0 are neglected, we find that if B_0 and T are kept constant:

$$S^{\max}(\nu = \nu_i) \propto |\gamma^3 N| \tag{3-24}$$

It is convenient to compare situations with the same atomic concentrations and same sample volumes, in which case the natural isotopic abundance, C, is the only variable contribution to N. The quantity $|\gamma^3 C|$ may be defined as the *receptivity* of a given nucleus, and represents roughly its suitability for NMR. Since ^{1}H and ^{13}C NMR are the most-studied nuclides, it is usual to quote receptivities relative to one or other of these nuclei. Table 3-1 gives such relative receptivities for the most suitable spin-$\frac{1}{2}$ isotopes of the elements, excluding the lanthanides and helium. It can be seen that ^{1}H, ^{19}F, ^{31}P and ^{205}Tl are the four easiest nuclei to study, whereas ^{187}Os and ^{57} Fe are at the other extreme, being more than six orders of magnitude less receptive than ^{1}H. In spite of its low receptivity, ^{15}N is frequently studied because of the importance of nitrogen in chemistry; however isotopic enrichment is often used in this case.

Table 3-1 Receptivities, relative to ^{1}H and ^{13}C (D^p and D^C respectively), of the major spin-$\frac{1}{2}$ nuclides[a]

isotope	^{1}H	^{13}C	^{15}N	^{19}F	^{29}Si
D^p	1	$1{\cdot}76\times10^{-4}$	$3{\cdot}85\times10^{-6}$	0·834	$3{\cdot}69\times10^{-4}$
D^C	$5{\cdot}67\times10^{3}$	1	$2{\cdot}19\times10^{-2}$	$4{\cdot}73\times10^{3}$	2·10
isotope	^{31}P	^{57}Fe	^{77}Se	^{89}Y	
D^p	0·0665	$7{\cdot}43\times10^{-7}$	$5{\cdot}30\times10^{-4}$	$1{\cdot}19\times10^{-4}$	
D^C	$3{\cdot}77\times10^{2}$	$4{\cdot}22\times10^{-3}$	3·01	0·675	
isotope	^{103}Rh	^{109}Ag	^{113}Cd	^{119}Sn	
D^p	$3{\cdot}16\times10^{-5}$	$4{\cdot}92\times10^{-5}$	$1{\cdot}35\times10^{-3}$	$4{\cdot}51\times10^{-3}$	
D^C	0·179	0·279	7·67	25·6	
isotope	^{125}Te	^{129}Xe	^{183}W	^{187}Os	
D^p	$2{\cdot}24\times10^{-3}$	$5{\cdot}69\times10^{-3}$	$1{\cdot}06\times10^{-5}$	$2{\cdot}00\times10^{-7}$	
D^C	12·7	32·3	$5{\cdot}99\times10^{-2}$	$1{\cdot}14\times10^{-3}$	
isotope	^{195}Pt	^{199}Hg	^{205}Tl	^{207}Pb	
D^p	$3{\cdot}39\times10^{-3}$	$9{\cdot}82\times10^{-4}$	0·140	$2{\cdot}01\times10^{-3}$	
D^C	19·2	5·57	$7{\cdot}91\times10^{2}$	11·4	

[a] A more comprehensive list is to be found in Appendix 1.

3-7 The effects of radiofrequency pulses

Modern sophisticated NMR spectrometers rely heavily on the use of short pulses of radiation, in contrast to the cheaper instruments which involve continuous wave methods. In a pulse experiment, the oscillating r.f. magnetic field is turned on for a time (usually in the range 1–50 μs) which is short compared to T_1 and T_2^*, and is then turned off again. The frequency (often termed the carrier frequency, ν_c) is chosen to be close to resonance for the nucleus of interest, so that $B_{eff} \simeq B_1$. However, as the Uncertainty Principle indicates, the pulse will contain, in effect, a range of frequencies centred on ν_c. It can be shown that the distribution of r.f. magnetic field amplitudes takes the form $\sin[\pi(\nu-\nu_c)\tau_p]/\pi(\nu-\nu_c)\tau_p$, where τ_p is the pulse duration. This distribution is illustrated in Fig. 3-6, which shows the *frequency domain* equivalent, $F(\nu)$, of a short pulse in the *time domain*, $f(t)$. The two domains are related by the mathematical procedure of *Fourier transformation* as follows

$$F(\nu) = \int_{-\infty}^{\infty} f(t)\exp(+i2\pi\nu t)\,dt \tag{3-25}$$

$$f(t) = \int_{-\infty}^{\infty} F(\nu)\exp(-i2\pi\nu t)\,d\nu \tag{3-26}$$

A pulse of duration 10 μs gives an effective range of frequencies[3] of the order of 10^5 Hz, which is adequate for the excitation of all protons in a sample, whatever their chemical shifts (or, correspondingly, of all ^{13}C nuclei). However, a rather flat distribution of B_1 over the frequencies of interest is required, so it is desirable to operate within the central portion of Fig. 3-6, i.e. to have τ_p^{-1} one or two orders of magnitude greater than the chemical shift range under study.

In the rotating frame of reference the nuclear magnetization, **M**, will precess about the x' direction at a rate $-\gamma\mathbf{B}_1$ during the pulse (see Fig. 3-3(b)). At the end of a pulse of duration τ_p it will have precessed by $\gamma B_1\tau$ in angular units. Now τ_p may be set at any pre-selected value; if it is $\pi/2\gamma B_1$, then M will have been turned through 90° into the y′ direction. If $\tau_p = \pi/\gamma B_1$, then at the end of the pulse M will lie in the $-z$ direction. These two situations are said to be the results of a 90° (or

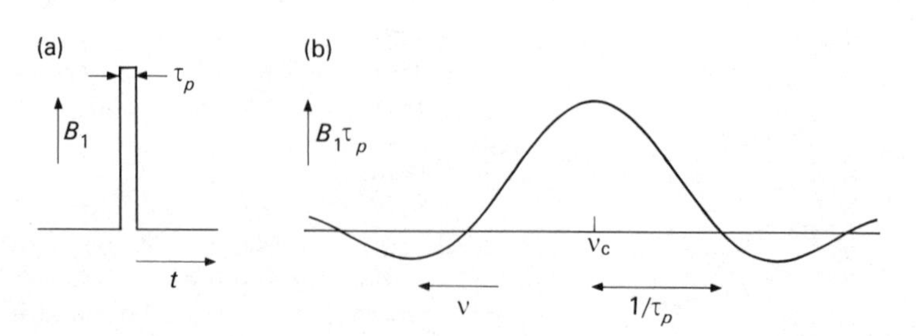

Fig. 3-6 A pulse of apparently monochromatic radiofrequency radiation as depicted (a) in the time domain, and (b) in the frequency domain. In each case the ordinate is in arbitrary units. Note that the nature of Fourier transformation implies that it is $B_1\tau_p$ which is plotted in (b) rather than B_1 itself, so that the maximum amplitude is τ_p times that in the time domain. Figure (a) shows the 'on' or 'off' status of the r.f.

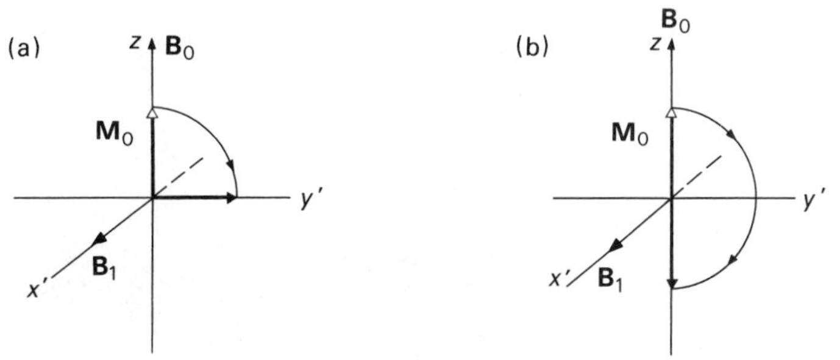

Fig. 3-7 The effects of (a) a 90° pulse, and (b) a 180° pulse (in the frame of reference rotating at $-\omega_1$).

$\pi/2$)-pulse and a 180° (or π)-pulse respectively (see Fig. 3-7). Thus the use of pulses enables the experimenter to place the magnetization in any chosen direction, without any loss in magnitude of M (provided $\tau_p \ll T_1, T_2^*$).

Following the pulse, Boltzmann equilibrium is slowly restored by relaxation. A 180° degree pulse is succeeded by purely spin–lattice relaxation, normally exponential as in Eqn 3-27:

$$M_z(t) - M_0 = [M_z(0) - M_0] \exp(-t/T_1) \tag{3-27}$$

Similarly, in the absence of B_0 inhomogeneities, a 90° pulse is followed by transverse relaxation, usually exponential as in Eqn 3-28:

$$M_y(t) = M_y(0) \exp(-t/T_2) \tag{3-28}$$

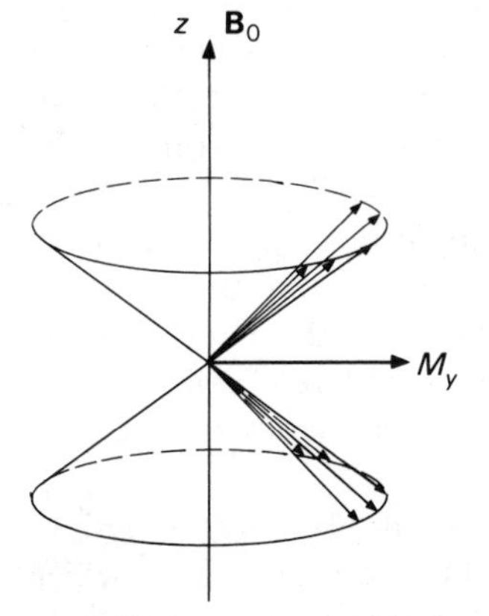

Fig. 3-8 Phase coherence of individual nuclear spins following a 90° pulse. Most spins are still randomly distributed around the two cones; only the small fraction that are 'bunched' are indicated schematically here.

Note that the times T_1 and T_2 are exactly those used in the Bloch equations, being the spin–lattice and transverse relaxation times respectively. In fact Eqns 3-27 and 3-28 are simply the integrated forms of Eqns 3-5 and 3-6. The reader should note that relaxation is being treated as a first-order kinetic process, so that T_1^{-1} and T_2^{-1} are relaxation rate constants.

The simple physical picture described in Sections 1-6 and 1-7 makes it easy to see that a 180° pulse simply inverts the populations of the α and β states, thus inverting the net magnetization. In terms of individual spins, a 90° pulse can be viewed as causing them to 'bunch' as indicated schematically in Fig. 3-8. Such 'bunching' is more properly termed *phase coherence.*

3-8 The Free Induction Decay and Fourier transform NMR spectra

Following a 90° pulse the magnetization is placed in the y' direction in the rotating frame of reference. Consequently a signal will be observed which will be attenuated as transverse relaxation occurs. The plot of signal intensity vs. time is referred to as a *free induction decay* (FID). If all the spins being monitored are equivalent, and if the carrier frequency is exactly on resonance, a simple exponential decay will be observed. Detection itself is usually based on the carrier frequency, so any deviation from resonance is manifested as a modulation of the FID, as in Fig. 3-9. If the sample contains many different spins of the type being monitored, each with its own resonance frequency, the FID will be a complicated interferogram, as in Fig. 3-10.

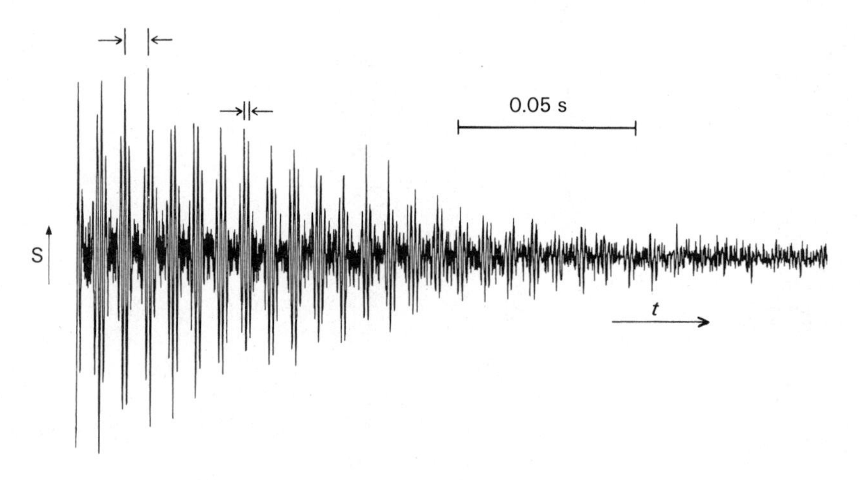

Fig. 3-9 Carbon-13 free induction decay for methyl iodide. The coarse modulation occurs at time intervals $|^1J_{CH}|^{-1}$; the faster modulation occurs at time intervals $|\nu_0 - \nu_c|^{-1}$ where ν_0 is the ^{13}C resonance frequency for CH_3I and ν_c is the carrier frequency (ca. 25 MHz in this case).

Now the information content of the FID must be the same as that in a conventional spectrum under appropriate conditions, and in 1966 Ernst and Anderson[4] demonstrated in practice what had been previously suggested, namely that Fourier transformation of a FID produced an acceptable spectrum. In other words, the spectrum is the frequency-domain equivalent of the time-domain FID. The Fourier transformation process is mathematically simple but tedious; an ideal situation for a computer. It is scarcely a coincidence that the commercial establishment of NMR spectrometers based on the Fourier transform (FT) principle occurred just at the time that reasonably-priced minicomputers, which could be dedicated to a spectrometer, became available.

For many experiments it is necessary to calibrate B_1. The strength of B_1 is usually reported in terms of the duration of a 90° pulse, which secures the maximum signal height obtainable from a single pulse. In fact, measurement of peak height as a function of τ_p should yield a sine curve (Fig. 3-11), the first null point (after $\tau_p = 0$) corresponding to a

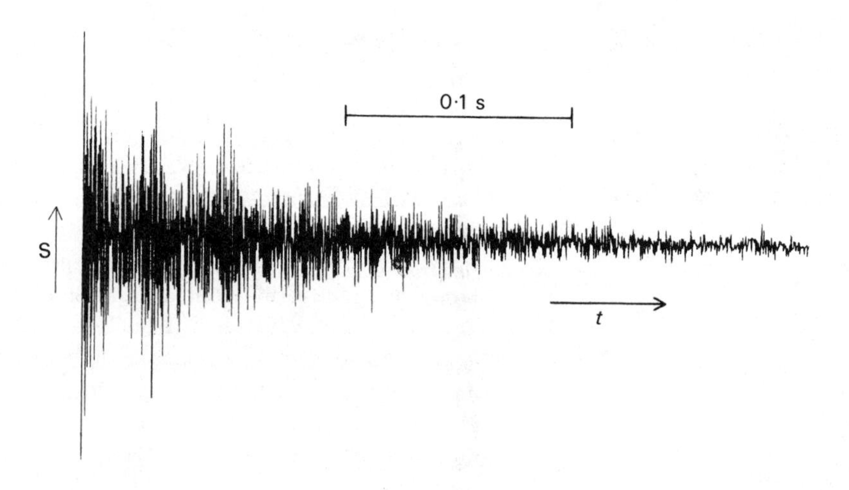

Fig. 3-10 Carbon-13 free induction decay for cholesterol (saturated solution in $CDCl_3$) at 25 MHz under conditions of proton noise decoupling.

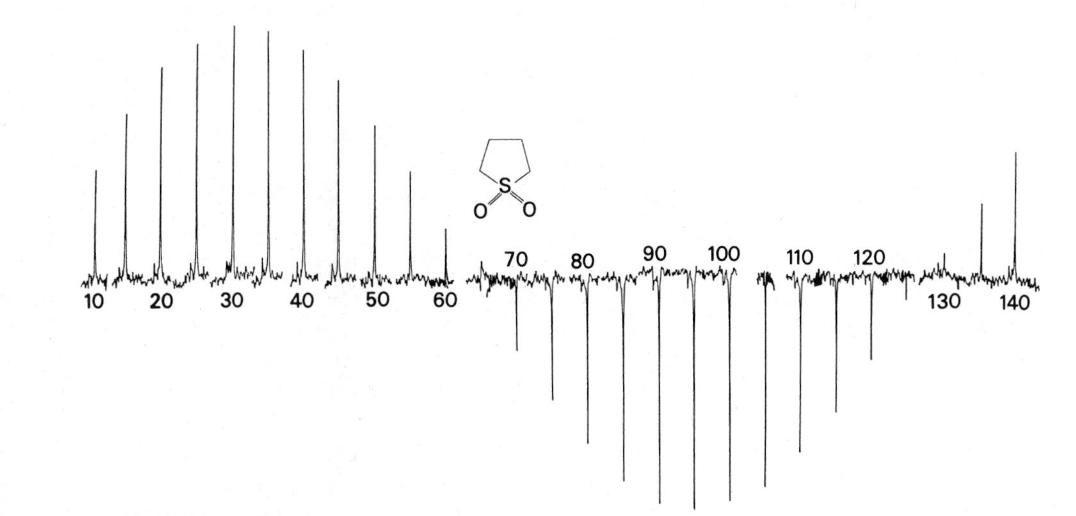

Fig. 3-11 Calibration of radio-frequency pulse power, carried out using a Bruker CXP 300 spectrometer for the ^{33}S signal of sulpholane. The pulse duration was incremented by 5 μs between experiments and the values are indicated on the figure. The first null point, ca. 65 μs, indicates the duration required for a 180° pulse.

180° pulse. Distortions from a sine curve often imply that the r.f. magnetic field B_1 is inhomogeneous over the volume of the sample, a condition it is generally desirable to avoid (e.g. by limiting the sample volume), particularly when measuring relaxation times.

3-9 Multipulse operation

There are two great advantages of FT operation for NMR. As will be seen later, one of these is that new experiments become possible. The more basic (albeit more trivial) advantage is that there is a substantial gain in sensitivity over CW operation. The essential reason for this is that CW spectroscopy wastes time by detecting only one frequency at a time and ignoring all others; this appears particularly undesirable when one realizes that most of the time the spectrometer is scanning baseline! In contrast, in FT NMR all frequencies of interest are being monitored all the time. Moreover, recording a typical ^1H spectrum by CW methods typically takes 1000 s, whereas monitoring a FID may be complete in 1 s. As always, time can be traded for sensitivity by repeating the experiment in a manner analogous to the multiscan principle discussed in Section 1-19. For FT NMR it is *multipulse* operation that is required.

Thus, following a pulse, the FID is monitored digitally and stored in the channels of a computer during the *acquisition time*. At a suitable point in time a new pulse is applied and the succeeding FID is again monitored and co-added to the first (in the same channels of the computer). This process is repeated as often as required to obtain the necessary signal intensity, and the accumulated FID is Fourier transformed at the end of the process to yield the desired spectrum. The number of transients acquired in this fashion may run into tens of thousands. For Figs. 3-9 and 3-10 1000 transients were accumulated. Note that only one Fourier transformation is required. It can be shown that the improvement in S/N in a given time of the FT method over the CW mode is of the order of $[F/\Delta\nu_{1/2}]^{1/2}$ where F is the spectral width examined and $\Delta\nu_{1/2}$ is the width of the individual lines in the spectrum.

However, so far the question of equilibration of the magnetization has not been discussed. If each successive FID from a train of pulses is to be identical to the first, then the interpulse time must suffice for full relaxation to occur, which implies a waiting period of $\geqslant 5T_1$. On the other hand the FID actually decays in a characteristic time T_2^* which is usually substantially less than T_2 because of the effects of inhomogeneities in B_0, and there is no point in having acquisition times longer than ca. $3T_2^*$. This implies a waiting period before the next pulse which is entirely unproductive, thus losing (at least partially) the big advantage of FT operation. If accurate relative signal intensities are required this may be necessary (see Section 4-12), but in general it is undesirable. If such a waiting period is *not* left, the result is attenuation of the signal due to saturation. This is, in effect, the same phenomenon as discussed in Section 3-5 but is manifested only as a reduction in peak heights in the spectrum and not as the line broadening which occurs for CW operation. In FT NMR saturation can be avoided in an alternative way to leaving delay periods, namely by reducing the pulse duration so that the pulse angle (through which precession of M about B_1 occurs) is less than 90°. In fact the recommended method of operation is to leave no delay period, but to initiate a new pulse immediately after the acquisition time, T_{ac}, and to have a pulse angle, α_E (known as the Ernst angle)[4] such that

$$\cos\alpha_E = \exp(-T_{ac}/T_1) \qquad (3\text{-}29)$$

Under such conditions multipulse operation produces a steady-state situation of partial saturation after the first few pulses. Of course, if T_1 varies for different peaks in the spectrum, as often occurs, difficult choices must be made.

It is the combination of the Fourier transform principle with multipulse operation and noise decoupling that has enabled ^{13}C NMR spectra to be obtained routinely. Under optimum operating conditions, including use of the Ernst angle, and with the assumption that $T_1 = T_2^*$ it can be shown that the definition of receptivity given in Eqns 3-24 and 5-29 also holds for FT operation.

3-10 Other operational facets of Fourier transform NMR

Normally the FID is detected *in quadrature*, that is the components of magnetization which are in-phase and out-of-phase with the irradiation (u and v respectively) are both observed simultaneously so that pairs of points are recorded in the computer. The interval between pairs of data points is known as the *dwell time*, τ_d. If N is the total number of computer channels used to store the FID, then $\tau_d = 2T_{ac}/N$ by definition. True frequency information is only obtained within a spectral width, F, given by

$$F = N/2T_{ac} = \tau_d^{-1} \tag{3-30}$$

with the carrier frequency, ν_c, at the centre. Any peak in the spectrum which falls outside this range, at $\nu_c + (F/2) + f$, say, appears as a ghost *fold-back* peak at $\nu_c - (F/2) + f$. It is therefore important to use an

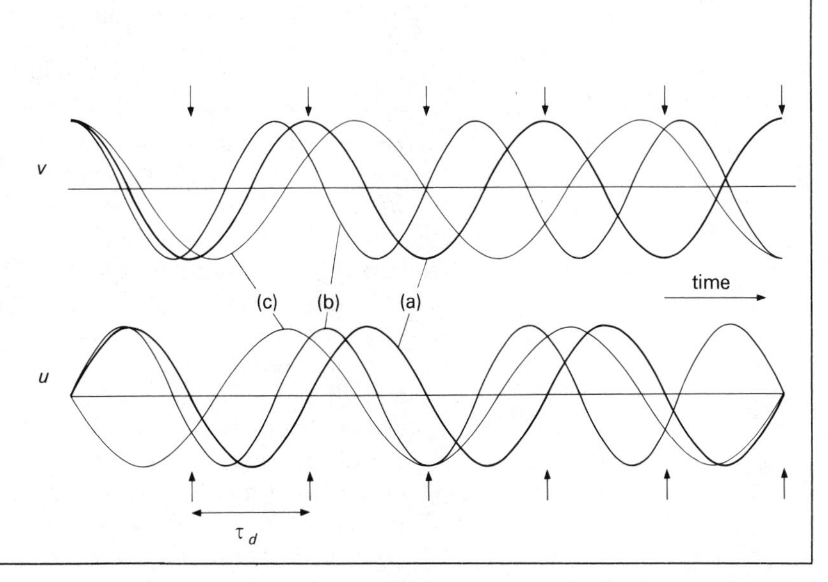

Fig. 3-12 Waveforms to indicate the origins of fold-back. The magnetization components v and u are drawn as a function of time for three different values of $\nu - \nu_c$. The vertical arrows indicate the digitization times, separated by τ_d. Traces (a) are for a frequency difference $F/2$ which matches half the digitization rate, so that there are exactly two pairs of digitization points per cycle. This is the highest frequency difference that will be properly represented in the spectrum following Fourier transformation. Traces (b) are of a slightly higher frequency difference than (a)—in the case illustrated the excess frequency difference, f, is $(1/6) \times (F/2)$, whereas traces (c) have a correspondingly lower frequency difference. It may be seen that the digitized points for traces (b) and (c) are identical and therefore the experiment cannot distinguish between the two frequencies. The magnetization u shows that this situation is only obtained when the phase of trace (c) is 180° different from that of trace (b); this is recognized by the computer as due to a *negative* frequency difference $\nu - \nu_c$. Therefore the frequency $\nu_c + (F/2) + f$ is recorded as equivalent to $\nu_c - (F/2) + f$.

appropriate electronic filter to attenuate the intensity of peaks (and also of noise!!) outside the range $\nu_c \pm F/2$. The reason for the appearance of fold-back is given in Fig. 3-12. If it becomes necessary to check whether a peak is folded back or not, the carrier frequency should be altered. The limiting frequency $F/2$ is often known as the *Nyquist frequency*. Clearly, the operator may choose f to cover the whole spectrum of interest by suitably adjusting N or T_{ac} (or the rate of digitization, τ_d^{-1}, directly).

It is of importance to realize that Fourier transformation yields absorption-mode and dispersion-mode spectra separately (usually the former is wanted), with $N/2$ points in each. Thus the digitization of the spectrum (points per Hz) is, by Eqn 3-30, equal to T_{ac}. Clearly, to reproduce lineshapes properly one requires $T_{ac}^{-1} < \Delta\nu_{1/2}$. One may regard T_{ac}^{-1} as a measure of the *resolution* implicit in the experiment. Therefore T_{ac} must be adjusted to give the resolution required by the spectrum in question. Unfortunately, with fixed N, increase in T_{ac} to achieve better resolution implies a narrowing of F. It also means the experiment takes longer (or sensitivity is lost). So the criteria for good S/N and good resolution are incompatible and a compromise must be reached depending on the objectives of the experiment.

3-11 Cosmetic improvements by computer

The fact that a computer is interfaced to the spectrometer allows the operator to process the data in many different ways before the final spectrum is produced. Four such ways will be mentioned at this point:

(i) *Sensitivity enhancement.* The form of a FID implies that in the time domain the S/N near the end is substantially less than at the beginning. An improvement in S/N in the resulting spectrum after Fourier transformation should therefore result if the FID is attenuated in some way. This is usually done by multiplying it by an exponential function $\exp(-t/T_2^*)$. Unfortunately, the increase in S/N is accompanied by some line broadening, and the value T_2^* for the exponential time constant is chosen so as to optimize the appearance of the spectrum.

(ii) *Resolution enhancement.* The end of the FID mediates the ability to distinguish between two signals which are close in frequency. Therefore to improve the resolution the end of the FID must be augmented, i.e. we must use the reverse of the sensitivity enhancement routine. Naturally, any improvement in resolution will be accompanied by a deterioration in S/N.

(iii) *Digitization improvement.* It has been shown above that to achieve a required spectral width the value of τ_d must be fixed by Eqn 3-30. On the other hand a given resolution requires a definite value of T_{ac}. It may be that these criteria involve the use of less than the available computer store. The extra channels may be used by filling each with a zero and incorporating them on the end of the FID before Fourier transformation. This process of *zero-filling* will improve the digitization of the spectrum (though not its intrinsic resolution), giving it a smoother and more realistic appearance and enabling peak maxima to be located more accurately.

(iv) *Difference spectra.* Occasions arise when it is important to study changes or differences in spectra, e.g. when there is an interfering signal. The computer-based nature of a FT NMR spectrometer allows this to be done readily by storing an FID from the experiment and one from a blank run of the interfering signal only. The two FIDs can be subtracted prior to Fourier transformation. Such an operation (but actually involving only one original FID) can be used to obtain a good spectrum of sharp peaks when broad peaks partially obscure the information. Sensitivity enhancement is first carried out, such that the sharp lines are greatly broadened but the broad lines are scarcely affected. The resulting FID is subtracted from the original FID and the difference is Fourier transformed to give a spectrum exhibiting the sharp lines only. This procedure is known as *convolution difference spectroscopy.*

3-12 Measurement of T_1 by the inversion recovery method

As stated in Section 3-7, a 180° pulse is followed by purely spin–lattice relaxation, frequently obeying Eqn 3-27, but no signal is seen because no magnetization is produced in the y direction. However, at any point in time following the pulse the state of the magnetization M_z may be monitored by applying a 90° pulse (sometimes referred to as a 'read' pulse). This situation forms the basis of a method for measuring T_1, known as the inversion-recovery method because the first step inverts the magnetization. The pulse sequence is

$$[180° - \tau - 90°(\text{FID}) - T_\text{d}]_n \tag{3-31}$$

where a delay time T_d, longer than the longest T_1 to be measured, must be left for the Boltzmann populations to be re-established between 180° pulses. The accumulated FID is Fourier transformed, and the parameter τ is varied. The natural logarithm of the peak height, S, for each peak may be plotted against τ, and thus T_1 found according to the following equation (see Eqn 3-27; $M_z(0) = -M_0$ at inversion):

$$\ln[S(\infty) - S(t)] = \ln 2 + \ln S(\infty) - \tau/T_1 \tag{3-32}$$

At short recovery times τ the signal will appear negative, since in effect a 270° pulse has been applied. At long times, full recovery is obtained so that $S_0 \equiv S(\infty)$. At some intermediate time the signal will be zero; this time, τ_null, can be used to give a measure (generally of low accuracy) for T_1 according to

$$T_1 = \tau_\text{null}/\ln 2 \tag{3-33}$$

An example of stacked plots from a $T_1(^{13}\text{C})$ determination for chlorobenzene is given in Fig. 3-13. It can be seen that the different

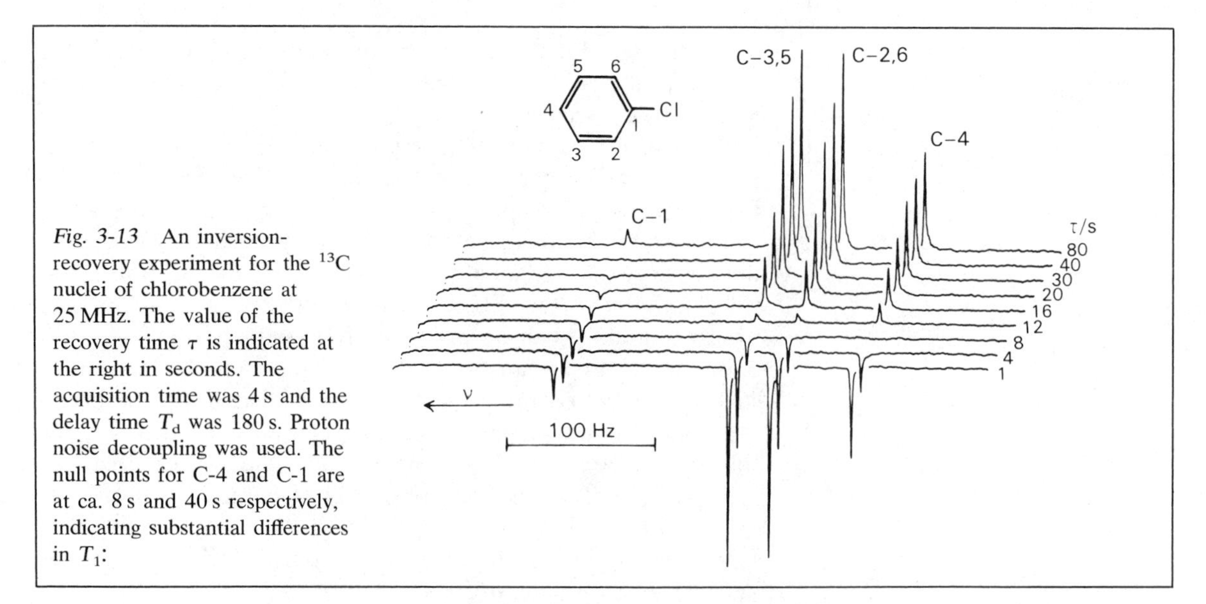

Fig. 3-13 An inversion-recovery experiment for the ^{13}C nuclei of chlorobenzene at 25 MHz. The value of the recovery time τ is indicated at the right in seconds. The acquisition time was 4 s and the delay time T_d was 180 s. Proton noise decoupling was used. The null points for C-4 and C-1 are at ca. 8 s and 40 s respectively, indicating substantial differences in T_1:

carbon nuclei differ in their values of T_1. This example illustrates the power of the FT techniques, since the T_1 determination is clearly *selective*—a single set of experiments gives the separate spin–lattice relaxation times for each carbon.

In order to minimize errors due to spectrometer drift, the Freeman–Hill modification is often used. This uses an extended pulse sequence, 3-34:

$$[180° - \tau - 90°(\text{FID}_1) - T_d - 90°(\text{FID}_2) - T_d]_n \qquad (3\text{-}34)$$

to monitor the equilibrium magnetization after each measurement of $M_z(\tau)$. The FID_1 is subtracted from FID_2 before Fourier transformation. The resulting peaks are all positive and range from $2S_0$ at $\tau = 0$ to zero at $\tau \to \infty$. Further improvements in accuracy can be obtained by cycling τ instead of accumulating the free induction decays for different τ values blockwise.

3-13 Spin-echoes and the measurement of T_2

As has been suggested earlier there are two factors which contribute to the FID following a single 90° pulse. Firstly, the Larmor frequencies of different portions of the sample differ slightly as a result of field inhomogeneities, and, in axes rotating at the mean Larmor frequency, the magnetization vectors of these different portions slowly fan out. Secondly, random processes tend to realign the nuclei within each portion to re-establish thermal equilibrium; this is true transverse relaxation. Both these processes normally contribute to the linewidth, but they can be measured separately by applying a succession of pulses and observing spin-echoes. The production of these echoes, first studied by Hahn,[5] is shown in Fig. 3-14.

An initial 90° pulse turns M_0 into the y direction in the system of rotating axes (Fig. 3-14(a)). The magnetization vectors from different

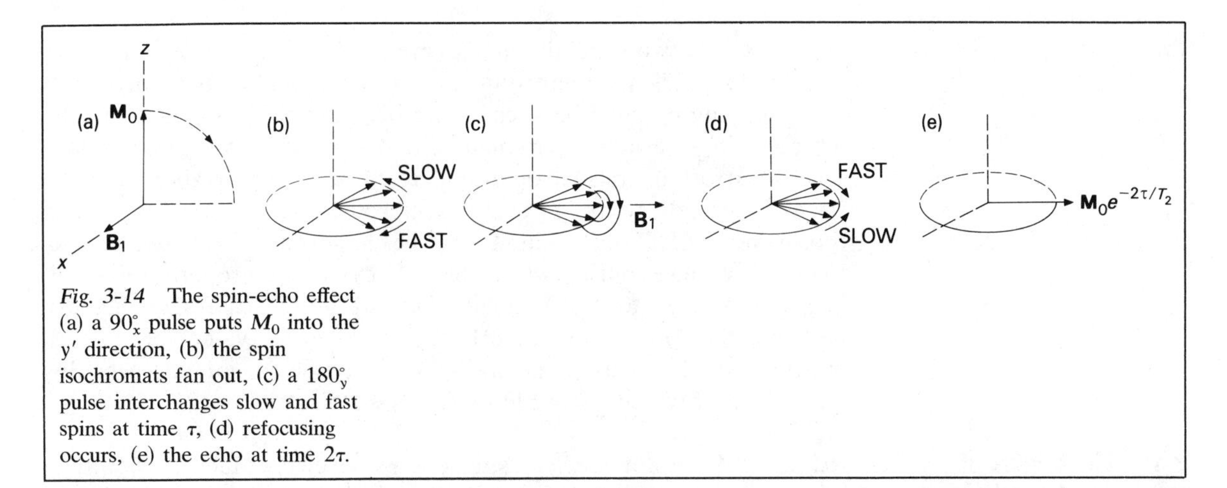

Fig. 3-14 The spin-echo effect (a) a 90°_x pulse puts M_0 into the y' direction, (b) the spin isochromats fan out, (c) a 180°_y pulse interchanges slow and fast spins at time τ, (d) refocusing occurs, (e) the echo at time 2τ.

portions of the sample then begin to fan out (b) and the signal decays. A 180°_y pulse (note the phase change) is then applied after time τ(c), which has the effect of rotating all the magnetization vectors by 180° about y', or in other words reflecting them in the $y'z$ plane. They continue to move in the same direction, and after a further time τ they are again in phase in the y' direction. The process can then be repeated by applying successive 180°_y pulses (expression 3-35), a sequence known as the Carr–Purcell–Meiboom–Gill[6] (CPMG) experiment. The amplitudes of successive echoes decay exponentially (see Fig. 3-15), due to relaxation, and the true value of T_2 can be found from the envelope of the echoes.

$$90^\circ_x\{\tau - 180^\circ_y - \tau\text{-echo}\}_n \qquad (3\text{-}35)$$

In appearance a spin-echo is like two free induction decays back-to-back, and it is possible to record the FID which follows the echo maximum and Fourier transform it to obtain the spectrum. In the absence of modulation (see Chapter 7) due to coupling (such a situation would be normal for a proton-decoupled ^{13}C system) the spectrum will be almost identical to that obtained from a FID following a simple

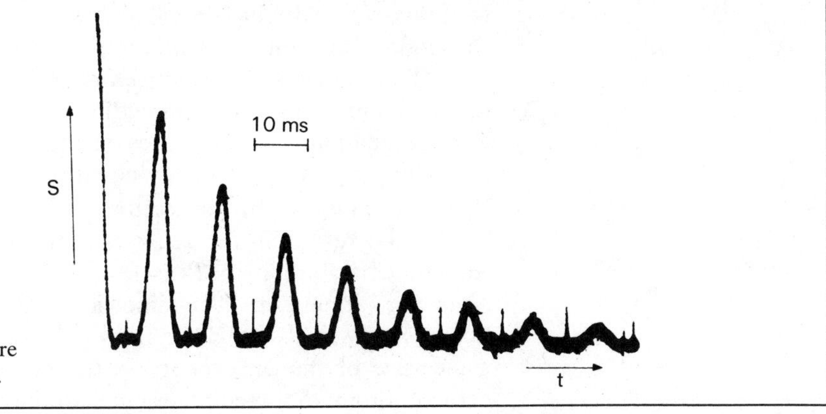

Fig. 3-15 The results of a CPMG experiment (see Eqn 3-33). Responses of the electronics to the 180° pulses are just visible between the echoes.

90° pulse. The two small differences are (a) there will be some intensity loss due to transverse relaxation, and (b) there will be no distortions at the beginning of the FID such as are occasioned by proximity to the 90° pulse in a normal experiment. The second point can cause serious spectral errors in normal operation if T_2^* is relatively short and the signal inherently weak. In such circumstances the spin-echo Fourier transform (SEFT) experiment can be important. The 'dead time' following a pulse, during which the FID cannot be properly recorded, is of the order of tens of μs. The 180° pulse of a SEFT sequence can be applied before the initial FID is completed, so values of τ of ca. 50–100 μs can be used, giving an echo which is not unreasonably attenuated provided $T_2^* \geqslant 200$ μs ($\Delta\nu_{1/2} \leqslant 1{\cdot}5$ kHz).

3-14 The origins of relaxation for spin-$\frac{1}{2}$ nuclei

Relaxation between nuclear spin energy levels is much slower than relaxation of rotational or vibrational levels. The distortion of molecules during collisions often results in changes of rotational and vibrational states, but changing the orientation of a nucleus is a more difficult matter. Consequently, relaxation times of seconds are common in nuclear resonance, while rotational and vibrational relaxation times are more often of the order of 10^{-9} and 10^{-6} s respectively. A common temperature of vibrational, rotational and translational degrees of freedom is quickly established.

As the only property of a nucleus of spin-$\frac{1}{2}$ which depends on orientation is its magnetic moment, transitions between nuclear spin levels can only be induced by magnetic fields. Thermal relaxation to give populations of the nuclear spin levels equal to those predicted by Boltzmann's distribution law with the same temperature as the bulk of the sample can only occur via random magnetic fields within the sample. A nucleus in a liquid will experience a fluctuating field, due to the magnetic moments of nuclei in other molecules as they execute Brownian motion. This random fluctuating field may be resolved by Fourier analysis into terms oscillating at different frequencies (see Section 3-15) and may be further subdivided into components perpendicular to B_0 and those parallel to B_0. The component perpendicular to the static field which oscillates with the Larmor frequency induces transitions between the levels in a similar way to an electromagnetic field. This gives rise to a *non-adiabatic* (or *non-secular*) contribution to relaxation of both the longitudinal and transverse components of **M**. The populations of the states change until they reach the values predicted by the Boltzmann equations for the temperature of the Brownian motion (the 'lattice' temperature). This process is described by T_1, and results in the relaxation of the longitudinal component of **M**. The contribution to T_2 is less obvious, but can be seen from the following argument. The linewidth (excluding the effect of inhomogeneity in B_0), which is inversely proportional to T_2 (Fig. 3-5), is a measure of the uncertainty in the energies of the two states concerned. From the Heisenberg uncertainty principle this is inversely proportional to the lifetimes of these states, which we have seen are reduced by random fluctuations of the local magnetic field. Thus, fluctuations which cause transitions between states result in both the changes of population associated with longitudinal relaxation (T_1) and the increase in linewidth associated with transverse relaxation (T_2). Since both x and y components of a fluctuating field affect T_1 but M_y can only be affected by the x components the non-secular contribution to the relaxation rate T_2^{-1} is half that of T_1^{-1} for mobile liquids. However, there is a second contribution to T_2^{-1} arising from fluctuating fields in the z direction. This is the *adiabatic* (or *secular*) contribution and involves no energy exchange with the lattice but is related directly to variations in the total magnetic field in the z direction, and hence to linewidths. For a mobile liquid this contribution is usually equal to $\frac{1}{2}T_1^{-1}$ so that $T_1 = T_2$. However, this is not normally the case in other circumstances (see the following section and Chapter 6).

3-15 The theory of relaxation

The general requirement for relaxation, as stated in the preceding section, is a magnetic field fluctuating at the appropriate frequency. The theory of relaxation is not altogether straightforward and requires considerable background knowledge. This section will attempt a derivation for a particularly simple case. A system of isolated spins of a single type is assumed, with an isotropic random magnetic field of unspecified origin acting on it in a time-dependent fashion. The x component of this local field may be written

$$B_{xL}(t) = B_{xL}^0 \mathrm{f}(t) \tag{3-36}$$

where $\mathrm{f}(t)$ has a zero average and a root-mean-square average of unity, while B_{xL}^0 is the r.m.s. average amplitude of the field. This field contributes a perturbation $\mathscr{H}'(t)$ to the Hamiltonian, by interacting with the x component of the nuclear magnetic moment:

$$\hat{\mathscr{H}}'(t) = -\gamma\hbar\hat{I}_x B_{xL}^0 \mathrm{f}(t) \tag{3-37}$$

Perturbation theory shows (see Appendix 3) that this induces transitions at a rate

$$\begin{aligned} W_1 &= \hbar^{-2}\langle\alpha| -\gamma\hbar\hat{I}_x B_{xL}^0 |\beta\rangle^2 J(\omega_0) \\ &= \gamma^2[B_{xL}^0]^2\langle\alpha|\hat{I}_x|\beta\rangle^2 J(\omega_0) \\ &= \tfrac{1}{4}\gamma^2[B_{xL}^0]^2 J(\omega_0) \end{aligned} \tag{3-38}$$

where $J(\omega_0)$ is the power available from the fluctuations at the relevant transition frequency ω_0; $J(\omega_0)$ is known as the *spectral density*, and is clearly a frequency-domain function which must be linked with $\mathrm{f}(t)$. Since an isotropic random field has been assumed, B_{yL} will contribute an equal term to W_1, though B_{zL} will not cause any transitions. Thus *in toto:*

$$W_1 = \tfrac{1}{2}\gamma^2[B_{xL}^0]^2 J(\omega_0) \tag{3-39}$$

However, each transition alters the population difference by 2, so:

$$T_1^{-1} = 2W_1 = \gamma^2[B_{xL}^0]^2 J(\omega_0) \tag{3-40}$$

Further progress can only be made by discussing the nature of f(t). The important facet about f(t) is its 'memory', expressed in terms of an *auto-correlation function* $G(\tau)$, defined[7] as:

$$G(\tau) = \overline{f(t)f(t+\tau)} \tag{3-41}$$

where the horizontal bar indicates an *ensemble average* (i.e. an average over all the spins). This function is independent of t but indicates how f(t) changes. Clearly it must decay with τ, as 'memory' is lost, and this decay is often assumed to be exponential:

$$G(\tau) = \exp(-|\tau|/\tau_c) \tag{3-42}$$

where τ_c is a characteristic of the decay known as the *correlation time* for the particular motion involved. For random molecular tumbling, τ_c roughly corresponds to the average time for a molecule to progress through one radian. Notice that a more mobile solution has a *lower* τ_c (shorter memory).

In fact $G(\tau)$ is the time-domain function corresponding to $J(\omega)$, related to it by Fourier transformation:

$$J(\omega) = \int_{-\infty}^{\infty} G(\tau)\exp(i\omega t)\,d\tau \tag{3-43}$$

Substitution of Eqn 3-42 followed by explicit integration gives

$$J(\omega) = 2\tau_c/(1+\omega^2\tau_c^2) \tag{3-44}$$

Substitution into Eqn 3-38 gives

$$T_1^{-1} = \gamma^2[B_{xL}^0]^2 \frac{2\tau_c}{1+\omega_0^2\tau_c^2} \tag{3-45}$$

It is instructive to look at the form of a plot of $J(\omega)$ vs. $\log \omega$ (see Fig. 3-16). Notice the flat portion and the sudden drop at $\omega\tau_c \sim 1$. The flat portion occurs when $\omega^2\tau_c^2 \ll 1$, known as the *extreme narrowing* condition. In such circumstances Eqn 3-45 reduces to

$$T_1^{-1} = 2\gamma^2[B_{xL}^0]^2\tau_c \tag{3-46}$$

For mobile solutions, molecular tumbling typically gives $\tau_c \sim 10$ ps, so that for the usual resonance frequencies (ω_0 being hundreds of Mrad s^{-1}) the extreme narrowing condition holds. This implies that the spin–lattice relaxation rate increases as τ_c increases, i.e. as mobility decreases. This occurs, for example, for molecular tumbling when the temperature is lowered.

However, if τ_c increases sufficiently so that it becomes of the order of ω_0^{-1}, a new regime is entered. The full form of $J(\omega)$ (Eqn 3-44) predicts a maximum in T_1^{-1} (a minimum in T_1) as a function of τ_c (see

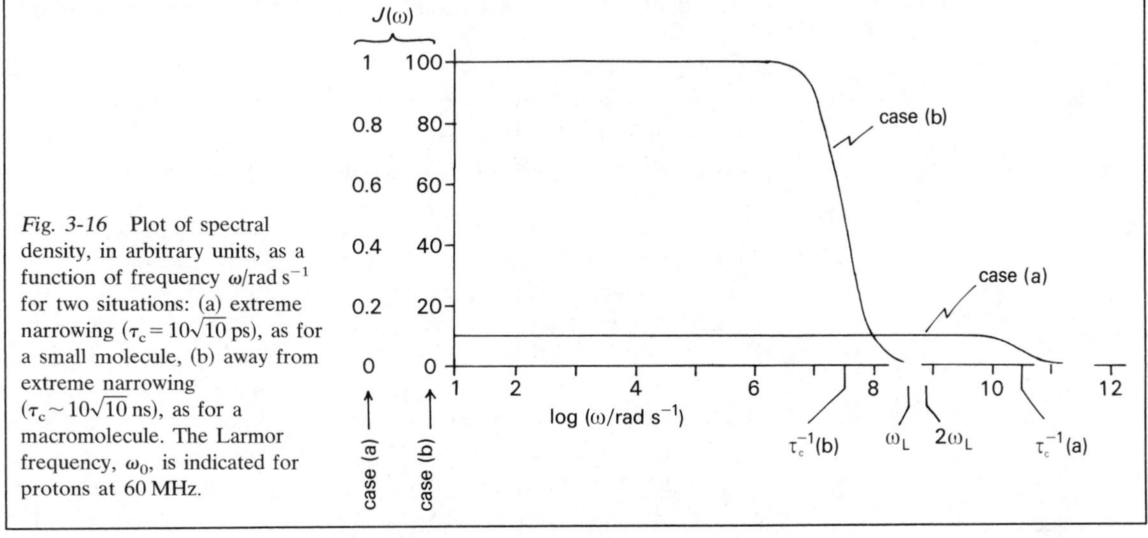

Fig. 3-16 Plot of spectral density, in arbitrary units, as a function of frequency ω/rad s^{-1} for two situations: (a) extreme narrowing ($\tau_c = 10\sqrt{10}$ ps), as for a small molecule, (b) away from extreme narrowing ($\tau_c \sim 10\sqrt{10}$ ns), as for a macromolecule. The Larmor frequency, ω_0, is indicated for protons at 60 MHz.

Fig. 3-17). The T_1 minimum occurs when the sudden drop in $J(\omega)$ as a function of $\log \omega$ (see Fig. 3-16) corresponds to the resonance frequency, i.e. when $\tau_c = \omega_0^{-1}$. In such circumstances, which can obtain for macromolecules in solution:

$$[T_1(\min)]^{-1} = \frac{\gamma^2}{\omega_0}[B_{xL}^0]^2 \tag{3-47}$$

Equation 3-47 makes it clear that at higher applied magnetic fields the minimum value of T_1 increases, as is shown in Fig. 3-17 and as may be deduced from Fig. 3-16.

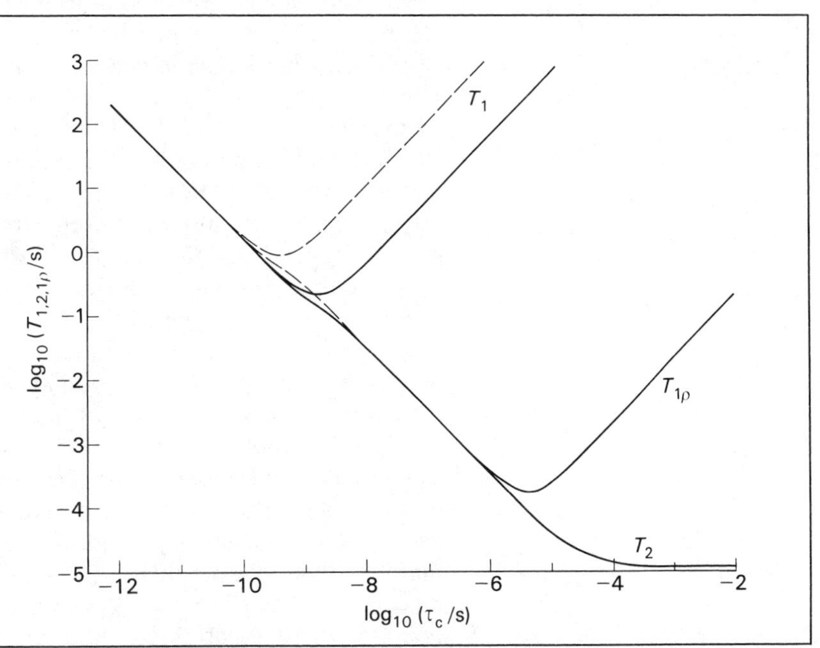

Fig. 3-17 The dependence of T_1, T_2 and $T_{1\rho}$ on the correlation time τ_c and the spectrometer operating frequency ν_0 for relaxation by isotropic random magnetic fields (see Eqns 3-46, 3-48 and 3-54). Note the log/log scale. The extreme narrowing regime is to the left. The solid lines are calculated for $\nu_0 \sim 100$ MHz and the dashed lines show the deviations for $\nu_0 \sim 400$ MHz. The curves for $T_{1\rho}$ are calculated for $\nu_1 = \gamma B_1/2\pi \sim 40$ kHz. The rigid lattice value for T_2 (as $\tau_c \to \infty$) is somewhat arbitrary since it depends on the distribution of the random fields—the theory of Section 3-15 is not valid in this region. A value of 0·2 mT has been used for the r.m.s. random field.

As has been pointed out earlier, factors affecting spin–lattice relaxation will also cause transverse relaxation. However, whereas the non-secular contribution requires a fluctuating magnetic field, and thus depends on $J(\omega_0)$, the secular contribution involves no energy change, and the appropriate spectral density is $J(0)$. In fact the general expression for T_2 in the random-field model discussed in this section is

$$T_2^{-1} = (2T_1)^{-1} + \tfrac{1}{2}\gamma^2[B_{zL}^0]^2 J(0) \tag{3-48}$$

which is equal to T_1^{-1} provided $B_{zL}^0 = B_{xL}^0$ and $J(0) = J(\omega_0)$. The influence of the $J(0)$ term causes T_2 to decrease monotonically as τ_c increases in contrast to T_1, as is shown in Fig. 3-17 (but see Section 6-3). In fact at the T_1 minimum $T_2 = \frac{2}{3}T_1$. The concepts regarding relaxation, as given in the present section and as further elaborated for the dipolar relaxation mechanism in Section 4-3 and Appendix 3, were first described by Bloembergen, Purcell and Pound[8] and are frequently referred to as BPP theory. This is only valid in the so-called 'weak collision' case, where $\tau_c < T_2$. The situation in relatively rigid materials, where the extreme narrowing condition is usually not valid and where, in the limit, BPP theory is also not applicable, is further discussed in Chapter 6.

3-16 Relaxation mechanisms for spin-$\frac{1}{2}$ nuclei

As noted in Section 3-14, the general requirement for spin–lattice relaxation is a magnetic interaction fluctuating at the resonance frequency. A nearly static component of the magnetic interaction is also effective for T_2. There are a number of physical mechanisms which provide the appropriate conditions, and at this point these are summarized, and the relevant equations given (Table 3-2), though in some cases more detail is supplied later.

(a) *Dipole–dipole interactions with other nuclei* (DD)

Dipolar interactions are of the same type as those observed macroscopically between two small bar magnets. They may be modulated by molecular tumbling (affecting intermolecular and intramolecular interactions) or by translational diffusion (intermolecular interactions), thus causing relaxation, and the equations for the two cases, given in Table 3-2, differ. There is also a difference of a factor of 3/2 between the case where the two nuclei under consideration both contribute to the magnetization being monitored (the homonuclear case) and that where one magnetization is under observation independently of the other (the heteronuclear case). Dipolar interactions are discussed in more detail in Chapter 4.

(b) *Shielding anisotropy* (SA)

The shielding at a nucleus, and therefore the magnetic field acting on it, varies with the molecular orientation in the static field B_0 (see Section 6-6), except for sites of very high symmetry. Molecular tumbling therefore modulates the local magnetic field, and can cause relaxation. The relevant equations are given in Table 3-2. It should be noted that the correlation time (in the case of isotropic motion) is

Table 3-2 Equations for spin–lattice relaxation rates, T_1^{-1}, of a spin-$\frac{1}{2}$ nucleus, A, for various mechanisms in the extreme narrowing approximation

Mechanism	T_1^{-1}	*Notes*
dd(intra)(homo)	$\left(\frac{\mu_0}{4\pi}\right)^2 \frac{3}{2}\gamma_A^4\hbar^2\tau_c/r^6$	for a single pair of spin-$\frac{1}{2}$ nuclei of separation r
dd(intra)(hetero)	$\left(\frac{\mu_0}{4\pi}\right)^2 \gamma_A^2\gamma_X^2\hbar\tau_c/r^6$	for a single pair of spin-$\frac{1}{2}$ nuclei AX of separation r
dd(inter)(hetero)	$\left(\frac{\mu_0}{4\pi}\right)^2 \frac{2}{15} N_X\gamma_A^2\gamma_X^2\hbar^2/Da$	for relaxation by a spin-$\frac{1}{2}$ nucleus X
ue(intra)(dipolar)	$\left(\frac{\mu_0}{4\pi}\right)^2 \frac{4}{3}\gamma_A^2\gamma_e^2\hbar^2 S(S+1)\tau_c/r^6$	for relaxation by unpaired electrons of total spin S at distance r
ue(scalar)	$\frac{8}{3}\pi^2 a_N^2 S(S+1)\tau_e$	for relaxation by unpaired electrons of total spin S
sr	$2I_r kTC^2\tau_{sr}/\hbar^2$	for isotropic molecular inertia
sa	$\frac{2}{15}\gamma_A^2 B_0^2\Delta\sigma^2\tau_c$	for cylindrical symmetry of shielding
sc	$\frac{8}{3}\pi^2 J_{AX}^2 I_X(I_X+1)\frac{\tau_{sc}}{1+(\omega_X-\omega_A)^2\tau_{sc}^2}$	relaxation by coupling to spin, X, of quantum number I_X

Meaning of symbols (where not otherwise defined in this Table or Section):

- τ_c Correlation time for molecular tumbling
- τ_e Correlation time related, *inter alia*, to the spin–lattice relaxation time for the unpaired electrons
- τ_{sc} For scalar relaxation of the first kind this is the exchange lifetime; for scalar relaxation of the second kind this is T_{1X}
- N_X Concentration of spins X (per unit volume)
- D Mutual translational self-diffusion coefficient of the molecules containing A and X
- a Distance of closest approach of A and X
- γ_e Magnetogyric ratio for the electron
- a_N Nucleus-electron hyperfine splitting constant (in frequency units)
- C Spin–rotation interaction constant (assumed to be isotropic)
- $\Delta\sigma$ Shielding anisotropy ($\sigma_\parallel - \sigma_\perp$; see Section 6-6)

exactly the same as that entering into the equations for dipolar relaxation caused by molecular tumbling. As shown by the equations in Table 3-2, the relaxation rate arising from shielding anisotropy is proportional to the square of the applied magnetic field B_0. This effect can cause unacceptable line broadening for some nuclei (e.g. ^{205}Tl) at high applied fields—for instance, the ^{205}Tl linewidth for Me_2TlNO_3 in D_2O in a magnetic field of 9·40 T is 140 Hz, whereas in a field of 1·41 T it is 4 Hz.[9] The SA mechanism is unusual in that, even in the extreme narrowing situation, T_1 and T_2 are unequal—in fact $6T_1 = 7T_2$.

(c) *Spin-rotation interactions* (SR)

Coherent molecular rotation generates a magnetic field which can couple with nuclear spin. Such coupling of angular momenta involves a magnetic field at the nucleus, and interruption of the coupling (e.g. by collisions) provides a relaxation mechanism. The correlation time, τ_{sr}, is therefore related to the time between collisions and clearly differs from the correlation time for molecular tumbling. Indeed, there is an *inverse* relationship—for spherical molecules with diffusion-controlled processes, Hubbard's Equation[10] (3-49) applies, where I_r is the molecular moment of inertia.

$$\tau_c\tau_{sr} = I_r/6kT \qquad (3\text{-}49)$$

(d) *Scalar interactions* (SC)

As noted in Chapters 1 and 2, two nuclear spins, A and X, can couple indirectly (via electrons) leading to a Hamiltonian term of the form:

$$hJ_{AX}\hat{\mathbf{I}}_A \cdot \hat{\mathbf{I}}_X \qquad (3\text{-}50)$$

Such an interaction, which involves a magnetic field produced by X acting at A (and vice versa), can lead to relaxation of A if a time-dependence occurs. This can happen in one of two ways:

(i) J_{AX} is time-dependent as a result of exchange
(ii) I_X is time-dependent as a result of relaxation

These two possibilities give what are known as scalar relaxation of the first kind and of the second kind respectively. The correlation times are the exchange rate and the X relaxation rate respectively. It is much more common for scalar relaxation to affect T_2 than T_1. This matter will be discussed further in Section 5-14.

(e) *Interactions with unpaired electrons* (UE)

Such interactions may be either of the dipolar or the scalar type. Modulation may either occur by exchange of the electron between different molecules or by spin–lattice relaxation of the electron itself. Dipolar interactions with electrons are much larger than those with nuclei because the magnetic moment of an electron is so high. Consequently the presence of paramagnetic impurities can have very severe consequences for nuclear relaxation. Traces of paramagnetics dissolved from the glass of NMR tubes by alkaline aqueous solutions can significantly alter relaxation times. So can dissolved oxygen, and it is essential to degass NMR samples by the freeze/pump/thaw method if meaningful relaxation measurements are required. Conversely, it is sometimes desirable to add traces of paramagnetic compounds such as $Cr(acac)_3$ (in ca. 0·03M concentration) to act as *relaxation agents*, which assist the efficiency of multipulse FT NMR operation by allowing interpulse times to be shorter or pulse angles larger.

Of these five mechanisms, only spin-rotation can strictly be treated using the isotropic random-field model,[11] though for most practical purposes the model can also be used for relaxation by interactions with unpaired electrons and for scalar relaxation of the second kind.

Relaxation rates due to different mechanisms are additive so that Eqn 3-51 applies:

$$T_1^{-1} = T_{1dd}^{-1} + T_{1sa}^{-1} + T_{1sr}^{-1} + T_{1sc}^{-1} + T_{1ue}^{-1} \qquad (3\text{-}51)$$

where dd, sa, sr, sc and ue stand for the five mechanisms discussed above. Indeed, further breakdown within these groups is also frequently possible. For example, dipolar relaxation can be apportioned to intra- and inter-molecular contributions:

$$T_{1dd}^{-1} = T_{1dd}^{-1}(\text{intra}) + T_{1dd}^{-1}(\text{inter}) \qquad (3\text{-}52)$$

However, care must sometimes be taken if there is correlation between different relaxation contributions, as for ^{13}C relaxation by (C, H) dipolar interactions in a CH_2 group, for which the two C–H vectors experience correlated motion.

It is frequently desirable to separate relaxation contributions from different mechanisms. Inter- and intra-molecular terms can be distinguished by dilution studies, but since this normally involves substantial viscosity changes, which affect relaxation through τ_c, it is best to use *isotopic* dilution. The commonest case is that of proton relaxation, when separation of inter- and intra-molecular contributions can be achieved by diluting with the corresponding perdeuterated material. Relaxation due to heteronuclear dipolar interactions can be separated from all the other terms (except, occasionally, scalar relaxation of the first kind) by measurement of the nuclear Overhauser effect, as will be fully discussed in Chapter 4. The temperature dependence of spin-rotation relaxation is opposite to that of all the other contributions, allowing a clear distinction. The dependence of T_{1sa}^{-1} on B_0^2 provides a means of distinguishing this mechanism, though scalar relaxation also depends on B_0 through the term $\omega_X - \omega_A$. Several of these separations can only be simply applied when the extreme narrowing condition is assumed.

3-17 Spin-locking and $T_{1\rho}$

As discussed earlier, a 90° pulse is followed by transverse relaxation. However, a different situation is obtained if the r.f. is not switched off, but its phase is changed by 90°, as illustrated in Fig. 3-18. In such a case the magnetization is aligned with the r.f. field, which is the only effective field in the rotating frame of reference when on resonance (Section 3-3). This situation is reminiscent of that with M_z in the laboratory frame with B_1 off. The magnetization is said to be *spin-locked* by B_{1y}. It will undergo no precession in the rotating frame. However, the magnitude of the magnetization is far larger than can be maintained by B_1, since it was developed in B_0, which is several orders of magnitude greater than B_1. Therefore M_y will decay with time to a value $(B_1/B_0)M_0$, which is very small. This is a relaxation phenomenon of a type not previously discussed in this book. It is known as

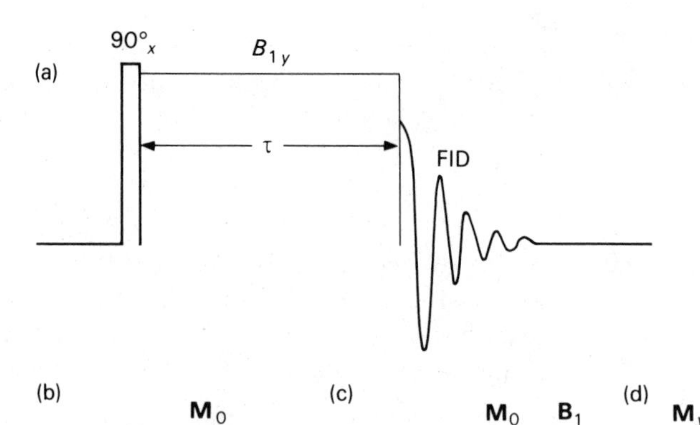

Fig. 3-18 Spin–lattice relaxation in the rotating frame (a) the pulse sequence, involving spin locking, (b) the situation at the end of the 90°_x pulse, (c) the situation at the start of spin-locking, (d) the situation at the end of spin-locking.

spin–lattice relaxation in the rotating frame, and is frequently exponential, characterized by a time designated $T_{1\rho}$.

$$M_y(\tau) = M_0 \exp(-\tau/T_{1\rho}) \tag{3-53}$$

Equation 3-53 applies if the equilibrium value of M_y is ignored. Values of $T_{1\rho}$ may be obtained selectively by Fourier transformation of the FID following the end of spin-locking and repetition of the experiment with variation of τ. Clearly $T_{1\rho}$ can be different from T_1 or T_2. In some ways it partakes of the character of both T_1 and T_2. For mobile solutions, $T_{1\rho} = T_1 = T_2$ in most circumstances, though not necessarily when chemical exchange occurs. For viscous solutions and solids $T_{1\rho}$ is likely to differ substantially from both T_1 and T_2. In fact it depends on the spectral density governed by $\omega_1 = \gamma B_1$, (which is of the order of tens of kHz), just as T_1 is governed by $J(\omega)$ at $\omega_0 = \gamma B_0$ (which is of the order of tens of MHz). The full expression for $T_{1\rho}$ is

$$T_{1\rho}^{-1} = \tfrac{1}{6}\gamma^2\{[B^0_{xL}]^2 + [B^0_{yL}]^2 + [B^0_{zL}]^2\}\{J(\omega_1) + J(\omega_0)\} \tag{3-54}$$

which is equal to T_1^{-1} and T_2^{-1} provided the random field is isotropic and $J(\omega_1) = J(0) = J(\omega_0)$. Variation of $T_{1\rho}$ with τ_c is shown in Fig. 3-17. When T_1 deviates from T_2, $T_{1\rho}$ follows T_2 until $\omega_1^2\tau_c^2 \gtrsim 1$, but the theory becomes questionable as τ_c decreases further (see Chapter 6).

When measuring $T_{1\rho}$ for ^{13}C nuclei in the solution state there are two situations to avoid in the proton-decoupling procedure, viz. (a) *noise*-decoupling must not be used since it can reduce the effective $T_{1\rho}$, so single-frequency decoupling must be employed, and (b) levels of r.f. magnetic fields such that Hartmann–Hahn matching [i.e. $\gamma_C B_1(^{13}C) = \gamma_H B_1(^1H)$ (see Section 6-5)] occurs must be avoided.

Notes and references

[1] For positive γ Larmor precession is anticlockwise about the applied field, so the relationship would be written vectorially as $\boldsymbol{\omega}_i = -\gamma \mathbf{B}_0$. In vector terms the rotating frame of reference is moving with angular velocity $\boldsymbol{\omega} = -\omega \mathbf{k}$.

[2] Although derived here for a CW experiment, Eqn 3-23 is also valid (in certain circumstances) for the FT mode.

[3] For $\nu_c \sim 100$ MHz there are ca. 1000 periods of the radiation in a pulse of this duration.

[4] R. R. Ernst & W. A. Anderson, *Rev. Sci. Instr.*, **37,** 93 (1966).

[5] E. L. Hahn, *Phys. Rev.*, **80,** 580 (1950).

[6] H. Y. Carr & E. M. Purcell, *Phys. Rev.*, **94,** 630 (1954).
S. Meiboom & D. Gill, *Rev. Sci. Instr.*, **29,** 688 (1958).

[7] Some authors include B_L in G.

[8] N. Bloembergen, E. M. Purcell & R. V. Pound, *Phys. Rev.* **73,** 679 (1948).

[9] F. Brady, R. W. Matthews, M. J. Forster & D. G. Gillies, *Inorg. Nucl. Chem. Lett.*, **17,** 155 (1981).

[10] P. S. Hubbard, *Phys. Rev.*, **131,** 1155 (1963).

[11] This implies that some of the relationships derived in Section 3-15 will not be exact in real situations.

Further reading

Fourier Transform NMR Techniques: A Practical Approach, K. Müllen & P. S. Pregosin, Academic Press (1976).

'Carbon-13 nuclear spin relaxation', J. R. Lyerla & G. C. Levy, Ch. 3 of *Topics in Carbon-*13 *NMR Spectroscopy*, Vol. 1, Ed. G. C. Levy, John Wiley and Sons (1974).

'Carbon-13 nuclear spin relaxation', J. R. Lyerla & D. M. Grant, Ch. 5 of *MTP International Rev. of Science: Physical Chemistry*, Series 1, Vol. 4, Ed. C. A. McDowell, Butterworths (1972).

Fourier Transform NMR Spectroscopy, D. Shaw, Elsevier Scientific Publishing Co. (1976).

Pulse and Fourier Transform NMR: Introduction to Theory and Methods, T. C. Farrar & E. D. Becker, Academic Press (1971).

Experimental Pulse NMR: A Nuts and Bolts Approach, E. Fukushima & S. B. W. Roeder, Addison-Wesley Publishing Co. (1981).

[See also the elementary texts listed at the end of Chapter 1.]

Problems

3-1 Use Eqn 3-19 to prove that the width of the absorption line at half-height is equal to $(\pi T_2)^{-1}$ in frequency units.

3-2 In the presence of partial saturation the full shape of a CW absorption signal as a function of frequency is defined by the expression 3-18. Show that the magnetic field, B_1(opt), which gives the maximum value of the signal at the Larmor frequency is expressed by Eqn 3-21. Using this result to simplify Eqn 3-18, derive an expression for the width of the NMR line at half maximum signal when the r.f. magnetic field is B_1(opt).

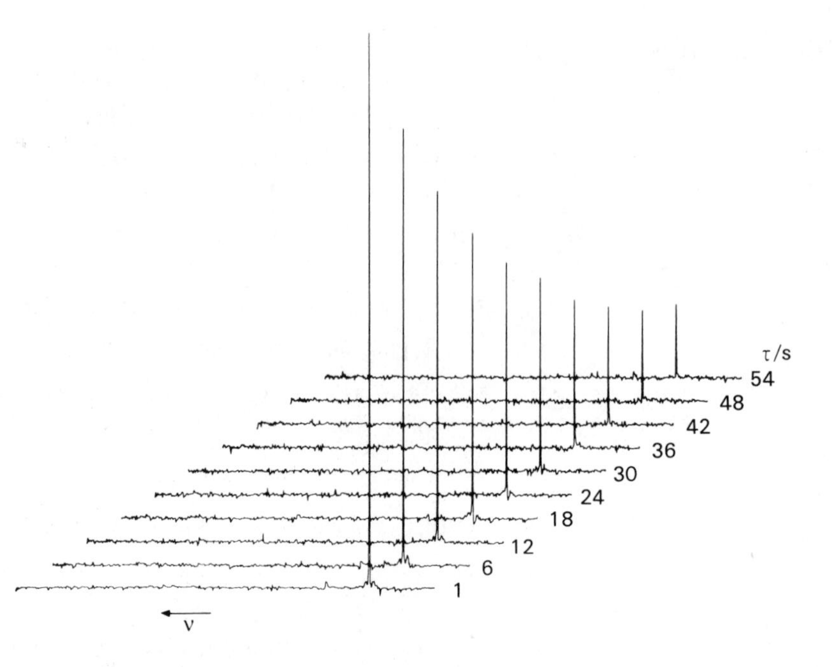

Fig. 3-19 Partially-relaxed ^{29}Si NMR spectra of diphenylsilane (see Problem 3-5).

3-3 In a Fourier transform NMR experiment the signal observed following the pulse was found to be a maximum for $t = 50\ \mu s$. Over what range of NMR transition frequencies would this pulse be effective? What would be the observed effect of a pulse of duration 100 μs?

3-4 A NMR signal is scanned (in CW mode) at time $t = 0$ using a high-power radio-frequency magnetic field. It is then scanned repetitively at intervals of 5 seconds with a low-power radiofrequency magnetic field. The maximum signal height S, of each signal is given as follows:

t/s	5	10	15	20	25	30	1000
S/cm	4·6	6·8	8·4	9·8	10·9	11·8	14·9

Explain the physical process occurring and derive a numerical value for the parameter which determines the rate of this process.

3-5 Fig. 3-19 shows ^{29}Si NMR spectra of diphenylsilane, Ph_2SiH_2, obtained with proton-decoupling. A pulse sequence of the form $180° - \tau - 90° -$ FID was used, and the free induction decays (FID) were recorded, subtracted from the corresponding FID obtained with τ very large, and the results transformed to give the spectra. The separate spectra are for different times, τ, as given on the diagram. Calculate the spin–lattice relaxation time.

3-6 Show that, if the observed signal is ωv, Eqns 3-18 and 3-19 are consistent with the expression for signal intensity given in Section 1-7.

3-7 Show, using Eqns 3-11, 3-12 and 3-18 that on resonance in the absence of relaxation the statement in Section 3-4 that the out-of-phase signal is the steady-state value of ωv is consistent with the statement in Section 1-7 that NMR spectrometers detect dM_y/dt.

4 Dipolar interactions and double resonance

4-1 Dipole–dipole coupling

As already noted, a spinning nucleus acts as a small magnet and therefore generates a local magnetic field. For a nuclear magnetic moment μ at an angle ξ to the z direction, this field at a distance $\mathbf{r}$ is given (see Fig. 4-1(a)) by

$$B_r = \frac{\mu_0}{4\pi}\frac{2\mu}{r^3}\cos\psi \tag{4-1}$$

$$B_\psi = \frac{\mu_0}{4\pi}\frac{\mu}{r^3}\sin\psi \tag{4-2}$$

where μ_0 is the permeability constant ($4\pi \times 10^{-7}$ kg m s^{-2} A^{-2}). The angle ξ is, in the model of Section 3-16, related directly to the quantum number m_I if the z direction is defined by an applied field B_0, but at a deeper level of quantum mechanics a general angle needs to be considered. The local field, B_L, is clearly of the order of $(\mu_0/4\pi)$ (μ/r^3), which at typical molecular distances ($r \sim 0{\cdot}2$ nm) for protons is ca. 0·2 mT. This is obviously far less than B_0 but it is much greater than coupling constants or proton chemical shifts expressed in field units (a chemical shift of 10 ppm at an applied field of 2·35 T is only 23·5 μT). When precession of the nuclear moment is considered it is clear that the local field contains a static component in the z direction and a fluctuating component at right angles to it. Both components will be modulated by any overall motion of the nucleus in question. The total local field at a given point in space will receive contributions from many nuclei.

At a molecular level a given nucleus will be affected by the local dipolar field of its neighbours. There is therefore a direct through-space interaction between nuclei in the same way as for macroscopic bar magnets, as mentioned in Section 3-16. Classically, the energy of interaction between two magnetic point dipoles $\boldsymbol{\mu}_1$ and $\boldsymbol{\mu}_2$ is given by Eqn 4-3, where the vector $\mathbf{r}$ is the distance between $\boldsymbol{\mu}_1$ and $\boldsymbol{\mu}_2$.

$$U = \left\{\frac{\boldsymbol{\mu}_1 \cdot \boldsymbol{\mu}_2}{r^3} - 3\,\frac{(\boldsymbol{\mu}_1 \cdot \mathbf{r})(\boldsymbol{\mu}_2 \cdot \mathbf{r})}{r^5}\right\}\frac{\mu_0}{4\pi} \tag{4-3}$$

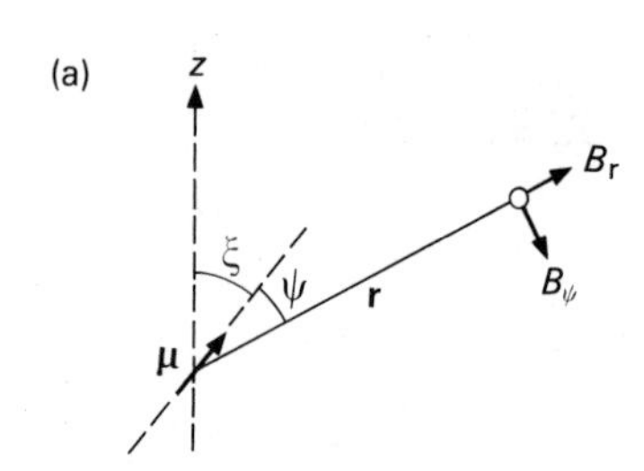

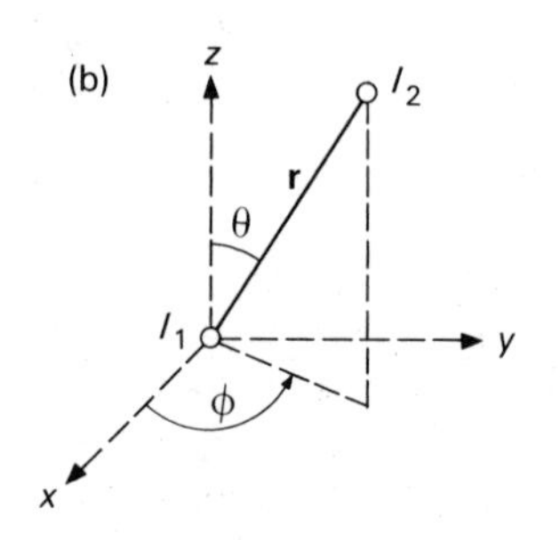

Fig. 4-1 Spatial effects of nuclear magnetic moments.
(a) The local field.
(b) Definitions of geometry for the dipole–dipole interaction of nuclei.

The appropriate quantum mechanical expression, Eqn 4-4, is obtained by using $\hat{\boldsymbol{\mu}} = \gamma\hbar\hat{\mathbf{I}}$.

$$\hat{\mathscr{H}}_{\mathrm{dd}} = \gamma_1\gamma_2\hbar^2\left\{\frac{\hat{\mathbf{I}}_1 \cdot \hat{\mathbf{I}}_2}{r^3} - 3\,\frac{(\hat{\mathbf{I}}_1 \cdot \mathbf{r})(\hat{\mathbf{I}}_2 \cdot \mathbf{r})}{r^5}\right\}\frac{\mu_0}{4\pi} \tag{4-4}$$

When the scalar products are expanded and the expression is put into polar coordinates (see Fig. 4-1(b) and Problem 4-1), Eqn 4-5 is obtained, where the five terms A to F are given by Eqns 4-6 to 4-11.

$$\hat{\mathscr{H}}_{\mathrm{dd}} = r^{-3}\gamma_1\gamma_2\hbar^2[A+B+C+D+E+F]\frac{\mu_0}{4\pi} \tag{4-5}$$

$$A = -\hat{I}_{1z}\hat{I}_{2z}(3\cos^2\theta - 1) \tag{4-6}$$

$$B = \tfrac{1}{4}[\hat{I}_{1+}\hat{I}_{2-} + \hat{I}_{1-}\hat{I}_{2+}](3\cos^2\theta - 1) \tag{4-7}$$

$$C = \tfrac{3}{2}[\hat{I}_{1z}\hat{I}_{2+} + \hat{I}_{1+}\hat{I}_{2z}]\sin\theta\cos\theta\exp(-i\phi) \tag{4-8}$$

$$D = -\tfrac{3}{2}[\hat{I}_{1z}\hat{I}_{2-} + \hat{I}_{1-}\hat{I}_{2z}]\sin\theta\cos\theta\exp(i\phi) \tag{4-9}$$

$$E = -\tfrac{3}{4}\hat{I}_{1+}\hat{I}_{2+}\sin^2\theta\exp(-2i\phi) \tag{4-10}$$

$$F = -\tfrac{3}{4}\hat{I}_{1-}\hat{I}_{2-}\sin^2\theta\exp(2i\phi) \tag{4-11}$$

Each of the terms A to F contains a spin factor and a geometrical factor, the effects of which can be appreciated separately. The common factor $(\mu_0/4\pi)r^{-3}\gamma_1\gamma_2(\hbar/2\pi)$ (in frequency units) is sometimes referred to as the dipolar coupling constant, and in this book it will be given the symbol R. Dipolar coupling should be clearly distinguished from the scalar coupling discussed in Chapters 1, 2 and 8, which has the symbol J for the coupling constant and is an indirect interaction, being mediated by the electronic framework of the molecule.

4-2 Averaging by molecular tumbling

When intramolecular dipolar interactions are considered for liquids and solutions, the way in which the geometrical factors in Eqns 4-4 and 4-9 are affected by molecular motion becomes important. Rapid isotropic tumbling will cause averaging of these factors. It is immediately apparent that the average of each of the terms C to F is zero since $\langle\exp(\pm i\phi)\rangle = \langle\exp(\pm 2i\phi)\rangle = 0$. Averaging of terms A and B each involves $\langle 3\cos^2\theta - 1\rangle$, the value of which is not so obvious. Since this average is extremely important a formal proof will be given here using elementary calculus.

Consider the distribution of the angle θ. For a sphere at unit distance from the origin (which is at I_1, say) the solid angle subtended by an annulus dependent on $\mathrm{d}\theta$ (see Fig. 4-2) is $2\pi\sin\theta\,\mathrm{d}\theta$. Since the total solid angle over a sphere is 4π, the fraction of internuclear interactions at angle θ for a random distribution is

$$\frac{\mathrm{d}n}{n} = \frac{2\pi\sin\theta\,\mathrm{d}\theta}{4\pi} = \tfrac{1}{2}\sin\theta\,\mathrm{d}\theta \tag{4-12}$$

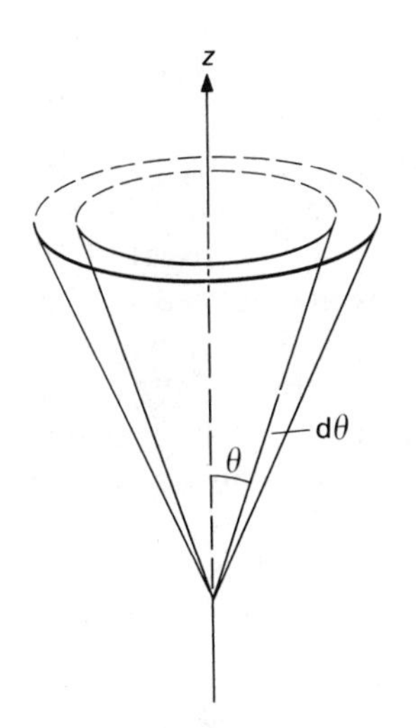

Fig. 4-2 Isotropic averaging (see the text).

Now the average value of any function of θ, $f(\theta)$ say, is

$$\langle f(\theta)\rangle = \int_0^{\pi} F(\theta) f(\theta)\, \mathrm{d}\theta \tag{4-13}$$

where $F(\theta)$ is the distribution function $(1/n)(\mathrm{d}n/\mathrm{d}\theta)$. Thus, for terms A and B of the dipolar Hamiltonian:

$$\begin{aligned}\langle(3\cos^2\theta - 1)\rangle &= \int_0^{\pi} \tfrac{1}{2}\sin\theta(3\cos^2\theta - 1)\, \mathrm{d}\theta \\ &= \tfrac{1}{2}[\cos^3\theta]_0^{\pi} + \tfrac{1}{2}[\cos\theta]_0^{\pi} = 0\end{aligned} \tag{4-14}$$

Thus the average value of all the terms A to F for isotropic tumbling is zero, and this means that dipolar interactions do not affect the transition energies or intensities for the high-resolution NMR spectra of isotropic fluids. This is fortunate, since the chemically-important chemical shifts and splittings due to scalar couplings, both of which are usually much smaller than dipolar interaction strengths, are thus revealed. Note that for these conclusions to hold there must be no preferred orientation during molecular tumbling, a situation which does not occur, for instance, for solutions in liquid crystals. Note also that even for isotropic solutions the dipolar interactions affect relaxation times, as will be shown below.

4-3 Relaxation induced by dipolar interactions

Since dipolar interactions occur mutually between spins they cannot strictly be treated by the random-field model of Section 3-15, and a more plausible theory will be discussed here.

For an ensemble of isolated two-spin systems of the AX (heteronuclear) type the basic product spin states are as shown in Table 4-1. Examination of the spin factors in terms A to F of the dipolar interaction indicate that these only link states (see Chapter 2) indicated by the matrix of Table 4-1. Thus, for example, B contains the 'flip-flop' operator, $\hat{I}_{1+}\hat{I}_{2-} + \hat{I}_{1-}\hat{I}_{2+}$, which links the $\alpha\beta$ and $\beta\alpha$ states only, whereas terms C and D, which contain one step-up or step-down operator, link states differing by 1 in the total component quantum number, m_T.

Table 4-1 The mixing of AX spin states by the various terms of the dipolar Hamiltonian

Spin state	$\beta\beta$	$\alpha\beta$	$\beta\alpha$	$\alpha\alpha$
$\beta\beta$	A	D	D	F
$\alpha\beta$	C	A	B	D
$\beta\alpha$	C	B	A	D
$\alpha\alpha$	E	C	C	A

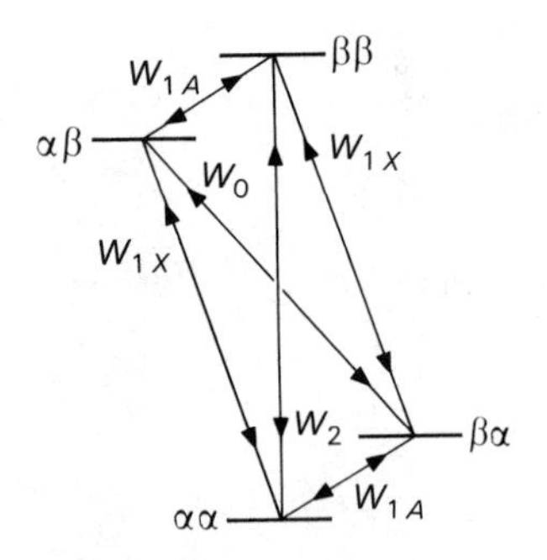

Fig. 4-3 Definition of transition rates for an AX spin system.

If intramolecular interactions are considered, molecular tumbling will modulate terms B to F, implying there are fluctuating magnetic fields. If there are components of these at the appropriate frequencies, spin–lattice relaxation will result. The appropriate frequencies are precisely those needed for transitions according to the usual Bohr condition $\Delta U = \hbar\omega$. The possible transition rates are defined in Fig. 4-3, where it is assumed that the $\alpha\beta \leftrightarrow \beta\beta$ and $\alpha\alpha \leftrightarrow \beta\alpha$ pathways are equivalent and have the same rate W_{1A} (and similarly for W_{1X}). The subscript numerals to W_0, W_1 and W_2 indicate the change in m_T involved. Table 4-2 summarizes the spectral densities and transitions concerned (ω_A and ω_X are the appropriate Larmor frequencies). Note that the form of the dipolar spin operators causes all six paths of Fig. 4-3 to be allowed for relaxation, whereas the selection rule $\Delta m_T = \pm 1$ limits those allowed for r.f.-excited transitions.

It will be shown in Section 4-9 that if the X spins are strongly irradiated during the experiment, the relaxation of the z component of the total A magnetization following a perturbation is exponential with rate:

$$T_{1A}^{-1} = W_0 + 2W_{1A} + W_2 \tag{4-15}$$

In order to calculate the dipolar contribution to T_1^{-1} it is necessary to obtain expressions for W_0, W_{1A} and W_2 separately. This is done in Appendix 3, giving

$$W_0 = \tfrac{1}{20}(2\pi R)^2 J(\omega_X - \omega_A) \tag{4-16}$$

$$W_{1A} = \tfrac{3}{40}(2\pi R)^2 J(\omega_A) \tag{4-17}$$

$$W_2 = \tfrac{3}{10}(2\pi R)^2 J(\omega_X + \omega_A) \tag{4-18}$$

where R is the dipolar interaction constant, $(\mu_0/4\pi)\gamma_A\gamma_X(\hbar/2\pi)r_{AX}^{-3}$ and the frequency-dependence of the spectral densities has been explained above. Note that

$$W_0 : W_1 : W_2 = \tfrac{1}{6} : \tfrac{1}{4} : 1 \tag{4-19}$$

Insertion of Eqns 4-16 to 4-18 into Eqn 4-15 gives

$$T_{1dd}^{-1} = \tfrac{1}{20}(2\pi R)^2[J(\omega_X - \omega_A) + 3J(\omega_A) + 6J(\omega_X + \omega_A)] \tag{4-20}$$

Table 4-2 Transitions induced by the dipolar Hamiltonian

Transition	*Dipolar term*	*Transition rate*	*Spectral density*
$\alpha\beta \leftrightarrow \beta\beta$, $\alpha\alpha \leftrightarrow \beta\alpha$	C,D	W_{1A}	$J(\omega_A)$
$\beta\alpha \leftrightarrow \beta\beta$, $\alpha\alpha \leftrightarrow \alpha\beta$	C,D	W_{1X}	$J(\omega_X)$
$\alpha\beta \leftrightarrow \beta\alpha$	B	W_0	$J(\omega_X - \omega_A)$
$\alpha\alpha \leftrightarrow \beta\beta$	E,F	W_2	$J(\omega_X + \omega_A)$

If a single-exponential correlation function is assumed, this becomes

$$T_{1dd}^{-1} = \tfrac{1}{10}\tau_c(2\pi R)^2\left[\frac{1}{1+(\omega_X-\omega_A)^2\tau_c^2}+\frac{3}{1+\omega_A^2\tau_c^2}+\frac{6}{1+(\omega_X+\omega_A)^2\tau_c^2}\right] \tag{4-21}$$

where τ_c is the correlation time. When the extreme narrowing condition is obeyed, Eqn 4-21 reduces to the expression given in Chapter 3. Otherwise the three spectral densities are unequal.

As usual, the terms causing spin–lattice relaxation lead to an uncertainty in the transition intensities and hence to line broadening. This in turn implies a non-secular contribution to T_2^{-1}, equal to $\frac{1}{2}T_1^{-1}$. In addition the dipolar term A modulates the energy levels directly, leading to a secular contribution to linewidths and T_2^{-1} which depends on the zero-frequency spectral density $J(0)$. The full equation for T_2^{-1} is therefore

$$T_{2dd}^{-1} = \tfrac{1}{40}(2\pi R)^2[4J(0)+J(\omega_X-\omega_A)+3J(\omega_A)+6J(\omega_X) + 6J(\omega_X+\omega_A)] \tag{4-22}$$

which, for a single-exponential correlation function, becomes

$$T_{2dd}^{-1} = \tfrac{1}{20}(2\pi R)^2\tau_c\left[4+\frac{1}{1+(\omega_X-\omega_A)^2\tau_c^2}+\frac{3}{1+\omega_A^2\tau_c^2} + \frac{6}{1+\omega_X^2\tau_c^2}+\frac{6}{1+(\omega_A+\omega_X)^2\tau_c^2}\right] \tag{4-23}$$

The corresponding equations for $T_{1\rho dd}^{-1}$ (in the limit $B_1 \to 0$) are

$$T_{1\rho dd}^{-1} = \tfrac{1}{40}(2\pi R)^2[4J(\omega_1)+J(\omega_X-\omega_A)+3J(\omega_A) + 6J(\omega_X)+6J(\omega_X-\omega_A)] \tag{4-24}$$

$$T_{1\rho dd}^{-1} = \tfrac{1}{20}(2\pi R)^2\tau_c\left[\frac{4}{1+\omega_1^2\tau_c^2}+\frac{1}{1+(\omega_X-\omega_A)^2\tau_c^2}+\frac{3}{1+\omega_A^2\tau_c^2} + \frac{6}{1+(\omega_X+\omega_A)^2\tau_c^2}+\frac{6}{1+\omega_X^2\tau_c^2}\right] \tag{4-25}$$

For *homo*nuclear two-spin dipolar interactions, term B is secular, and the corresponding equations to those above are

$$T_{1dd}^{-1}(\text{homo}) = \tfrac{3}{20}(2\pi R)^2[J(\omega_0)+4J(2\omega_0)] \tag{4-26}$$

$$T_{1dd}^{-1}(\text{homo}) = \tfrac{3}{10}(2\pi R)^2\tau_c\left[\frac{1}{1+\omega_0^2\tau_c^2}+\frac{4}{1+4\omega_0^2\tau_c^2}\right] \tag{4-27}$$

$$T_{2dd}^{-1}(\text{homo}) = \tfrac{3}{40}(2\pi R)^2[3J(0)+5J(\omega_0)+2J(2\omega_0)] \tag{4-28}$$

$$T_{2dd}^{-1}(\text{homo}) = \tfrac{3}{20}(2\pi R)^2\tau_c\left[3+\frac{5}{1+\omega_0^2\tau_c^2}+\frac{2}{1+4\omega_0^2\tau_c^2}\right] \tag{4-29}$$

$$T_{1\rho dd}^{-1}(\text{homo}) = \tfrac{3}{40}(2\pi R)^2[3J(2\omega_1)+5J(\omega_0)+2J(2\omega_0)] \tag{4-30}$$

$$T_{1\rho dd}^{-1}(\text{homo}) = \tfrac{3}{20}(2\pi R)^2\tau_c\left[\frac{3}{1+4\omega_1^2\tau_c^2}+\frac{5}{1+\omega_0^2\tau_c^2}+\frac{2}{1+4\omega_0^2\tau_c^2}\right] \tag{4-31}$$

where the factor R is now $(\mu_0/4\pi)\gamma^2(\hbar/2\pi)r^{-3}$. Under extreme narrowing conditions these equations reduce to that given in Chapter 3, with $T_{1dd}^{-1}(\text{homo}) = T_{2dd}^{-1}(\text{homo}) = T_{1\rho dd}^{-1}(\text{homo}) = \frac{3}{2}T_{1dd}^{-1}(\text{hetero})$.

For most molecules there are many spin-pairs to consider, and the above equations must be modified by summing over all relevant interactions with the spin under consideration. Complications sometimes arise because motions of different spin pairs in the same molecule are not independent, but if this is ignored the above equations need only be modified by replacing r_{AX}^{-6} by the sum over all X nuclei, $\sum_X r_{AX}^{-6}$ (assuming a single correlation time).

4-4 Double resonance—introduction

In a double resonance experiment two radiofrequency fields are applied to the sample. Several different phenomena can be observed, depending on the frequencies and amplitudes of these oscillating fields, ranging from changes in intensity to completely altered spectra. The first radiofrequency field, $\mathbf{B}_1 \cos \omega_1 t$, is known as the observing field and is analogous to the radiofrequency field in an ordinary single-resonance experiment. The second, or perturbing, radiofrequency field, $\mathbf{B}_2 \cos \omega_2 t$, is also applied perpendicular to $\mathbf{B}_0$, the static field. The effects of $\mathbf{B}_2 \cos \omega_2 t$ on the spin system are observed using the field $\mathbf{B}_1 \cos \omega_1 t$ to monitor the signal at angular frequency ω_1. Different double-resonance spectra are obtained by changing the parameters of the perturbing field, which may be either CW or pulsed (gated). The notation A-$\{X\}$ is used to indicate that nucleus A is being monitored while nucleus X is being perturbed.

All the effects observed in double-resonance experiments may be placed in one of two categories. The first type encompasses all phenomena involving population changes caused by $\mathbf{B}_2$ (or by $\mathbf{B}_2$ together with $\mathbf{B}_1$), usually resulting in intensity changes in the spectra. Such effects are not instantaneous but take a time of the order of T_1 to build up and decay. The second category concerns experiments dependent on the additional term contributed to the Hamiltonian by $\mathbf{B}_2$. Usually such experiments involve changes in transition frequencies; these observed effects occur immediately $\mathbf{B}_2$ is switched on and cease immediately it is switched off. Of course, in many cases both categories of effect will be observed in the same experiment.

4-5 Selective population transfer

Consider an experiment in which a perturbing r.f. field $\mathbf{B}_2$ is continuously applied and the system is monitored at equilibrium by means of the r.f. ν_1 (which may be either CW or pulsed). If the amplitude of the perturbing field $\mathbf{B}_2$ is small and its frequency corresponds to a transition of the spin system, the double-resonance spectrum will have lines at the same frequencies as the single-resonance spectrum but will show changes of intensity. This is illustrated in Fig. 4-4(a), where P, Q, R and S are nuclear spin energy levels with allowed transitions from R to Q, from Q to P and from S to Q. If the transition from R to Q is saturated by the field $\mathbf{B}_2$, then the population of Q is increased and

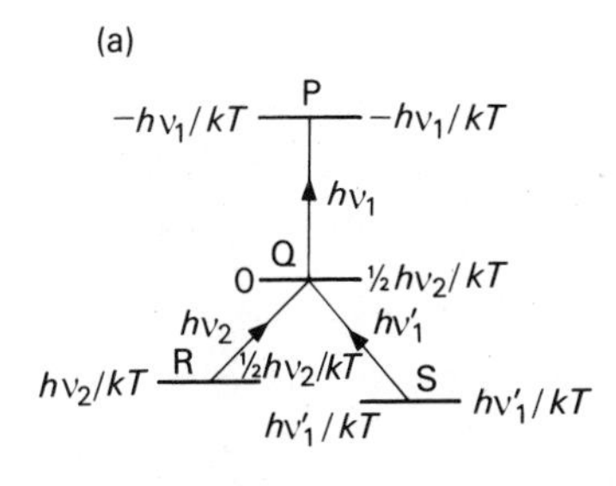

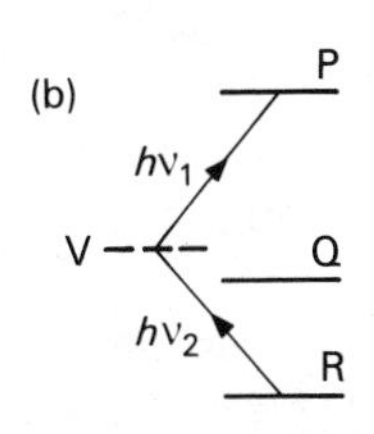

Fig. 4-4 Selective population changes and double quantum transitions. (a) If B_2 is relatively weak and is applied at resonance it changes the populations of the levels Q and R. The excess Boltzmann populations (relative to that of Q) are given to the left of each level and the populations if transition R → Q is saturated (ignoring relaxation) are given to the right of each level. (b) If ν_2 is off-resonance a two-quantum transition occurs when $h(\nu_1 + \nu_2)$ equals the difference in energy between levels P and R. It may be considered as involving a virtual level V.

that of R decreased. As the intensity of the transition Q to P depends on the difference in populations of levels Q and P, it is enhanced by the increase in the population of Q. For similar reasons transition S → Q is reduced in intensity by the population transfer due to $\mathbf{B}_2$. Other transitions involving one of the perturbed levels have enhanced or lowered intensities. Transitions R → Q and Q → P are said to be in a *progressive* relationship, since the total spin component, m_T changes monotonically in the series R → Q → P. If one transition of any such pair is subjected to selective population transfer, the intensity of its companion is *in*creased. The pair of transitions R → Q and S → Q are *regressively* related (m_T for R equals that for S) and selective population transfer causes a *de*crease of intensity. It should be noted that even with the low levels of B_2 needed to cause such effects, the monitoring field B_1 in a CW experiment is much weaker.

Two-quantum transitions can be observed when ω_2 does not correspond exactly to a transition. This is shown in the energy level diagram 4-4(b). The two-quantum transition is a transition from state R to state P in which two quanta are absorbed, one from $\mathbf{B}_1$ and one from $\mathbf{B}_2$. The double-resonance spectrum in this case will contain all the single-resonance lines and in addition these two-quantum transitions. The intensity of the latter are small, and depend both on the value of $B_2^2 B_1$ and on the difference in energy between the virtual level V and the state Q.

Selective population transfer may be readily seen in FT NMR if the perturbing field is pulsed and is of suitable power. Figure 3-8 shows that for a pulse to be selective all that is necessary is that its duration is relatively long. Thus a pulse of duration 1 s will only affect a frequency range of ca. 1 Hz. Moreover a 180° pulse corresponds to population inversion and a 90° pulse achieves population equalization (saturation). It is therefore feasible, by monitoring a spectrum (using a pulse in ν_1) at intervals of time following a selective ν_2 pulse, to study the complicated processes of relaxation in a multi-spin system. A more detailed discussion of the effects of selective population transfer is to be found in Section 7-4.

4-6 The double resonance Hamiltonian

Most double resonance experiments using CW perturbing and monitoring radiofrequencies are of neither of the two types described in the preceding section, but employ a much larger perturbing field so that the spin energy levels themselves are altered. Similar effects have been observed with lasers in optical spectroscopy, but in the radio region it is easier to obtain electromagnetic radiation with sufficient amplitude and coherence to perturb molecular energy levels. In these circumstances it is no longer possible to describe the spectrum in terms of one- or two-quantum transitions between the energy levels of an isolated molecule in a field B_0; instead, we must determine the energy levels of the spin system in the presence of two fields, B_0 and $B_2 \cos \omega_2 t$, one of which is time-dependent. In this section the problem is discussed theoretically, using similar techniques to those described in Chapter 2. Succeeding sections describe the types of observations made and discuss their applications with examples. Those sections should be comprehensible by the reader who does not wish to examine the theoretical detail of this section.

Double resonance spectra which are more complex than the simple two-quantum transitions described in the previous section can most easily be interpreted as one-quantum spectra observed with the measuring field $B_1 \cos \omega_1 t$, in which transitions occur between the levels of the combined system of spins and fields B_0 and $B_2 \cos \omega_2 t$. The problem is to determine the energy levels and wave functions of the latter system. In order to do this, we first remove the time dependence of the perturbing field by using axes which rotate with the latter field.

In Section 3-3, transformation to rotating axes was discussed for a classical system and it was found that the equations of motion in axes rotating about the z direction with angular frequency $-\omega$ contain extra terms which have the same form as an additional magnetic field of $-\omega\gamma^{-1}$ in the z direction. Similarly we can write an effective Hamiltonian in rotating axes (x, y, z) (see Problem 4-7)

$$h^{-1}\hat{\mathscr{H}}_{\text{eff}} = \sum_j -(2\pi)^{-1}\gamma_j\{[(1-\sigma_j)B_0 - \omega_2\gamma_j^{-1}]\hat{I}_{jz} + B_2\hat{I}_{jx}\} + \sum_{j<k} J_{jk}\hat{\mathbf{I}}_j \cdot \hat{\mathbf{I}}_k \quad (4\text{-}32)$$

whose eigenstates are stationary in the system of rotating axes. This Hamiltonian differs from that given in Chapter 2 in two respects. First, there is a change from B_0 to $(B_0 - \omega_2\gamma_j^{-1})$, and secondly there is an additional term involving the spin operators $\hat{I}_{jx}$. This latter term mixes states with different values of m, which is no longer a good quantum number.

The method of finding the eigenstates and eigenvalues of this double resonance Hamiltonian is to set up the secular equations and to solve them by diagonalizing the Hamiltonian matrix as described in Section 2-9.

The AX system The spin Hamiltonian for the AX system in rotating axes is

$$h^{-1}\hat{\mathscr{H}}_{\text{eff}} = -(\nu_A - \nu_2)\hat{I}_{Az} - (\nu_X - \nu_2)\hat{I}_{Xz} + (2\pi)^{-1}B_2(\gamma_X\hat{I}_{Xx} + \gamma_A\hat{I}_{Ax}) + J\hat{\mathbf{I}}_A \cdot \hat{\mathbf{I}}_X \quad (4\text{-}33)$$

where $\nu_A = \gamma_A(1-\sigma_A)B_0$, $\nu_X = \gamma_X(1-\sigma_X)B_0$ and $2\pi\nu_2 = \omega_2$. If ν_2 is in the X region, we may omit the term in $B_2\hat{I}_{Ax}$ as the latter mixes states with m_A values which differ by unity. As the difference in diagonal elements of such states $|\nu_A - \nu_2|$ is large compared with $\gamma_A B_2/2\pi$, the term in $\hat{I}_{Ax}$ has a negligible effect. Similarly, we may make the X approximation and omit the terms $\frac{1}{2}J\hat{I}_{A+}\hat{I}_{X-}$ and $\frac{1}{2}J\hat{I}_{A-}\hat{I}_{X+}$ (see Section 2-10), leaving an effective Hamiltonian

$$h^{-1}\hat{\mathscr{H}}_{\text{eff}} = -(\nu_A - \nu_2)\hat{I}_{Az} - (\nu_X - \nu_2)\hat{I}_{Xz} + (2\pi)^{-1}B_2\gamma_X\hat{I}_{Xx} + J\hat{I}_{Az}\hat{I}_{Xz} \quad (4\text{-}34)$$

for which the value of m_A is a good quantum number as $\hat{\mathscr{H}}_{\text{eff}}$ commutes with $\hat{I}_{Az}$.

This Hamiltonian, like the normal AX Hamiltonian, has four eigenvalues. However, m_T is no longer a good quantum number and there are now six possible transitions, two in the X and four in the A region. The frequencies and intensities of these transitions depend on the values of ν_2 and B_2. Referred to the set of states ($\alpha\alpha$, $\alpha\beta$, $\beta\alpha$, $\beta\beta$) defined in the rotating frame, the Hamiltonian matrix is

$$\begin{bmatrix} \frac{1}{2}(2\nu_2 - \nu_A - \nu_X) + \frac{1}{4}J & \gamma_X B_2/4\pi & 0 & 0 \\ \gamma_X B_2/4\pi & \frac{1}{2}(\nu_X - \nu_A) - \frac{1}{4}J & 0 & 0 \\ 0 & 0 & \frac{1}{2}(\nu_A - \nu_X) - \frac{1}{4}J & \gamma_X B_2/4\pi \\ 0 & 0 & \gamma_X B_2/4\pi & \frac{1}{2}(\nu_A + \nu_X - 2\nu_2) + \frac{1}{4}J \end{bmatrix} \quad (4\text{-}35)$$

The eigenvalues and eigenfunctions are given in Table 4-3. If the double resonance spectrum is observed in the A region, transitions in which $\Delta m_A = -1$ are seen. There are four such transitions whose frequencies can be determined from Table 4-3. However, it must be remembered that this table refers to rotating axes. An applied field with frequency ν_1 in laboratory axes appears to have frequency $\nu_1 - \nu_2$ in rotating axes, so the appropriate resonance condition for transition between states $|r\rangle$ and $|s\rangle$ is

$$|\nu_1 - \nu_2| = U_r - U_s \quad (4\text{-}36)$$

The relative intensities depend on

$$\left|\langle r| \sum_j \gamma_j \hat{I}_{j-} |s\rangle\right|^2 \quad (4\text{-}37)$$

where $\hat{I}_-$ refers to the system of rotating axes.

Table 4-3 Eigenvalues and eigenfunctions of the double-resonance Hamiltonian for the AX system

State[a]	*Eigenvalue*[b]
$\cos\theta_+ \|\alpha\alpha\rangle + \sin\theta_+ \|\alpha\beta\rangle$	$\frac{1}{2}(\nu_2 - \nu_A) + \frac{1}{2}[(\nu_2 - \nu_X + \frac{1}{2}J)^2 + (\gamma_X B_2/2\pi)^2]^{1/2}$
$-\sin\theta_+ \|\alpha\alpha\rangle + \cos\theta_+ \|\alpha\beta\rangle$	$\frac{1}{2}(\nu_2 - \nu_A) - \frac{1}{2}[(\nu_2 - \nu_X + \frac{1}{2}J)^2 + (\gamma_X B_2/2\pi)^2]^{1/2}$
$\cos\theta_- \|\beta\alpha\rangle + \sin\theta_- \|\beta\beta\rangle$	$-\frac{1}{2}(\nu_2 - \nu_A) + \frac{1}{2}[(\nu_2 - \nu_X - \frac{1}{2}J)^2 + (\gamma_X B_2/2\pi)^2]^{1/2}$
$-\sin\theta_- \|\beta\alpha\rangle + \cos\theta_- \|\beta\beta\rangle$	$-\frac{1}{2}(\nu_2 - \nu_A) - \frac{1}{2}[(\nu_2 - \nu_X - \frac{1}{2}J)^2 + (\gamma_X B_2/2\pi)^2]^{1/2}$

[a] $\sin 2\theta_+ = \gamma_X B_2/2\pi[(\nu_2 - \nu_X + \frac{1}{2}J)^2 + (\gamma_X B_2/2\pi)^2]^{1/2}$,
$\sin 2\theta_- = \gamma_X B_2/2\pi[(\nu_2 - \nu_X - \frac{1}{2}J)^2 + (\gamma_X B_2/2\pi)^2]^{1/2}$.
[b] In frequency units.

4-7 Tickling experiments

If one particular transition of a multispin system is irradiated by the perturbing field, the two levels connected by that transition are perturbed, and other transitions involving these levels will be the first to be affected. We have already seen that the intensities of these transitions are altered by relatively weak perturbing fields. As the strength of the perturbing field is increased, these transitions are split in the double-resonance spectrum by an amount $\gamma B_2/2\pi$. This allows one to identify transitions which share a common energy level, and so to assign lines in a complex spectrum and in many cases to find the relative signs of coupling constants. Such an experiment, in which one transition is selectively irradiated and the resulting line splittings are observed, is known as a tickling experiment.

Figure 4-5 shows the double-resonance spectrum of the M and A parts of an AMX system in which the lowest frequency line of the X group is irradiated. The molecule is $CCl_3CHClCH_2Cl$, in which the two methylene protons are non-equivalent because of restricted internal rotation about the C—C bond and intrinsic asymmetry.

Figure 4-6 shows the energy level diagram and the single-resonance spectrum of this molecule. The lines are labelled to correspond to the transitions on the energy level diagram for four possible choices of signs of coupling constants. In the double-resonance spectrum shown in Fig. 4-5, the line marked with an arrow in Fig. 4-6 is being tickled, and the lines marked with asterisks are observed to be affected. From the energy level diagram, we see that irradiating any X transition will affect two A transitions; if X_2 is tickled, A_3 and A_4 are affected, while tickling X_4 also affects A_3 and A_4. The third assignment of Fig. 4-6 is consistent with the observations, while the others are ruled out. The reader should ascertain for himself that it can be deduced that the two- and three-bond coupling constants in this molecule have opposite signs

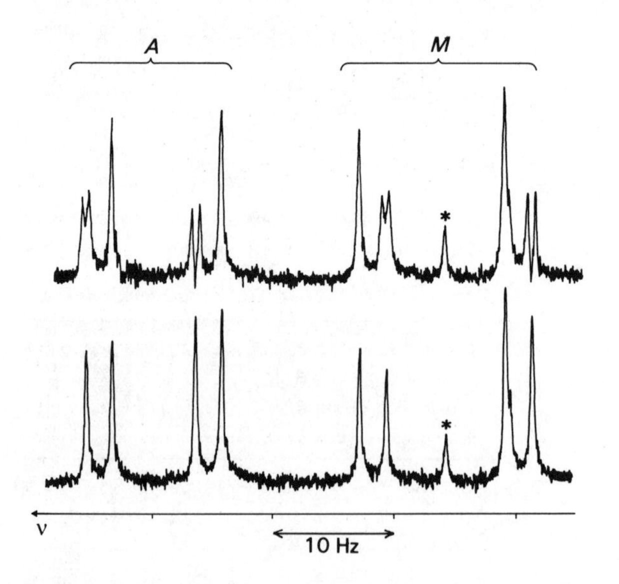

Fig. 4-5 Tickling experiment on an AMX system, $CCl_3CH_AClCH_MH_XCl$. The single-resonance spectrum is shown below a double-resonance spectrum in which the lowest frequency X line is irradiated. The line marked with an asterisk is due to an impurity. The AX and MX couplings can be seen to have opposite signs. A full analysis (see Fig. 4-6) shows that $^3J_{AM} = +2{\cdot}15$ Hz, $^3J_{AX} = +9{\cdot}4$ Hz, and $^2J_{MX} = -12{\cdot}0$ Hz. Note the better resolution of regressive transitions compared to progressive transitions. (A. B. Dempster, K. Price and N. Sheppard are thanked for this spectrum).

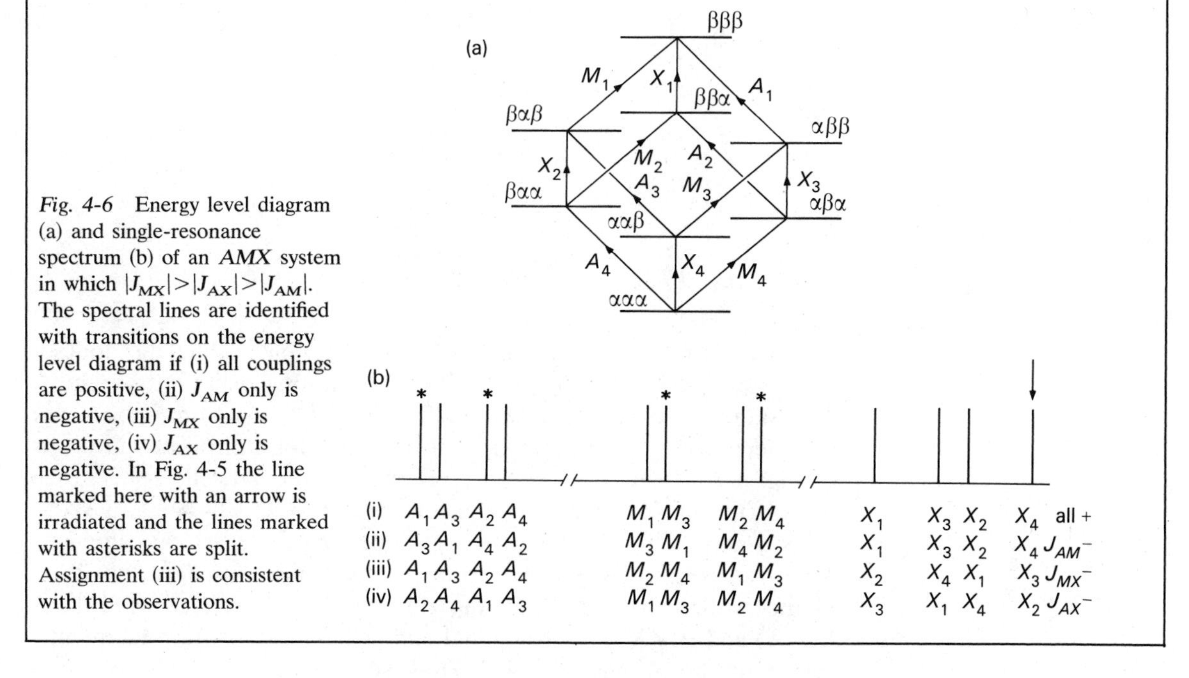

Fig. 4-6 Energy level diagram (a) and single-resonance spectrum (b) of an *AMX* system in which $|J_{MX}|>|J_{AX}|>|J_{AM}|$. The spectral lines are identified with transitions on the energy level diagram if (i) all couplings are positive, (ii) J_{AM} only is negative, (iii) J_{MX} only is negative, (iv) J_{AX} only is negative. In Fig. 4-5 the line marked here with an arrow is irradiated and the lines marked with asterisks are split. Assignment (iii) is consistent with the observations.

(Problem 4-6). It is not possible to determine *absolute* signs in a double-resonance experiment.

In Fig. 4-5 some of the splittings are better resolved than others. In fact, both progressive and regressive lines are split by an amount $\gamma_A B_2/2\pi$, but the effect of field inhomogeneity is such that progressive transitions (such as A_1 when X_3 is irradiated) are broader and therefore less well-resolved than regressive transitions (such as A_2). This fact is often an aid to line-assignment.

4-8 Spin decoupling

Figure 4-7 shows the predicted double-resonance spectra for the AX system as a perturbing field of increasing strength is applied at ν_X, the centre of the X doublet. A new line of increasing intensity (as B_2

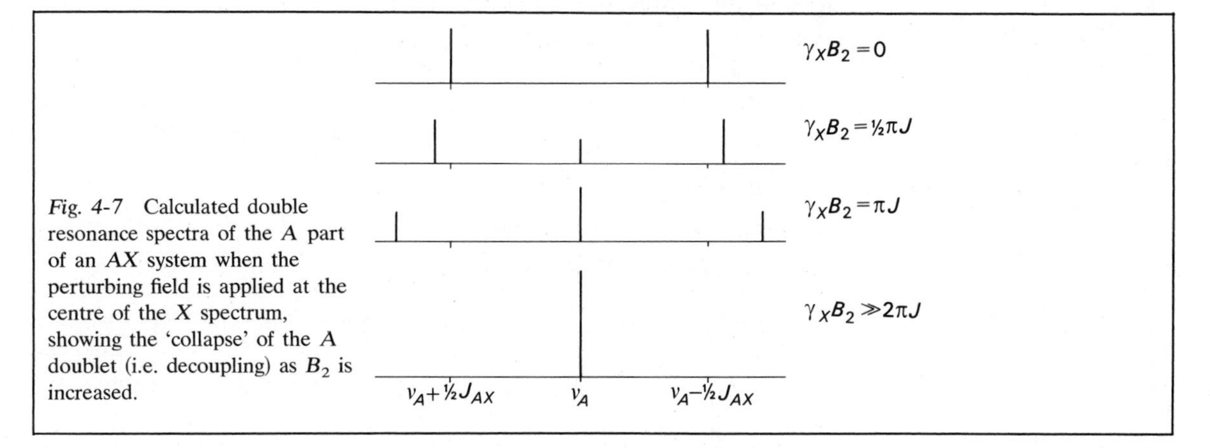

Fig. 4-7 Calculated double resonance spectra of the A part of an AX system when the perturbing field is applied at the centre of the X spectrum, showing the 'collapse' of the A doublet (i.e. decoupling) as B_2 is increased.

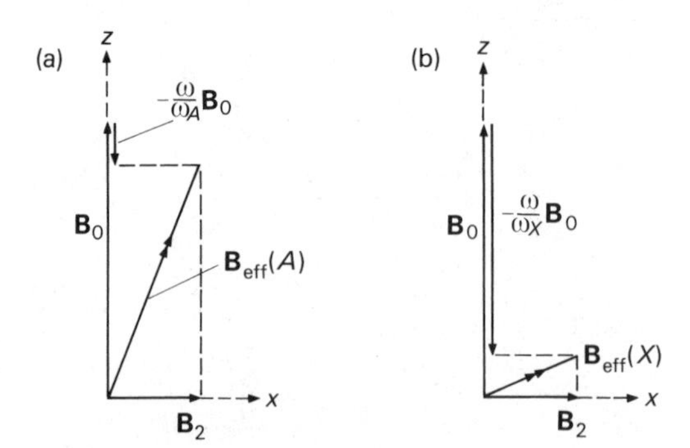

Fig. 4-8 Effective magnetic fields in rotating axes for (a) the A nuclei and (b) the X nuclei in an A-$\{X\}$ double-resonance experiment.

increases) appears at the centre of the A doublet, while the doublet lines move apart and lose intensity. The limiting result with high B_2 consists of a single line at ν_A, and a decoupled spectrum is obtained (see Section 1-18). Decoupling can be understood physically in terms of Fig. 4-8. If $\gamma B_2/2\pi \ll |\nu_X - \nu_A|$, as would automatically be the case for a heteronuclear decoupling experiment, $\mathbf{B}_{eff}(A)$ will lie along $\mathbf{B}_0$ to a good approximation, whereas, provided $\gamma B_2/2\pi$ is strong compared to all resonance offsets for ν_2 from X-type transitions, $\mathbf{B}_{eff}(X)$ will be in the x direction for X (in the rotating frame of reference). Thus $\mathbf{B}_{eff}(A)$ and $\mathbf{B}_{eff}(X)$ are orthogonal, and will lead to orthogonal quantization of the A and X spins. However, the coupling phenomenon depends on $\mathbf{I}_A \cdot \mathbf{I}_X$, which is zero for orthogonal spins. It therefore cannot affect the spectrum in these circumstances.

4-9 The Solomon equations[1]

Considerable enlightenment concerning the relaxation properties of a heteronuclear two-spin (AX) system can be obtained using elementary ideas of kinetics. In terms of the relaxation pathways of Fig. 4-3 the rate of change of the population of the $\alpha\alpha$ state, for instance, will be

$$\begin{aligned} \mathrm{d}n_{\alpha\alpha}/\mathrm{d}t = -(W_{1A} + W_{1X} + W_2)(n_{\alpha\alpha} - n^0_{\alpha\alpha}) + W_2(n_{\beta\beta} - n^0_{\beta\beta}) \\ + W_{1A}(n_{\beta\alpha} - n^0_{\beta\alpha}) + W_{1X}(n_{\alpha\beta} - n^0_{\alpha\beta}) \end{aligned} \quad (4\text{-}38)$$

where the superscript zero labels populations at Boltzmann equilibrium in the absence of any r.f. magnetic field. There will be similar equations for the other populations. The signals observed in NMR due to r.f.-induced transitions are proportional to the appropriate population differences. The total A signal is therefore proportional to the quantity N_A (Eqn 4-39), and similarly for the X signal (Eqn 4.40):

$$N_A = (n_{\alpha\alpha} - n_{\beta\alpha}) + (n_{\alpha\beta} - n_{\beta\beta}) \quad (4\text{-}39)$$

$$N_X = (n_{\alpha\alpha} - n_{\alpha\beta}) + (n_{\beta\alpha} - n_{\beta\beta}) \quad (4\text{-}40)$$

Thence, from Eqn 4-38 and its analogues (Problem 4-10):

$$\mathrm{d}N_A/\mathrm{d}t = -(W_0 + 2W_{1A} + W_2)(N_A - N^0_A) - (W_2 - W_0)(N_X - N^0_X) \quad (4\text{-}41)$$

$$\mathrm{d}N_X/\mathrm{d}t = -(W_2 - W_0)(N_A - N^0_A) - (W_0 + W_{1X} + W_2)(N_X - N^0_X) \quad (4\text{-}42)$$

This pair of coupled differential equations indicates that in general if the system is perturbed relaxation will not be single exponential, so there will be no simple definition of a T_1. Suppose a double-resonance experiment is performed such that the A signal is monitored while the X resonances are simultaneously irradiated strongly. There will, of course, be decoupling (Section 4-8), so the A signal will be a single line. At this point, however, interest is centred on intensity effects. If there is complete saturation of the X magnetization, the populations will be equalized ($n_{\alpha\alpha} = n_{\alpha\beta}$ and $n_{\beta\alpha} = n_{\beta\beta}$) so that $N_X = 0$. Moreover, at any steady state $dN_A/dt = 0$, so that Eqn 4-41 becomes

$$0 = (W_2 - W_0)N_X^0 - (W_0 + 2W_{1A} + W_2)(N_A^* - N_A^0) \tag{4-43}$$

where the asterisk is used to denote the particular conditions of the experiment. On rearrangement this gives

$$\frac{N_A^*}{N_A^0} = 1 + \frac{N_X^0}{N_A^0}\left(\frac{W_2 - W_0}{W_2 + 2W_{1A} + W_0}\right) \tag{4-44}$$

But the values of population differences at equilibrium are proportional to the appropriate magnetogyric ratios, so the integrated signal in the A-$\{X\}$ double-resonance situation S_A^*, is related to that for a Boltzmann-equilibrium situation, S_A^0, by

$$\frac{S_A^*}{S_A^0} = 1 + \frac{\gamma_X}{\gamma_A}\left(\frac{W_2 - W_0}{W_2 + 2W_{1A} + W_0}\right) \tag{4-45}$$

This intensity change is known as the nuclear Overhauser effect (NOE) and is discussed more fully in the next section.

Now suppose that a 180° pulse is applied to the A spins to invert their populations, and that the recovery of the A signal is monitored while the X spins are still being saturated. Then Eqn 4-41 gives

$$dN_A/dt = (W_2 - W_0)N_X^0 - (W_0 + 2W_{1A} + W_2)(N_A - N_A^0) \tag{4-46}$$

While irradiation of the X spins continues, relaxation will occur towards the steady-state condition 4-43, which, when substituted into Eqn 4-46 becomes

$$dN_A/dt = -(W_0 + 2W_{1A} + W_2)(N_A - N_A^*) \tag{4-47}$$

Thus relaxation is single-exponential (i.e. first order) with a rate constant

$$T_1^{-1} = W_0 + 2W_{1A} + W_2 \tag{4-48}$$

as was used in Section 4-3.

4-10 The nuclear Overhauser effect

As shown in the preceding section, decoupling is accompanied by intensity changes which are given by Eqn 4-45. This equation (and Eqn 4-48) are general and do not depend on the relaxation *mechanism*. However, it is easy to see what effect each mechanism has. For heteronuclear dipolar relaxation, Eqn 4-19 may be substituted into

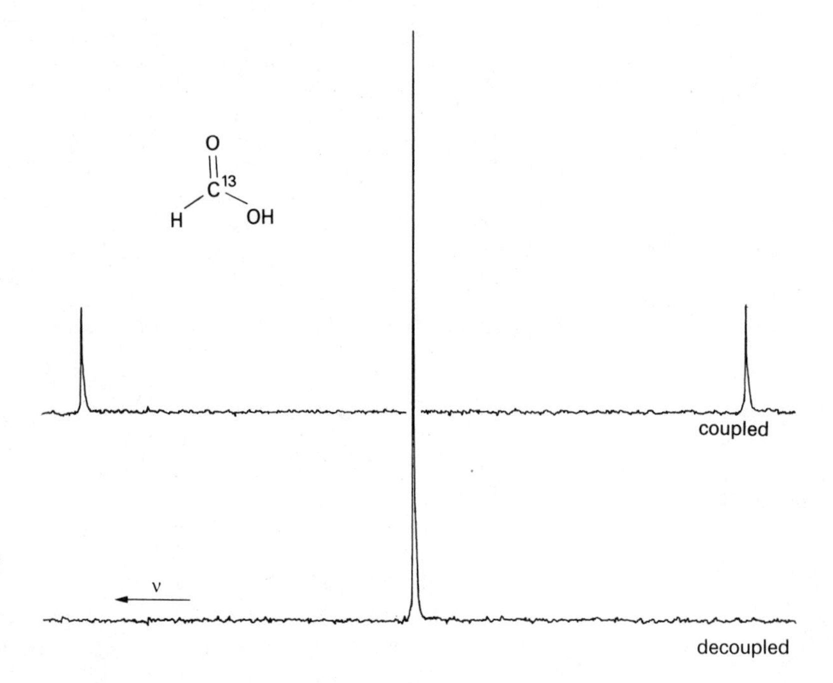

Fig. 4-9 The coupled and proton-decoupled ^{13}C spectra of formic acid (HCOOH) at 25 MHz, exhibiting a NOE near the maximum value of 2·98.

Eqn 4-45 to yield

$$NOE = 1 + (\gamma_X/2\gamma_A) \tag{4-49}$$

Usually, this is the maximum observable NOE, and for the common ^{13}C-$\{^1H\}$ situation it is 2·988. This gives a useful enhancement of the ^{13}C intensity (see Fig. 4-9).

Most mechanisms other than dipolar contribute only to W_1 and not to W_2 or W_0. An exception is modulated scalar coupling between the observed and perturbed spins—since this contains a 'flip-flop' spin operator it contributes to W_0 only, and if it forms the dominant mechanism the NOE will be $1-\gamma_A/\gamma_X$. If the SR or SA or UE mechanisms dominate then $W_2 = W_0 = 0$, and the NOE is 1 (i.e. no effect). Clearly the observed NOE depends on the balance of relaxation mechanisms. If scalar (A, X) relaxation is ignored, then $W_2 - W_0 = \frac{1}{2}T_{1dd}^{-1}$ and Eqn 4-45 becomes

$$NOE = 1 + \frac{\gamma_X}{2\gamma_A}\frac{T_{1dd}^{-1}}{T_1^{-1}} \tag{4-50}$$

where T_1^{-1} is the total A relaxation rate and T_{1dd}^{-1} is that from (A, X) dipolar interactions only. Note that T_{1dd} includes *inter*- as well as *intra*-molecular (A, X) relaxation but that dipolar interactions of A with nuclei *other* than X contribute to T_1^{-1} but not to T_{1dd}^{-1}. The factor $(\gamma_X/2\gamma_A)(T_{1dd}^{-1}/T_1^{-1})$ is sometimes called the nuclear Overhauser *enhancement* and is given the symbol η, while $\gamma_X/2\gamma_A$ is regarded as η_{max}.

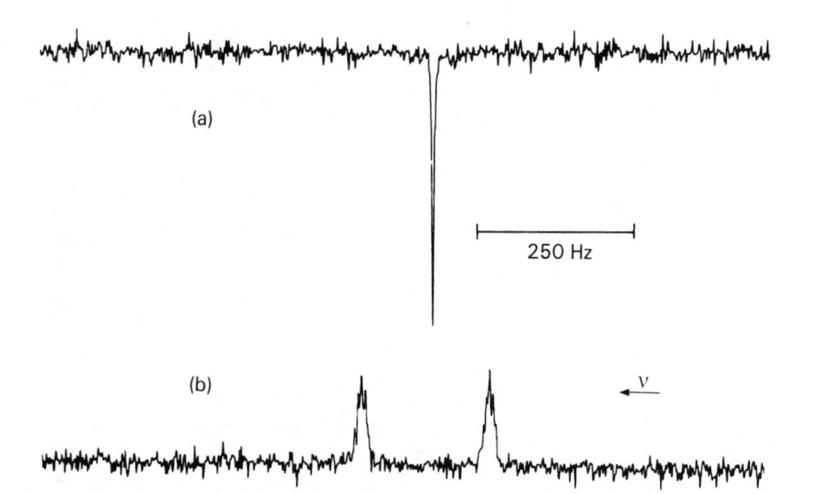

Fig. 4-10 The coupled and proton-decoupled ^{29}Si spectra of triphenylsilane, the latter giving a negative peak ($\eta = -2{\cdot}41$). [Reproduced, with permission, from R. K. Harris & B. J. Kimber, *Appl. Spectrosc. Rev.*, **10,** 117 (1975)].

There are three situations which require a little more comment at this stage:

(a) *Nuclei with negative magnetogyric ratios.* In such cases, η_{max} is negative, being $-4{\cdot}92$ for ^{15}N-{^{1}H} and $-2{\cdot}52$ for ^{29}Si-{^{1}H}. This means that if the relevant dipolar interactions dominate the relaxation, the observed *A* signal will be *negative* (see Fig. 4-10). Moreover, if $T_{1dd}/T_1 \sim \gamma_X/2\gamma_A$, the resonance will have a near-zero intensity—the 'null-signal' situation.

(b) *Homonuclear double resonance.* Equation 4-50 applies to homo- as well as heteronuclear situations, provided the spectrum is first order. In the former case $\gamma_A = \gamma_X$ and the value of η_{max} is $0{\cdot}5$, i.e. a 50% enhancement of the signal is the maximum achievable. Experiments of the ^{1}H-{^{1}H} type are common.

(c) *The electron-nucleus system.* Equation 4-50 also applies if the spin *X* is an unpaired electron irradiated with microwaves at its resonance frequency. Indeed such an experiment provided the original Overhauser effect. Since the electron has a high magnetic moment compared to all nuclei, the effect can be very large, with $\eta_{max} = -328$ if dipolar interactions dominate. The SC mechanism often contributes in such cases, however, leading to a positive Overhauser effect if dominant.

The NOE can be useful in a number of ways, in addition to providing additional sensitivity for ^{13}C-{^{1}H} experiments. In particular measurement of both T_{1A} and the NOE yields a value for the heteronuclear contribution to T_{1dd} (see Section 3-16) by solution of Eqns 4-50 and 3-49 (ignoring T_{1sc}). The contribution of all other mechanisms together can then be obtained from Eqn 3-49.

Because T_{1dd} is directly related to internuclear distance (Table 3-2), the NOE is often used to obtain information on conformations and configurations. Thus Ohtsuro *et al.*[2] have used the ^{1}H-{^{1}H} NOE to determine the configuration of the two isomers of citral. When the

CH_3 protons are irradiated and the ethylenic protons observed, there is signal enhancement of 18% for one isomer but no observable effect for the other. Clearly the former has structure 4-[II] and the latter 4-[I].

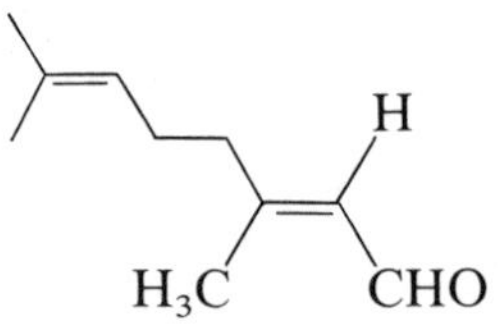

H, H_3C, CHO

4-[I] citral a

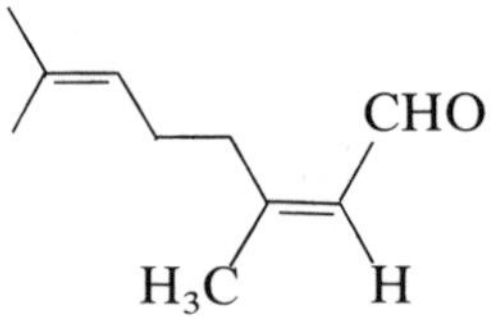

CHO, H_3C, H

4-[II] citral b

The above discussion has implicitly assumed the extreme narrowing condition, which may not hold for large molecules. However, Eqn 4-45 is more general, and it can be seen that since spectral density is first lost (as motion slows) from W_2, the NOE is decreased by departure from extreme narrowing. When motion is sufficiently slow that all terms of the type $\omega^2\tau_c^2$ are much greater than unity, the ^{13}C-{^1H} nuclear Overhauser enhancement can be shown to be $\eta = 0{\cdot}154$ (Problem 4-11).

4-11 Carbon-13 relaxation times and the ^{13}C-{^1H} NOE

Because ^{13}C is a dilute spin and usually occurs in molecules in which protons are the only other magnetically active nuclei, its relaxation properties are largely dominated by (C,H) dipolar interactions. Moreover the tetravalency of carbon implies it is not normally accessible to protons on other molecules, and only *intra*molecular (C,H) interactions need normally be considered. Since dipolar relaxation rates are inversely proportional to the sixth power of the appropriate distance, molecular geometry has a profound effect. For ^{13}C nuclei with directly-bonded protons T_1^{-1} is usually dominated by (C,H) dipolar interactions, whereas for quaternary carbons, this may not be the case, at least for small molecules. This implies that for quaternary carbons, values of T_1 are likely to be relatively long, as for 4-[III][3] and for the *ipso* carbon of chlorobenzene (Fig. 3-14). Moreover, quaternary carbons of small molecules may show low nuclear Overhauser enhancements (see Fig. 4-11(a)).

As indicated at the end of Section 4-3, T_{1dd}^{-1} actually depends on a summation over possible dipolar interactions of the nucleus in question. In the case of ^{13}C, when long-range effects are neglected and rigid isotropically-rotating molecules are considered, the dominant effect is

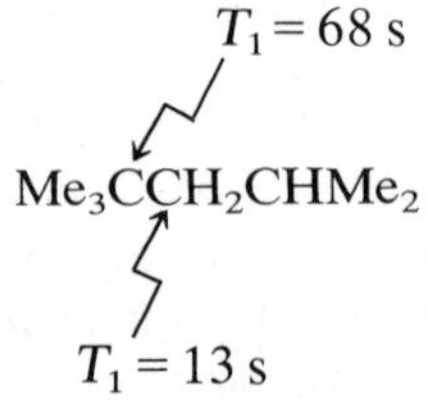

$T_1 = 68$ s

$Me_3CCH_2CHMe_2$

$T_1 = 13$ s

4-[III]

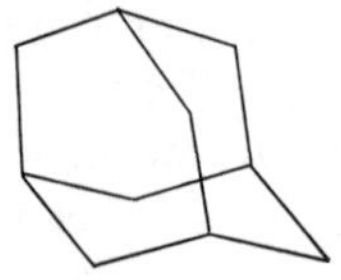

4-[IV]

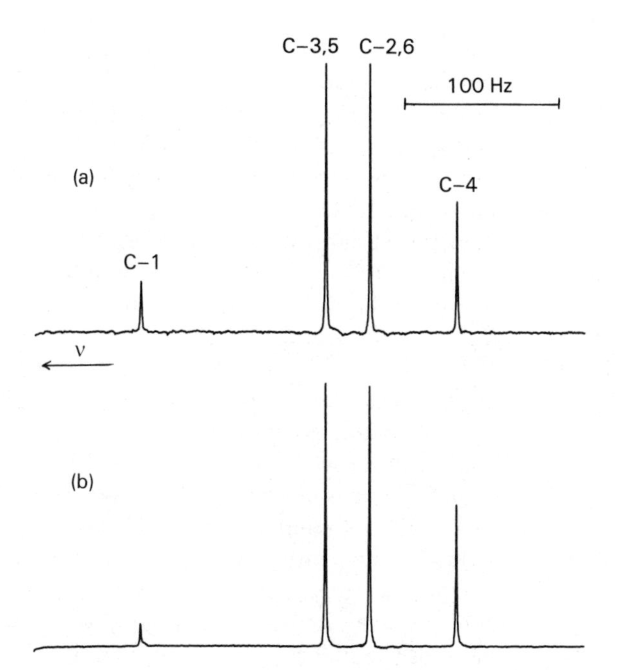

Fig. 4-11 Proton-decoupled ^{13}C spectra of chlorobenzene: (a) Using a single pulse—the relative intensities are affected by the NOE but not by T_1. The low intensity of the *ipso* carbon arises because the dipolar mechanism does not dominate its relaxation. (b) Using 10 pulses, with acquisition time 8 s and no further inter-pulse delay. The relative intensity of the *ipso* carbon is further reduced by its long T_1 (see Fig. 3-14).

from directly-bonded protons, so it is expected that $T_1^{-1}(CH) = \frac{1}{2}T_1^{-1}(CH_2) = \frac{1}{3}T_1^{-1}(CH_3)$ for carbons in a given molecule. The classic case is adamantane (4-[IV]), for which[3] $T_1(CH_2) = 11{\cdot}4$ s and $T_1(CH) = 20{\cdot}5$ s. In fact when the minor contributions of more distant hydrogens are considered the theoretical ratio of the relaxation times is very close to the experimental one.

Larger molecules move more sluggishly than small ones. Consequently, τ_c is longer in the former case and T_{1dd}^{-1} larger. For example, the total T_1 (dominated by the dipolar mechanism) for cyclohexane is 18 s whereas for the various non-quaternary carbons of cholesterol T_1 ranges from 0·2 to 2·0 s. Relaxation mechanisms other than the (C, H) dipolar effect are normally negligible for the carbons of large organic molecules, even (in the limit) for quaternary carbons.

Actually, most molecules possess some form of internal motion which affects relaxation, and the situation rapidly becomes more complicated as the number of motions increases. In general, however, the more mobile a particular CH_n grouping is, the longer its ^{13}C T_1 will be. Methyl and phenyl groups are invariably affected in this fashion, as will be discussed further in Section 4-13.

4-12 Quantitative ^{13}C NMR intensities

The fact that for CW ^1H NMR relative signal intensities are strictly proportional to the relevant proton concentrations for first-order spectra if saturation is avoided has long been of considerable importance. Unfortunately two factors complicate the situation for FT ^{13}C NMR, namely (a) the variable values of $T_1(^{13}C)$, and (b) variations in the NOE for different carbons. Quaternary carbons tend to have long T_1 values and low NOE's, so their intensities often appear much reduced. Figure 4-11 illustrates the difficulties. Of course the T_1 problem does not arise if only a single pulse is used or if adequate delay times (>5 times the longest T_1) are left between pulses. However, single pulses rarely give adequate S/N, and it is extremely inefficient to leave long delay times. Moreover, long delays will not help the variable NOE problem.

There are two ways over these difficulties. To circumvent the NOE, one can add a paramagnetic compound which decreases T_1 but leaves T_{1dd} unaffected, thus changing the ratio in Eqn 4-50. Suitable paramagnetics do not affect chemical shifts and are therefore called *shiftless relaxation reagents.* The commonest additive for this purpose is $Cr(acac)_3$, and a concentration of 0·03M is generally adequate (higher concentrations cause undesirable line broadening). It is usually assumed that such an additive reduces all carbon-13 T_1's for a given molecule to a common value. This is not so, but generally they are all reduced sufficiently that only modest inter-pulse delays are required.

Unfortunately it is frequently chemically undesirable to use such an additive. An alternative technique is to use *gated decoupling.* A suitable pulse sequence is shown in Fig. 4-12(a). The decoupler is switched on at the time of the pulse in the observe channel. Decoupling occurs instantaneously whereas the NOE builds up only slowly (and in any case cannot affect the FID since the NOE is a phenomenon of longitudinal, not transverse, magnetization). A delay time must be left following the FID for any NOE which has built up to decay, so the pulse sequence is not very efficient for sensitivity. It is, however, very effective, as shown in Fig. 4-13 (this depicts a ^{29}Si rather than ^{13}C example, but the principle is the same). Of course, the gated decoupling technique can be combined with use of a relaxation reagent so that only small delay times are used.

The gated decoupling sequence (Fig. 4-12(a)) is also valuable in allowing the NOE to be measured by recording decoupled spectra only, normally involving the comparison of the intensities of single lines obtained with continuous and gated decoupling. The inverse gated sequence shown in Fig. 4-12(b) may be used for obtaining coupled spectra *with* the intensity gains arising from the NOE.

There is another criterion to be fulfilled if ^{13}C intensities are to be quantitative: the irradiation power must be constant across the whole width of the spectrum. Thus pulse powers fulfilling the conditions

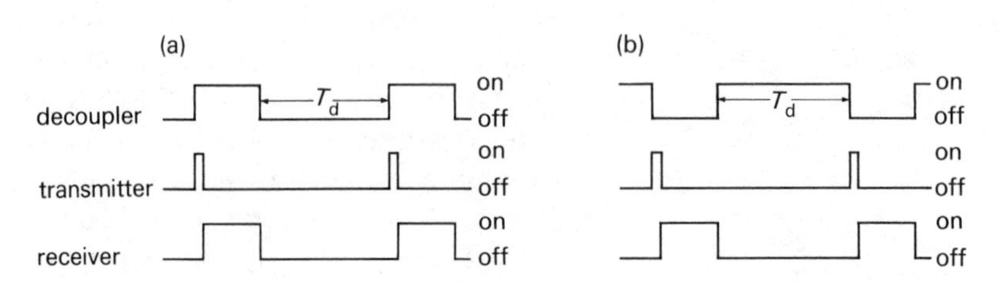

Fig. 4-12 Gated decoupler pulse sequences: (a) To obtain decoupling without the NOE. (b) To obtain a coupled spectrum with the NOE. The delay time T_d must be $>5T_1$.

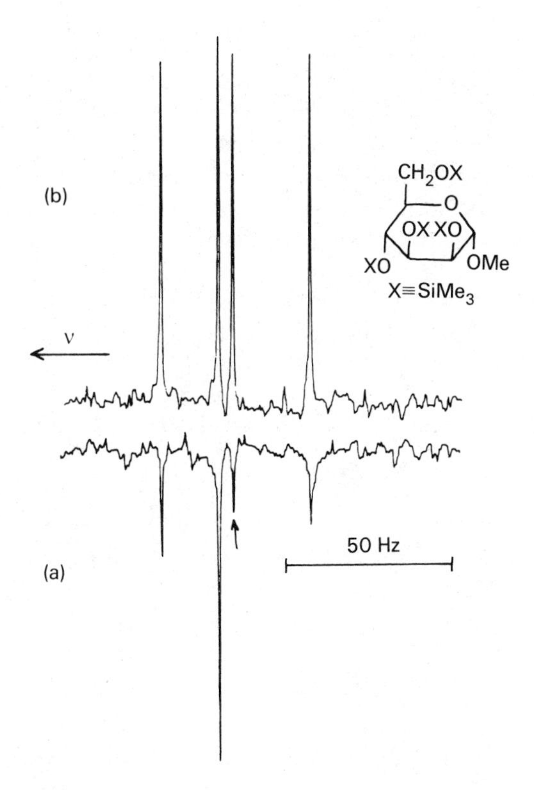

Fig. 4-13 Proton-decoupled ^{29}Si spectra of a trimethylsilylated mannose at 19.87 MHz:
(a) continuously decoupled,
(b) with gated decoupling. The inequality of signal intensities in (a) is due to variations in NOE and T_1 for the different silicons. The arrow indicates the signal due to the silicon of the —CH_2OSiMe_3 group. [Adapted from D. J. Gale, M. Sc. thesis, University of East Anglia (1975)].

$\gamma B_1/2\pi \gg \Delta$ must be used. Moreover, the digitization of the spectrum must be adequate to reflect true bandshapes. Finally it is best to use electronic integration of peak areas rather than to rely on simple measurement of peak heights.

It should be noted that the techniques mentioned in this section are also important to overcome the 'null-signal' problem for nuclei with negative magnetogyric ratios (see Section 4-10). Use of the gated decoupler sequence (Fig. 4-12(a)) with the decoupler switched on a time τ before the 'read' pulse enables T_1 values to be determined for null-signal cases. This is known as the method of NOE growth. The observed magnetization follows the equation:

$$\eta(\tau) = \eta(\infty)[1 - \exp(-\tau/T_1)] \tag{4-51}$$

4-13 Anisotropic molecular tumbling—the Woessner Equations

The equations (4-21, 4-23, 4-25 and the analogues for the homonuclear mechanism) developed earlier for relaxation which involves molecular tumbling assume that only one correlation time, τ_c, is involved, implying a rigid molecule undergoing isotropic motion (a very rare situation).

Real molecules usually tumble anisotropically, and equations have been developed for such situations. At this point equations for the special case of an axially symmetrical tumbling ellipsoid will be presented, subject to the extreme narrowing approximation. There will be

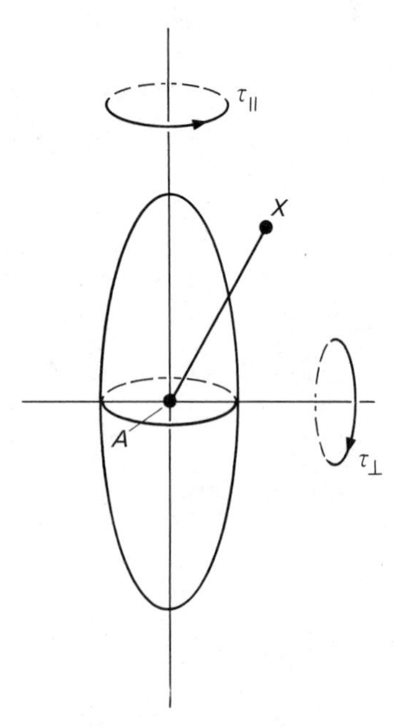

Fig. 4-14 Correlation times for an axially symmetrical tumbling ellipsoid.

two independent correlation times, which may be specified $\tau_{\parallel}$ and $\tau_{\perp}$ (see Fig. 4-14). Suppose that relaxation of spin A is dominated by heteronuclear dipolar interactions with spin X, and that the angle between the molecular axis and r_{AX} is specified Δ. Then it can be shown that Eqn 4-21 is still valid provided the isotropic tumbling correlation time τ_c is replaced by an effective correlation time τ_{eff} which is given by

$$\tau_{\text{eff}} = A(\Delta)\tau_A + B(\Delta)\tau_B + C(\Delta)\tau_C \tag{4-52}$$

where the three geometrical factors are

$$A(\Delta) = \tfrac{1}{4}(3\cos^2\Delta - 1)^2 \tag{4-53}$$

$$B(\Delta) = 3\sin^2\Delta\cos^2\Delta \equiv \tfrac{3}{4}\sin 2\Delta \tag{4-54}$$

$$C(\Delta) = \tfrac{3}{4}\sin^4\Delta \tag{4-55}$$

It may be noted that $A + B + C = 1$. The three correlation times are not independent, but are given by:

$$\tau_A = \tau_{\perp} \tag{4-56}$$

$$\tau_B^{-1} = \tfrac{5}{6}\tau_{\perp}^{-1} + \tfrac{1}{6}\tau_{\parallel}^{-1} \tag{4-57}$$

$$\tau_C^{-1} = \tfrac{1}{3}\tau_{\perp}^{-1} + \tfrac{2}{3}\tau_{\parallel}^{-1} \tag{4-58}$$

If the molecular geometry is known, then measurement of T_1 for two nuclear sites in a molecule will yield values of $\tau_{\parallel}$ and $\tau_{\perp}$ separately. For instance, analysis[4] of ^{13}C relaxation data for triptycene (4-[V]) yielded $\tau_{\perp} = 16 \pm 1$ ps and $\tau_{\parallel} = 48 \pm 8$ ps. (In this case three different CH carbons were measured and the correlation times are those which give the

CH_3
H_3C CH_3
CH_3

4-[V] 4-[VI]

best fit. The effects of distant protons in the molecule were taken into account.) It is interesting that reorientation about the symmetry axis is ca. three times slower than reorientation of the symmetry axis itself.

A second common occurrence for molecular motion is internal motion. Woessner's Eqn. 4-52 can again be used for the situation where a small rotor such as a methyl or phenyl group is attached to a large molecule, which is assumed to tumble isotropically with correlation time τ_R. The angle Δ is now that between the relevant dipolar interaction vector and the rotor axis, and the correlation times take on new definitions, it being assumed in effect that $\tau_{\perp} \rightarrow \tau_R$ and $\tau_{\parallel}^{-1} \rightarrow \tau_R^{-1} + \tau_G^{-1}$ where τ_G is the diffusional correlation time for rotation of the small group about its axis. Thus

$$\tau_A = \tau_R \qquad (4\text{-}59)$$

$$\tau_B^{-1} = \tau_R^{-1} + \tfrac{1}{6}\tau_G^{-1} \qquad (4\text{-}60)$$

$$\tau_C^{-1} = \tau_R^{-1} + \tfrac{2}{3}\tau_G^{-1} \qquad (4\text{-}61)$$

When internal rotation is fast, i.e. $\tau_R \gg \tau_G$, $\tau_{eff} \rightarrow A(\Delta)\tau_R$, except near the singularity $\Delta = \cos^{-1}(1/\sqrt{3})$ when $A \rightarrow 0$. For ^{13}C relaxation of methyl groups with a tetrahedral geometry, Eqn 4-52 reduces to

$$\tau_{eff} = \tfrac{1}{9}\tau_A + \tfrac{8}{27}\tau_B + \tfrac{16}{27}\tau_C \qquad (4\text{-}62)$$

If the methyl rotation is rapid $\tau_{eff} \rightarrow \tau_R/9$, and the relaxation rate will be one third of the value for a CH carbon rigidly fixed in the molecule (three C–H vectors contribute to relaxation for the methyl carbon). 1,2,3,5-tetramethylbenzene (4-[VI]), known as isodurene, provides[5] an example of this effect, though the molecule is not of the correct geometry. The 5-methyl group has a very low barrier to internal rotation and so, apparently, does the 2-methyl. The values of T_{1dd} are 24 and 25 s, respectively, roughly three times as long as for the ring CH carbon (8 s). The 1- and 3-methyl carbons, being in an asymmetrical steric environment, have higher barriers to internal rotation, and $T_{1dd} = 12$ s for these nuclei.

The theory is often applied to phenyl groups, although it is not strictly applicable. For *para*-carbons $\Delta = 0$ and $A = 1$, $B = C = 0$. Thus relaxation rates for *para* carbons yield values for τ_R^{-1} directly. For *ortho*- and *meta*-carbons $\Delta \simeq 60°$, so $A = \frac{1}{64}$, $B = \frac{9}{16}$, $C = \frac{27}{64}$. Very rapid internal rotation or extreme anisotropic behaviour should yield[6] a ratio $T_{1dd}^{-1}(p) : T_{1dd}^{-1}(o, m) = 64$, where complete rigidity would give a ratio of unity. The ratio therefore gives a qualitative indication of mobility anisotropy. For phenylsilane,[7] the *ortho* and *meta* carbons have $T_{1dd} = 22.9$ s whereas the *para* carbon has 14.9 s. For phenyltrimethylsilane the corresponding values are 11·5 s and 5·5 s. For large molecules only, the motion may be assumed to be entirely that of the phenyl group, so that information about internal rotation is derived. The equations in this section may be modified for homonuclear interactions (by the factor $\frac{3}{2}$), for the lack of extreme narrowing, and for a model of jumping rather than diffusional motion for internal rotation.

Notes and references

[1] I. Solomon, *Phys. Rev.*, **99**, 559 (1955).

[2] M. Ohtsuro, M. Teraoka, K. Tori & K. Takeda, *J. Chem. Soc. B.*, 1033 (1967).

[3] Page 187 of *Carbon-13 NMR Spectroscopy* (first edition), G. C. Levy & R. L. Lichter, Wiley & Sons (1972).

[4] K. F. Kuhlmann, D. M. Grant, & R. K. Harris, *J. Chem. Phys.*, **52**, 3439 (1970).

[5] R. K. Harris & R. H. Newman, *Mol. Phys.*, **38**, 1715 (1979).

[6] T. D. Alger, D. M. Grant & R. K. Harris, *J. Phys. Chem.*, **76,** 281 (1972).

[7] G. C. Levy, J. D. Cargioli & F. A. L. Anet, *J. Amer. Chem. Soc.*, **95,** 1527 (1973).

[8] R. K. Harris & B. J. Kimber, *Adv. Mol. Relaxation Processes,* **8,** 23 (1976).

Further reading

The Nuclear Overhauser Effect: Chemical Applications, J. H. Noggle & R. E. Schirmer, Academic Press (1971).

'High-resolution nuclear magnetic double and multiple resonance', R. A. Hoffman & S. Forsén, *Prog. NMR Spectrosc.*, **1,** 15 (1966).

'Multiple magnetic resonance', W. McFarlane & D. S. Rycroft, *Ann. Repts. NMR Spectrosc.*, **9,** 320 (1979).

'Nuclear magnetic relaxation spectroscopy', F. Noack, *NMR Basic Principles & Prog.*, **3,** 83 (1971).

Problems

4-1 By expanding the scalar products and expressing the result in polar coordinates, show that the dipolar Hamiltonian 4-4 may be put into the form 4-5.

4-2 Evaluate the nuclear Overhauser effect from first principles (by discussing energy level populations) for a system of two non-equivalent spin-$\frac{1}{2}$ nuclei of the same isotope, subject to the following model conditions:
(a) The X nuclei are completely saturated while the A spins are observed.
(b) The W_2 relaxation process (interconverting $\alpha\alpha \leftrightarrow \beta\beta$) is very efficient whereas W_{1A}, W_{1X} and W_0 are negligible.

4-3 For certain spin systems (e.g. observing ^{19}F spectra and irradiating the resonance of the unpaired electron in a free radical in the same solution as the fluorine-containing compound) modulation of a *scalar* coupling can give the dominant relaxation effect. If the spins are labelled I and S the appropriate Hamiltonian is:

$$h^{-1}\hat{\mathscr{H}} = A(t)\hat{\mathbf{I}}\cdot\hat{\mathbf{S}} = A(t)[\tfrac{1}{2}(\hat{I}_+\hat{S}_- + \hat{I}_-\hat{S}_+) + \hat{I}_z\hat{S}_z]$$

where the coupling constant A is indicated to be time-dependent. Suppose spin I is observed and spin S is irradiated. What would be the nuclear Overhauser effect?

4-4 Draw a diagram to illustrate approximately the variation in the direction and size of the macroscopic magnetic moment in rotating axes under steady state (CW) conditions as the applied frequency is increased through resonance. Assume $T_1 = T_2$. From the variation in u and v in your diagram deduce the approximate forms of the absorption and dispersion spectra.
(Ans. for diagram: see the book by Pople, Schneider, and Bernstein, Fig. 3-4, page 38; note the definition of v used here has the opposite sign.)

4-5 Assign the lines in Fig. 4-6 (an AMX spectrum) if (a) all couplings are negative and (b) J_{MX} only is positive, and find which assignment is consistent with the double-resonance spectrum, Fig. 4-5. Verify that changing the signs of all the coupling constants does not alter the double-resonance spectrum. Deduce that irradiating an AMX spectrum in the M region and observing in the X region will only give the relative signs of J_{AM} and J_{AX}.

4-6 If $\psi(x, y, z)$ is a wave function referred to a static set of axes and $\psi'(x', y', z')$ is the same function referred to a set of rotating axes then

$$\psi' = \exp(i\omega\hat{I}_z t)\psi$$

when the axes are rotating with angular velocity ω about the z direction.
Show that the function $\hat{A}\psi$ becomes $\hat{A}'\psi'$ in the rotating axes where

$$\hat{A}' = \exp(i\omega\hat{I}_z t)\hat{A}\exp(-i\omega\hat{I}_z t)$$

and that

$$\langle r|\hat{A}|s\rangle = \langle r'|\hat{A}'|s'\rangle.$$

By differentiating ψ' with respect to time find the time-dependent Schrödinger equation:

$$i\hbar\frac{\mathrm{d}\psi'}{\mathrm{d}t} = (-\omega\hbar\hat{I}_z + \hat{\mathcal{H}}')\psi'$$

Hence deduce Eqn 4-32 for the effective double-resonance Hamiltonian.

4-7 Explain the trend of the $T_1(^{13}C)$ data given below (in seconds) for n-decanol as a neat liquid:

C-1	C-2	C-3	C-4	C-5	C-6	C-7	C-8	C-9	C-10
0·65	0·77	0·77	0·84	0·84	0·84	1·1	1·6	2·2	3·1

4-8 For tri-n-propylsilane at ambient probe temperature the $^{29}Si\text{-}\{^1H\}$ NOE is $-0{\cdot}87$ and the ^{29}Si spin–lattice relaxation time is 45·4 s. Calculate the (Si, H) dipolar contribution to the rate $T_1^{-1}(^{29}Si)$.

4-9 Describe what you expect to observe in the A region of the spectrum for a double-resonance experiment on an A_2X spin system in which ν_2 is set to $\nu_X - |J_{AX}|$ with relatively low power.

4-10 Derive Eqns 4-41 and 4-42, using the preceding discussion.

4-11 Show that when $\omega^2\tau_c^2 \gg 1$ for all relevant frequencies, the $^{13}C\text{-}\{^1H\}$ NOE is 0·154.

4-12 Using the discussion of Section 4-6, calculate the frequencies and intensities for the A lines of an AX spin system with $J_{AX} = 5$ Hz subjected to an $A\text{-}\{X\}$ double resonance experiment with $\nu_2 = \nu_X$ and $\gamma B_2/2\pi = 15$ Hz.

5 Chemical exchange and quadrupolar effects

5A CHEMICAL EXCHANGE

5-1 Introduction

So far it has been assumed that the molecule under study is rigid and that none of the bonds are labile. No molecule fulfils such a criterion rigorously, since all have vibrations which modulate geometry. However, such motion is usually so rapid (frequencies $\sim 10^{12}$ to 10^{14} Hz) that it has little direct effect on NMR spectra. This chapter is concerned with somewhat slower motions which nonetheless cause *reversible* changes, i.e. dynamic equilibria are involved. In NMR parlance this is 'chemical exchange', whether it involves bond-breaking or not. Actually, a better phraseology might be *magnetic site exchange.* When bonds are broken and reformed, as for exchange of hydroxyl protons between methanol molecules, the exchange is said to be intermolecular, whereas a process such as hindered internal rotation may also lead to effects on NMR spectra but is definitely intramolecular. A further distinction which may be made is between mutual and non-mutual exchange. In *mutual exchange* two magnetic sites (i.e. two different environments for a given nuclide) are *interchanged* by the process, as is the case for the methyl protons of dimethylnitrosamine (5-[I]) when internal rotation occurs. For the static molecule there are two distinct

(A) H_3C–N–N=O, (B) H_3C PhH_2C–N–N=O, H_3C PhH_2C–N–N=O, H_3C

5-[I]   5-[IIa] *syn*   5-[IIb] *anti*

sites, A and B. Internal rotation about the central N—N bond interconverts these, and the process may be designated $A_3B_3 \rightleftharpoons B_3A_3$ for the protons for NMR purposes. There is only one distinguishable molecule involved. However, methylbenzylnitrosamine (5-[II]) provides an example of *non-mutual exchange.* In this case there are two distinguishable rotamers (5-[IIa] and [IIb]), and although there are still two methyl sites these are not *interchanged* by the internal rotation in a given molecule though one site is converted into the other. If the ethyl protons are ignored the exchange may be designated $A_3 \rightleftharpoons B_3$, and it is clear that rotamer populations give the relative weights of the two sites.

5-2 Slow-exchange and fast-exchange regimes

Magnetic site exchange of the types discussed above leads to NMR spectra which depend on the rate of the relevant process. In particular the rate needs to be compared with the 'NMR timescale', to be defined more closely below. The NMR timescale generally refers to lifetimes of the order of 1 s to 10^{-6} s. Two extreme types of NMR behaviour can be readily distinguished. If the exchange lifetime is greatly in excess of the NMR timescale, the system is in the slow-exchange regime, whereas if the lifetime is substantially less than the NMR timescale, the fast-exchange regime results. NMR observations in these two regimes are governed by simple considerations, viz:

(i) Slow-exchange spectra are the superposition of sub-spectra due to each species present.

(ii) In the fast-exchange regime the observed spectrum may be considered as due to a single species whose NMR parameters (chemical shifts and coupling constants) are the relevant *averages* of those for the individual species (suitably weighted to take account of differing populations if necessary)

The systems 5-[I] and 5-[II] are in the slow exchange regime at room temperature. Therefore in proton NMR the dimethyl compound 5-[I] gives two distinct methyl signals of equal intensity, separated by ca. 0·74 ppm. The higher frequency peak is attributed to the protons of the methyl group *trans* to the nitroso oxygen. Compound 5-[II] also gives two methyl proton signals but these are of different intensity, the low frequency signal being the more intense. The intensity ratio leads immediately to information about the relative concentrations of the rotamers and hence to the equilibrium constant. It may be noted that the timescale of normal chemical operations is substantially longer than the NMR timescale so that NMR information on dynamic equilibria may be obtained even when chemical separation of the species is not feasible. A case in point is illustrated later (see Fig. 8-2).

The fast exchange regime may be achieved for the nitrosamines 5-[I] and 5-[II] above about 180 °C (depending on the magnetic field B_0). For species 5-[I] a single proton peak is then observed, midway between the Larmor frequencies for the *trans* and *cis* methyl protons at *that temperature.* Since chemical shifts are in principle temperature-dependent this may not be at the average frequency of the peaks observed at room temperature. Similarly for 5-[II] a single peak will be observed for the methyl protons in the fast-exchange regime. It will occur at a frequency ν_{ave} given by

$$\nu_{\text{ave}} = p_s \nu_t + (1 - p_s)\nu_c \tag{5-1}$$

where p_s is the fractional population of the *syn* form and ν_t the Larmor frequency in that form (methyl *trans* to the nitroso group), whereas ν_c is the methyl proton Larmor frequency in the *anti* form. The parameters p_s, ν_t and ν_c refer to the temperature of measurement. If it can be assumed that ν_t and ν_c are independent of temperature (and are

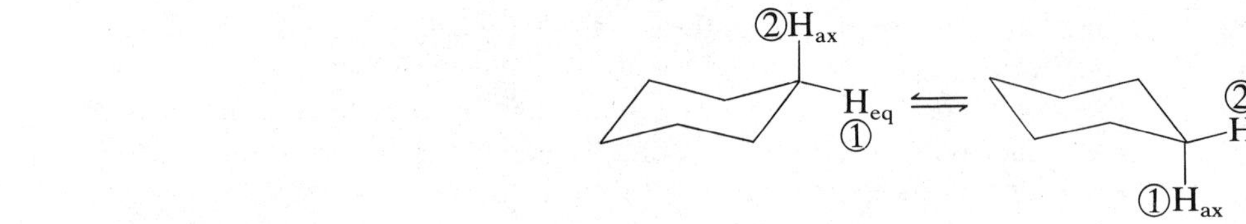

5-[III]

therefore given by room temperature data) then p_s can be determined, but the procedure is subject to big errors.

Inversion of six-membered rings is rapid on the NMR timescale at room temperature. In the case of cyclohexane itself (5-[III]) the inversion is of the mutual exchange type—as proton 1 changes from equatorial to axial so proton 2 changes in the reverse sense. On average all the protons in cyclohexane are equivalent and give rise to a single line. At low temperatures the spectrum is very complicated since there is an $[AB]_6$ spin system.

Coupling constants are also averaged by magnetic site exchange which is fast on the NMR timescale. However, in this case there are clear differences between intramolecular exchange and intermolecular exchange. The former may be illustrated using internal rotation in 1,1,2,2-tetrachlorofluoroethane (5-[IV]), $CFCl_2.CHCl_2$, as an example. The rotamers (essentially in staggered forms) are as follows:

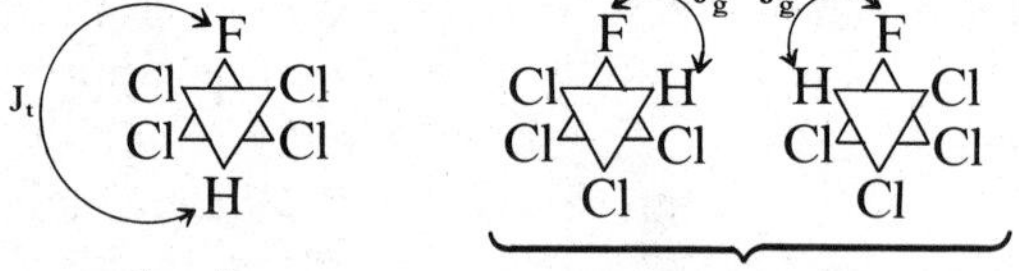

5-[IVa] *trans* 5-[IVb] *gauche*

The two *gauche* rotamers are equivalent and have a total fractional population $(1-p_t)$ if that of the *trans* form is p_t. Under fast-exchange conditions the proton will resonate at a chemical shift given by

$$\nu_{ave} = p_t\nu_t + (1-p_t)\nu_g \qquad (5\text{-}2)$$

and there will be a splitting due to coupling to fluorine with magnitude

$$J_{ave} = p_tJ_t + (1-p_t)J_g \qquad (5\text{-}3)$$

As p_t varies with temperature so will J_{ave}, and this fact may be used to monitor the variation of p_t, since individual coupling constants are less prone to temperature-dependence than chemical shifts.

In the case of intermolecular processes the fast-exchange regime results in the elimination of coupling across the bond which breaks.[1] Consider, for example, the methyl protons in methanol. Under slow-exchange conditions they give rise to a doublet due to coupling to the hydroxyl proton. The two methyl lines may be attributed to molecules in which the hydroxyl proton is α and β respectively. When intermolecular exchange occurs the incoming hydroxyl proton may be α or

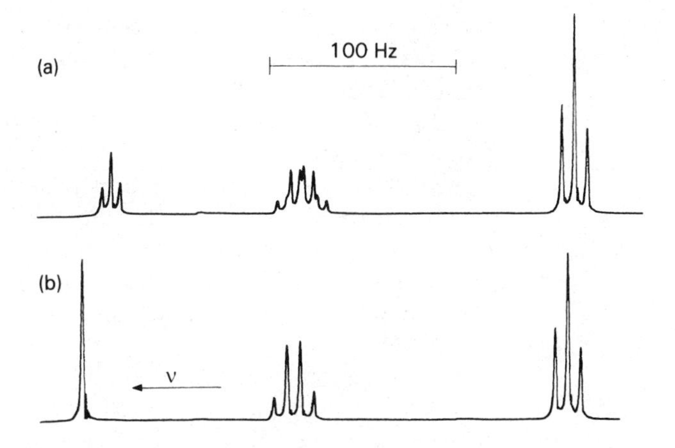

Fig. 5-1 Proton NMR spectra (60 MHz) of ethanol (a) pure liquid, (b) with the addition of a small amount of dilute hydrochloric acid. In addition to the 'decoupling' caused by intermolecular exchange of the hydroxyl protons in (b), there is a small shift in the Larmor frequency of the hydroxyl proton.

β at random. On average the methyl protons are influenced by α and β hydroxyl protons to an equal extent, and therefore resonate at the average frequency, i.e. at the chemical shift position. The effect of coupling is eliminated, as it is also in the hydroxyl region of the spectrum, and therefore only single lines will be observed for the CH_3 and OH protons. The analogous effect for ethanol is shown in Fig. 5-1. This may be referred to as *chemical exchange decoupling*. Observation as to whether coupling is eliminated or merely averaged therefore provides evidence as to the inter- or intramolecular nature of the exchange process. Rapid intermolecular exchange is frequently the rule for OH and NH protons in aqueous solution, so that a single line is seen for all such protons (including those of the H_2O solvent) at an average chemical shift.

A word of explanation as to why rapid exchange leads to *sharp* averaged lines is in order at this point. Consider a FT experiment in which the FID is recorded. An exchange process which occurs during T_2^* changes the frequency of the FID. If such an event occurs in a time short *compared to the difference between the two frequencies*, the FID will be indistinguishable from that given by a system with the average frequency, i.e. a sharp line results. This gives the clue to what is meant by the NMR timescale.

5-3 Near the exchange rate limits

When the rate of magnetic site exchange increases from the slow-exchange regime so as to begin to affect the NMR spectrum the situation can be treated very simply by invoking the Uncertainty Principle. If the lifetime of a nucleus in environment A is τ_A, then the uncertainty in the energy is $\Delta U = \hbar\tau_A^{-1}$. This will influence directly the widths of the relevant NMR line, giving a broadening contribution (at half-height) of

$$\Delta\nu_{1/2} = (\tau_A\pi)^{-1} \tag{5-4}$$

In fact in such a case the lifetime acts exactly like the non-secular contribution to T_2, giving a Lorentzian lineshape. This gives a further clue to the meaning of the NMR timescale, since effects of exchange on the spectra will only be noticeable if $\tau_A \lesssim T_2$ (or, in practice, T_2^*). Clearly, linewidths can be used to monitor lifetimes provided the widths appropriate to a static system are known. For two-site exchange, $A \rightleftharpoons X$ say, with unequal populations, the site with the smaller population will have the shorter lifetime and hence the broader line. The lifetimes are the reciprocals of the relevant pseudo-first-order rate constants, i.e. $\tau_A = k_r^{-1}$, where, for example the rate of internal rotation about the N—N bond in dimethylnitrosamine 5-[I] is written k_r [Me_2NNO].

As exchange rates approach the fast-exchange regime, the spectrum also takes on a simple appearance. For the case of simple (uncoupled) two-site exchange, $A \rightleftharpoons X$, with equal populations a single Lorentzian line is observed with a linewidth (in excess of that caused by relaxation and/or inhomogeneities in B_0) given by

$$\Delta\nu_{1/2} = \tfrac{1}{2}\pi(\nu_A - \nu_X)^2 k_r^{-1} \qquad (5\text{-}5)$$

In this region of exchange, information on k_r can only be obtained from the linewidth if the chemical shift difference $(\nu_A - \nu_X)$ can be reliably estimated. Equation 5-5 shows that observable broadening near the fast exchange regime only occurs when $\tau_A(\nu_A - \nu_X)^2 \gtrsim T_2^{-1}$ (or, in practice, $(T_2^*)^{-1}$).

5-4 Intermediate rates of exchange: the $A \rightleftharpoons X$ case

The preceding section has demonstrated that for

$$T_2^* \gtrsim k_r^{-1} \gtrsim (\tfrac{1}{2}\pi^2\,\Delta\nu^2 T_2^*)^{-1} \qquad (5\text{-}6)$$

where $\Delta\nu$ is the frequency difference being averaged by the exchange process (which may be a coupling constant rather than a chemical shift), the NMR spectrum will be affected. Equation 5-6, therefore, defines the NMR timescale. For spin-$\tfrac{1}{2}$ nuclei, T_2^* can usually be taken as ca. 1 s, so that the condition may be re-written

$$1 \gtrsim k_r^{-1} \gtrapprox 0{\cdot}2\,\Delta\nu^{-2} \qquad (5\text{-}7)$$

Of course, the relevant $\Delta\nu$ varies widely with the nature of the exchanging system and the magnetic field of the spectrometer, but it is clear that the lifetimes which can be studied are in the range 1 to 10^{-6} s, as already mentioned.

In the intervening region of rates, where Eqns 5-4 and 5-5 do not apply, complicated bandshapes are observed. In order to enhance understanding of such spectra the simple uncoupled equally-populated two-site exchange situation $A \rightleftharpoons X$ will be considered in more detail at this point. The approach of McConnell[2] will be used, but the influence of non-exchange contributions to T_2 will be ignored. For each type of environment a complex magnetization may be defined by

$$G = u - iv \qquad (5\text{-}8)$$

and the Bloch equations 3-14 and 3-15 used to obtain

$$\frac{dG}{dt} = -i\,\Delta\omega G - i\gamma B_1 M_z \tag{5-9}$$

where $\Delta\omega = \omega - \omega_i$. Now if there is a constant probability, τ_A^{-1}, per unit time of an exchange $A \to X$, and an equal probability for the reverse exchange, the complex magnetization for A must obey the equation

$$\frac{dG_A}{dt} = -i\,\Delta\omega_A G_A - i\gamma B_1 M_{zA} - \tau_A^{-1}(G_A - G_X) \tag{5-10}$$

and a corresponding equation may be written for X. If it is assumed that spin–lattice relaxation times in the two sites are equal, then $M_{zA} = M_{zX} = \frac{1}{2}M_z$. The total complex magnetization is clearly $G = G_A + G_X$, and according to Eqn 5-8 the absorption signal will be proportional to minus the imaginary part of G. Equation 5-10 and its analogue for X may be solved for G under conditions of equilibrium when $dG_A/dt = dG_X/dt = 0$. The result for the absorption mode is

$$v = \tfrac{1}{2}\gamma B_1 M_z \frac{\tau_A(\omega_A - \omega_X)^2}{(\Delta\omega_A + \Delta\omega_X)^2 + \tau_A^2\,\Delta\omega_A^2\,\Delta\omega_X^2} \tag{5-11}$$

In frequency units the signal shape is therefore given by

$$g(\nu) = \frac{2\tau_A(\nu_A - \nu_X)^2}{[\nu - \frac{1}{2}(\nu_A + \nu_X)]^2 + \pi^2\tau_A^2(\nu - \nu_A)^2(\nu - \nu_X)^2} \tag{5-12}$$

This equation may be recast into a form involving a dimensionless abscissa $x = \Delta\nu/\nabla$ and a parameter $a = \pi\tau_A\nabla$, where $\Delta\nu = \nu - \frac{1}{2}(\nu_A + \nu_X)$ and $\nabla = \frac{1}{2}(\nu_A - \nu_X)$ as follows

$$g(x) = 2\tau_A/[x^2 + a^2(x^2 - 1)^2] \tag{5-13}$$

The basic shape therefore depends only on a. Calculated shapes for various values of a are shown in Figure 5-2. Under conditions of slow exchange ($a \gg 1$) Eqn 5-12 can be shown to give rise to two Lorentzian

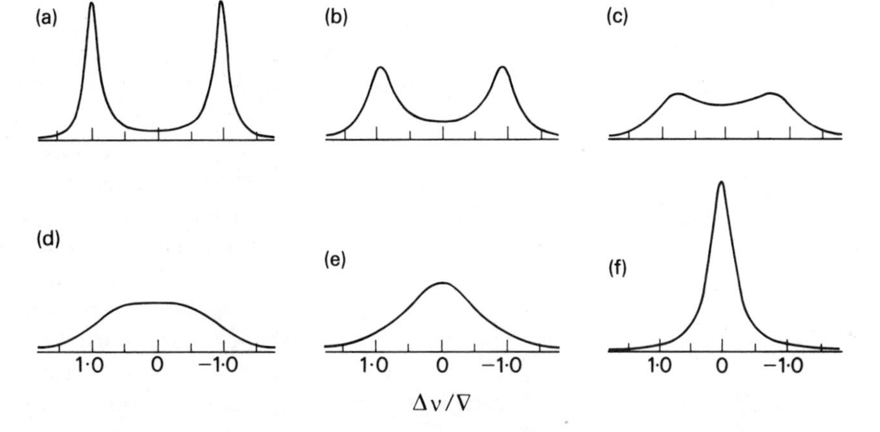

Fig. 5-2 Calculated NMR bandshapes for equally-populated uncoupled two-site exchange as a function of the parameter $a = \pi\tau_A\nabla$. The abscissa is the relative offset parameter $\Delta\nu/\nabla$. (a) $a = 4$; (b) $a = 2$; (c) $a = 1$; (d) $a = 1/\sqrt{2}$ (coalescence); (e) $a = 0.5$; (f) $a = 0.2$. Spectra (a) and (f) are near the slow- and fast-exchange limits respectively.

lines, centred at ν_A and ν_X respectively, with widths given by Eqn 5-4, whereas at the opposite extreme of fast exchange ($a \ll 1$) a single Lorentzian line of double intensity at $\frac{1}{2}(\nu_A + \nu_X)$, with width given by Eqn 5-5, is predicted. When $a = 2^{-1/2}$ a single peak with a flat top (Fig. 5-2d) occurs. This situation is referred to as coalescence. It is readily observable in practice and then the lifetime is given (Problem 5-5) by

$$\tau_A^c = \sqrt{2}/\pi(\nu_A - \nu_X) \tag{5-14}$$

This equation gives further meaning to the term 'NMR timescale' since it defines the critical lifetime for any given situation. When variation of τ_A is achieved by changing the temperature, that at which coalescence occurs is referred to as the *coalescence temperature.* For other exchange rates total bandshape fitting is required to give the required kinetic information. It would normally be necessary to incorporate into the calculations information about transverse relaxation times (which may or may not be equal) at the two sites. Such corrections are particularly important near the slow- and fast-exchange limits. Figure 5-3 shows a chemical example of how bandshapes for the $A \rightleftharpoons X$ case change as τ_A varies due to temperature variation.

The equations given above may be readily modified to deal with uncoupled two-sites exchange with *un*equal populations. Such cases are very common in ^{13}C NMR, the absence of coupling being due to the dilute nature of the carbon-13 isotope. Thus ring inversion of any monosubstituted cyclohexane will give rise to at least four $A \rightleftharpoons X$ exchange systems (assuming $C^2 \equiv C^6$ and $C^3 \equiv C^5$) with a common $A:X$ intensity ratio and common rate constants. Bandshape fitting will yield two rate constants (or one rate constant and an equilibrium constant).

5-5 More complicated exchange cases

It is not difficult to extend the equations of the previous section to deal with exchange between any number of uncoupled sites. The bandshape will then depend on a number of rate constants and equilibrium constants. Frequently it is possible to make simplifying kinetic assumptions. It should be stressed that it is only the magnetization transfer that matters (for uncoupled sites). The mechanism of exchange is only important in so far as it may determine relative lifetimes between sites. Thus equations derived in the manner discussed in the previous section on the implicit assumption that the sites differ in chemical shift are also applicable to cases where the site difference arises from coupling, as in the case of intermolecular exchange of hydroxyl protons for methanol. That is to say, such coupling is treated as giving rise to effective chemical shifts. Another situation in which coupling may be eliminated by 'exchange' is when one of the coupled nuclei relaxes rapidly (see Section 5-14), and once more the above approach to calculating intermediate bandshapes may be used.

Many intramolecular exchange cases, however, involve coupling which is retained (though, perhaps, subject to averaging) during the

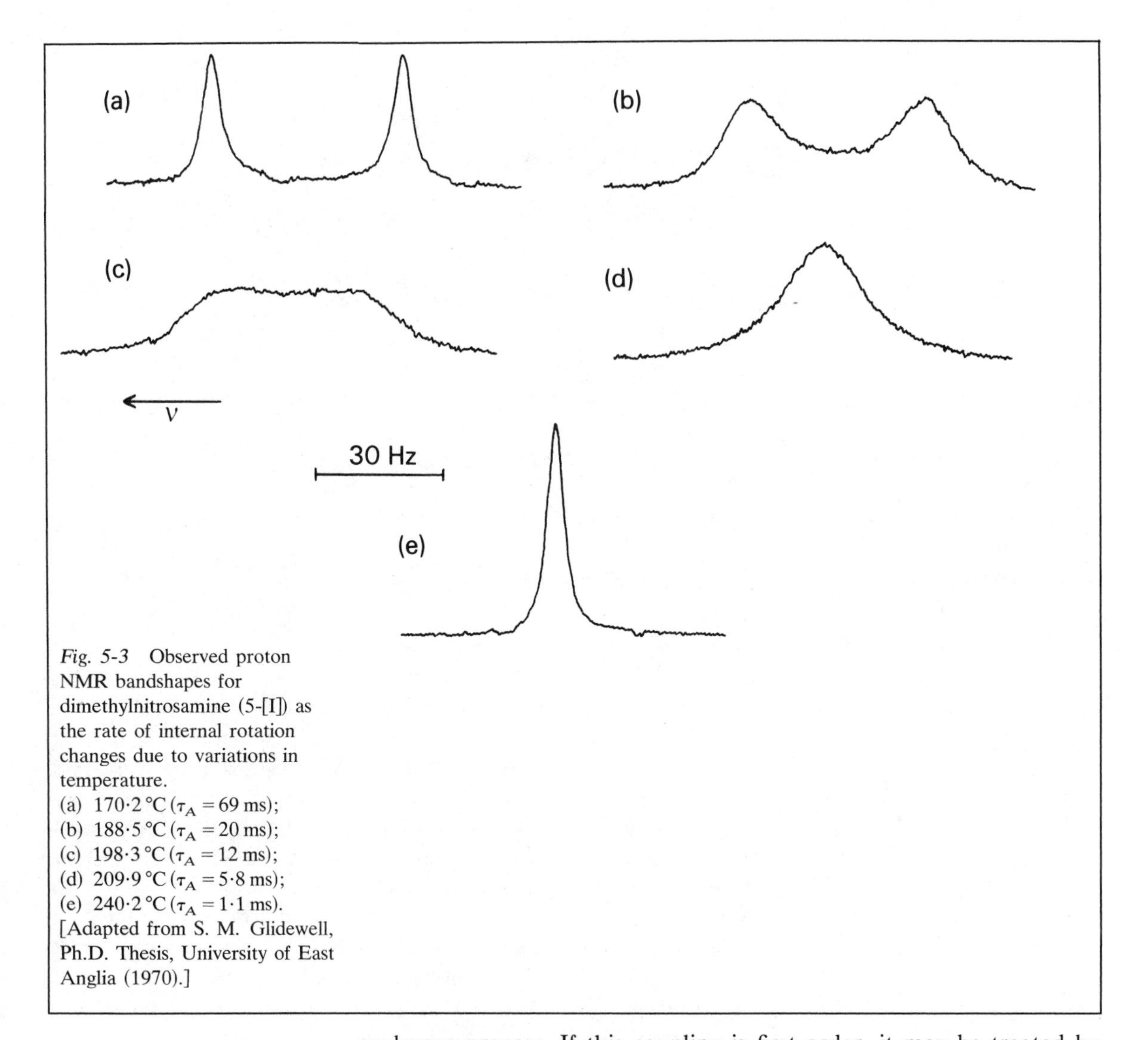

Fig. 5-3 Observed proton NMR bandshapes for dimethylnitrosamine (5-[I]) as the rate of internal rotation changes due to variations in temperature.
(a) 170·2 °C ($\tau_A = 69$ ms);
(b) 188·5 °C ($\tau_A = 20$ ms);
(c) 198·3 °C ($\tau_A = 12$ ms);
(d) 209·9 °C ($\tau_A = 5{\cdot}8$ ms);
(e) 240·2 °C ($\tau_A = 1{\cdot}1$ ms).
[Adapted from S. M. Glidewell, Ph.D. Thesis, University of East Anglia (1970).]

exchange process. If this coupling is first-order, it may be treated by the '*X* approximation' (see Section 2-9), and the bandshape for the exchanging system calculated as the sum of several simple cases. Consider, for instance, the methyl protons of *cis*-2,6-dimethyl-1-nitrosopiperidine 5-[V]), which differ in chemical shift at room temperature because of the orientation of the nitroso group but which are rendered equivalent on the NMR timescale by the internal rotation

Me Me N N O $\rightleftharpoons$ Me Me N N O

5-[Va] 5-[Vb]

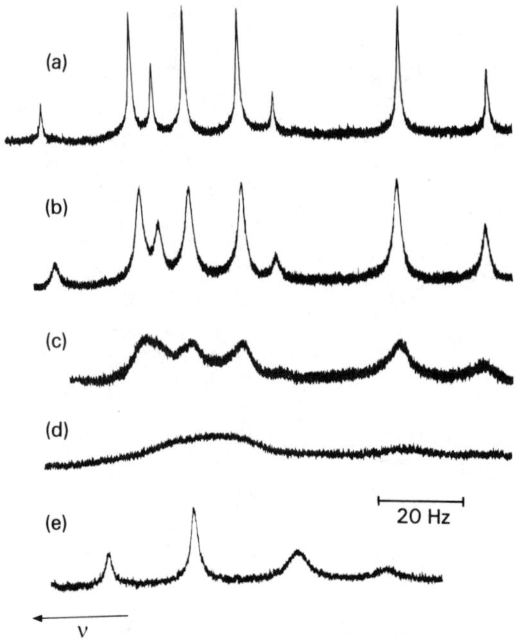

Fig. 5-4 Observed proton NMR bandshapes for the exchange case 5-[VI].
(a) 35 °C, $k_r < 0{\cdot}2\ s^{-1}$;
(b) 90 °C, $k_r = 2{\cdot}1\ s^{-1}$
(c) 105 °C, $k_r = 8{\cdot}0\ s^{-1}$;
(d) 120 °C, $k_r = 24\ s^{-1}$
(e) 160 °C, $k_r = 310\ s^{-1}$. The rate constants quoted are for the interconversion 5-[VIa] → 5-[VIb].
[Reproduced, with permission, from C. J. Creswell & R. K. Harris, *J. Magn. Reson.*, **4**, 99 (1971).]

about the N—N bond at temperatures above about 100 °C (mutual exchange).[3] The methyl proton NMR signals show a splitting due to coupling to the nearby vicinal protons in a manner not markedly influenced by the exchange, which therefore causes two doublets to coalesce to one doublet. The bandshape may be calculated on the basis of two uncoupled $A \rightleftharpoons X$ cases (one for each spin of the neighbouring methine proton) with all the populations equal and with the same rate constant but with differing effective chemical shifts.

In second-order cases, however, more sophisticated methods of calculation must be used. For example, [1,2,5]thiadiazolo[3,4-*e*]-benzofurazan oxide (5-[VI]) represents an exchange system (Figure 5-4) for the protons of the type $AB \rightleftharpoons CD$, so the slow-exchange spectrum shows two distinct *ab* sub-spectra. In the high-temperature limit there

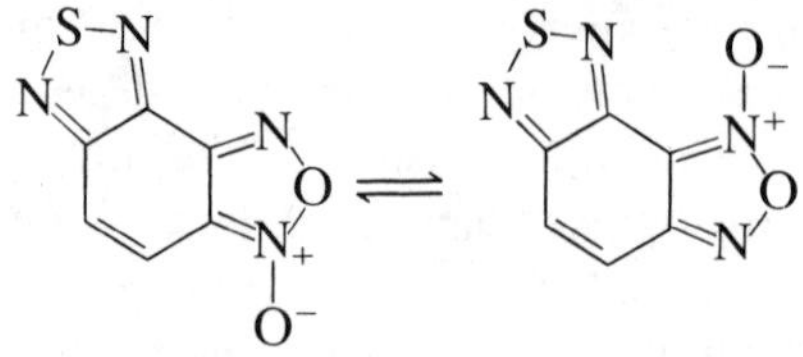

5-[VIa] ⇌ 5-[VIb]

is a single *AB* pattern in the spectrum. Computer programs[4] are available[5] for calculating the bandshapes of such exchanging systems.

5-6 Kinetic information

It has been shown above that in principle NMR bandshapes yield information about the kinetics of various types of magnetic site exchange, provided the lifetimes are of the order of the NMR timescale.

Variation of lifetimes, and hence of bandshapes, may be obtained by changing temperature, pH, catalyst concentrations, etc. For intramolecular processes, the NMR-derived rates may be interpreted in terms of energy barriers to internal rotation etc. by

$$k_r = \kappa \frac{kT}{h} \exp(-\Delta G^{\ddagger}/RT) \quad (5\text{-}15)$$

where $\Delta G^{\ddagger}$ is the free energy of activation and κ is a transmission coefficient which depends on the nature of the process but is often taken as unity. Barriers in the approximate range 20 to 100 kJ mol^{-1} may be studied by NMR bandshape methods. Coalescence temperatures provide approximate values of $\Delta G^{\ddagger}$, but further information requires a study of temperature variation according to

$$k_r = \kappa \frac{kT}{h} \exp(-\Delta H^{\ddagger}/RT)\exp(\Delta S^{\ddagger}/R) \quad (5\text{-}16)$$

Thus plots of $\log(k_r/T)$ vs. $1/T$ yield values of $\Delta H^{\ddagger}$ and $\Delta S^{\ddagger}$. Bandshape fitting provides the necessary data for k_r, but the procedure is fraught with difficulties. For example, estimation of the temperature variation of chemical shift differences and of transverse relaxation times is required. At intermediate exchange rates the bandshape is sensitive to the values of these parameters but an increase in the number of variables to be obtained is implied. The difficulties are greatest near the fast- and slow-exchange limits, and often lead to systematic errors in $\Delta H^{\ddagger}$ and $\Delta S^{\ddagger}$. It is important to extend the temperature range of observation as much as possible. This may be done in a number of ways, viz:

(a) Obtaining spectra at a variety of magnetic fields (B_0) so that coalescence occurs at different temperatures.
(b) Making use of complicated spectra containing a range of splittings to be averaged.
(c) Using a series of simple spectra with different splittings, as is naturally the case when studying ^{13}C (e.g. for ring inversion of a monosubstituted cyclohexane.
(d) Using NMR to study kinetics directly by isolating (if possible) one chemical form (in cases of non-mutual exchange) and running spectra as a function of time as it converts to the equilibrium situation. Clearly this is only feasible at temperatures very much lower than coalescence.
(e) Using the principles of magnetization transfer by selective pulse methods (see Section 7-4). This also gives kinetic information at exchange rates much slower than those needed for coalescence.
(f) Making use of spin–lattice relaxation in the rotating frame (see the next section).

It is not normal for NMR studies of chemical exchange to yield *mechanistic* information. However, different mechanistic models may in some cases give bandshapes at intermediate exchange rates which differ sufficiently to enable a distinction to be made.[6] As already pointed out, the effects on coupling enable inter- and intra-molecular exchange to be distinguished readily. Data on $\Delta H^{\ddagger}$ and $\Delta S^{\ddagger}$ can give information about the nature of the intermediate state in the exchange process. For example, NMR results for the ring inversion of cyclohexane ($\Delta H^{\ddagger} = 45$ kJ mol^{-1}, $\Delta S^{\ddagger} = 12$ J K^{-1} mol^{-1}) are consistent with the intermediate being in the half-chair form (four adjacent carbons in a single plane).

5-7 Chemical exchange information from spin-lattice relaxation in the rotating frame

Section 3-17 discussed the concepts related to $T_{1\rho}$, and concluded that for mobile isotropic fluids it is normal to find $T_1 = T_{1\rho} = T_2$. However, it was mentioned that $T_{1\rho}$ is affected by spectral densities at frequencies governed by γB_1, and hence lying in the region of tens of kHz. It is clear, then, that chemical exchange at such rates may affect $T_{1\rho}$ but not T_1. If both $T_{1\rho}^{-1}$ and T_1^{-1} are measured in such circumstances, their difference, $T_{1\rho}^{-1}(\text{ex})$, represents an exchange contribution to $T_{1\rho}$ given for the simple equally populated uncoupled $A \rightleftharpoons X$ exchange case, by

$$T_{1\rho}^{-1}(\text{ex}) = \tfrac{1}{2}\pi^2(\nu_A - \nu_X)^2\tau_A/(1 + \tfrac{1}{4}\tau_A^2\omega_1^2) \qquad (5\text{-}17)$$

where $\omega_1 = \gamma B_1$ measures the strength of the r.f. field. This equation may be re-arranged to give

$$T_{1\rho}(\text{ex}) = \frac{1}{\tfrac{1}{2}\pi^2(\nu_A - \nu_X)^2\tau_A} + \frac{\tau_A}{2\pi^2(\nu_A - \nu_X)^2}\,\omega_1^2 \qquad (5\text{-}18)$$

and thus a plot of $T_{1\rho}(\text{ex})$ vs. ω_1^2 will yield data on lifetimes and chemical shift differences independently. This method is relevant principally in the fast exchange regime, where only a single line is observed. Of course, it requires that $T_{1\rho}(\text{ex})$ is significant in relation to T_1. Figure 5-5 shows an example of a $T_{1\rho}(\text{ex})$ vs. B_1 plot for the ^{19}F

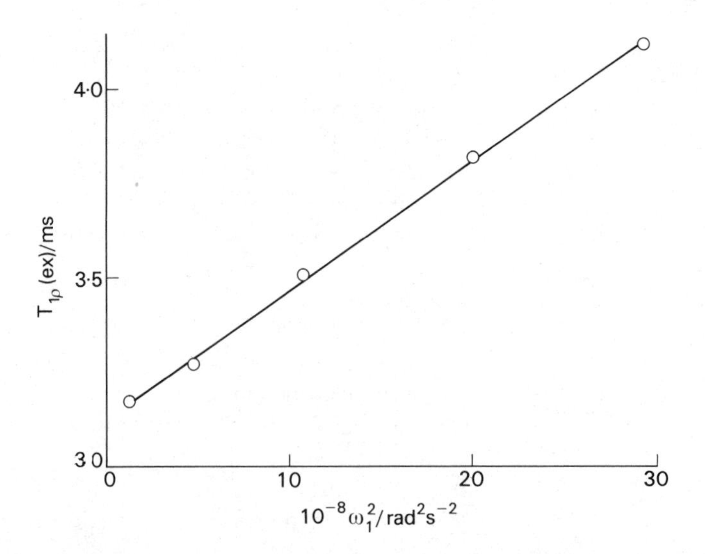

Fig. 5.5 Data for $T_{1\rho}(\text{ex})$ of ^{19}F in gas-phase C_6F_{12} at 36 °C. The plot leads to $|\nu_{ax} - \nu_{eq}| = 1760 \pm 40$ Hz and $\tau_A = 21$ μs.

resonance of perfluorocyclohexane, which is subject to ring inversion. Coupling between the fluorines was ignored in the analysis.

Measurements of T_1 and T_2 may also be used to obtain kinetic data, the former relating to considerably lower activation energies than those obtained by the methods discussed in this chapter.

5-8 Chemical exchange involving paramagnetic species

In the discussion of exchange situations so far, relaxation in the various sites has been relatively unimportant, being treated in terms of minor corrections to the bandshape. However, a modified approach is necessary when one or more of the chemical sites forms part of a paramagnetic species. This occurs very commonly for aqueous solutions containing a small amount of a paramagnetic metal ion with an excess of a ligand. A given ligand molecule may be in the primary coordination sphere of the metal ion for a small fraction of its time—the probability, p_M, depending on the various concentrations and the relative affinities of the metal ion for the various species present. There are several different effects on the observed linewidth depending on the relationships between three parameters, viz:

(i) The difference between the chemical shift, $\Delta\omega_M$ (in angular frequency terms), when coordinated to the metal and when free. This coordination shift, often referred to as the contact shift (see Section 8-14) may be very large, but often depends on the extent of delocalization of unpaired electron spin onto the nucleus in question.

(ii) The transverse relaxation time of the nucleus in the bound ligand state, T_{2M}. This is also dominated by the presence of unpaired electron spin.

(iii) The lifetime, τ_M, of the ligand in the primary coordination sphere.

Swift and Connick[7] have developed equations for such situations. Using an approach similar to that outlined in Section 5-4 they showed that in general the enhancement T_{2p}^{-1} of the transverse relaxation rate of the ligand due to the presence of the paramagnetic species is

$$T_{2p}^{-1} = \tau_\ell^{-1}[T_{2M}^{-2} - (T_{2M}\tau_M)^{-1} + \Delta\omega_M^2]/[(T_{2M}^{-1} + \tau_M^{-1})^2 + \Delta\omega_M^2] \qquad (5\text{-}19)$$

where τ_ℓ is the lifetime of the free ligand before becoming coordinated (detailed balancing shows that $\tau_\ell = \tau_M/p_M$). The four limiting types of behaviour are given in Table 5-1. Limiting conditions I and II correspond to the slow-exchange and fast-exchange regimes discussed in

Table 5-1 Summary of the results of Swift and Connick[7]

Regime	*Limiting conditions*	*Resulting equation for* T_{2p}^{-1}
I	$\Delta\omega_M^2 \gg T_{2M}^{-2}, \tau_M^{-2}$	$p_M/\tau_M \equiv \tau_\ell^{-1}$
II	$\tau_M^{-2} \gg \Delta\omega_M^2 \gg (T_{2M}\tau_M)^{-1}$	$p_M\tau_M\Delta\omega_M^2$
III	$T_{2M}^{-2} \gg \Delta\omega_M^2, \tau_M^{-2}$	$p_M/\tau_M \equiv \tau_\ell^{-1}$
IV	$(T_{2M}\tau_M)^{-1} \gg T_{2M}^{-2}, \Delta\omega_M^2$	p_M/T_{2M}

Section 5-3, and are independent of the relaxation rate in the bound state. Condition III relates to rapid relaxation in the bound state and gives the same expression as for condition I. Under condition IV, however, the observed linewidth enhancement depends only on the transverse relaxation time in the bound state—the total observed linewidth is the weighted average of those expected for the bound and free states.

Detailed study of the linewidth as a function of temperature through the various regimes yields considerable information both about the exchange and about the NMR parameters governing the bound state. It should be noted that $\Delta\omega_M$ and T_{2M} are likely to be affected by temperature.

5B QUADRUPOLAR EFFECTS

5-9 Nuclear electric quadrupole moments

A general distribution of point electric charges cannot be fully described simply in terms of the overall charge and a dipole. However, an adequate description can usually be obtained as a series of multipoles of the form:

$$\sum_i \varepsilon_i \mathbf{r}_i^n \tag{5-20}$$

where $\mathbf{r}$ is the distance to the charge ε_i from the electrical centre of gravity, and n is an integral power of two (1, 2, 4, 8 . . .). The term with $n=1$ is the electric dipole moment; that with $n=2$ is the electric quadrupole moment. This book will not be concerned with any higher terms. The molecule CO_2 is an example of a system with zero dipole moment but non-zero quadrupole moment. The latter can readily be seen as being constituted by two equal dipoles back-to-back (Fig. 5-6).

An electric monopole is affected by electric potential, V, whereas a dipole is only influenced by a potential *gradient*, i.e. by an electric field, $E=-\mathrm{d}V/\mathrm{d}z$ say. A quadrupole does not interact with a homogeneous electric field, as may be seen by consideration of the former as two dipoles back-to-back. However, there will be an energy of interaction of a quadrupole with an *in*homogeneous electric field, i.e. with an electric field gradient, $\mathrm{d}^2V/\mathrm{d}z^2$ say.

It is clearly feasible for nuclei to possess electric quadrupole moments, since, on a naive view, there is a distribution of positively charged protons in a nucleus. Of course, all nuclei have electric monopoles but they do not possess electric dipoles. It is found that those with spin quantum numbers greater than $\frac{1}{2}$ also possess electric quadrupole moments. It is usual to express such moments in terms of eQ where e is the charge on the proton and Q is the quantity that is normally referred to as the electric quadrupole moment of the nucleus.

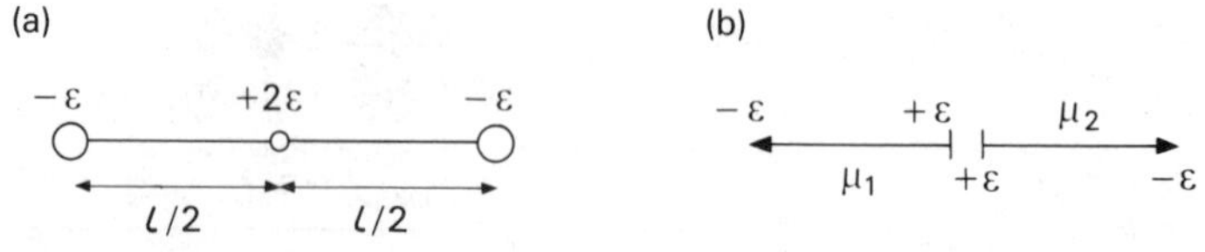

Fig. 5-6 (a) An electric quadrupole system as in the molecule CO_2. (b) A quadrupole as in (a) viewed as two equal dipoles back-to-back

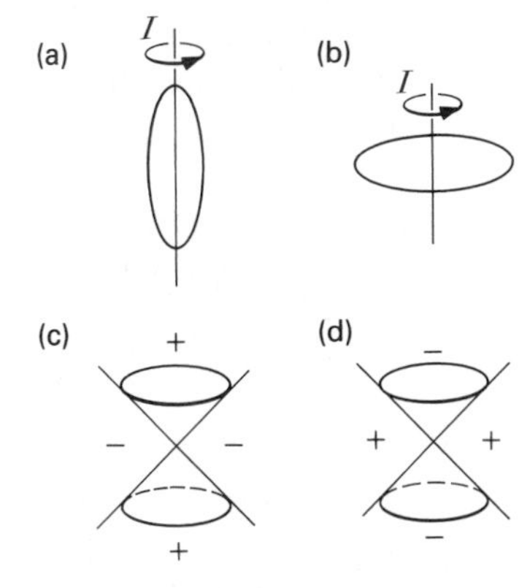

Fig. 5-7 Nuclear electrical properties (a) a prolate charge distribution; (b) an oblate charge distribution; (c) and (d) represent the quadrupole moments of (a) and (b), respectively, when the monopole (spherically symmetrical) charge distribution is subtracted. Note that cylindrical symmetry about the nuclear spin direction is preserved.

The dimensions of Q are clearly those of length squared. Since nuclear radii are ca. 10^{-14} m, the magnitudes of Q are of the order of 10^{-28} m^2. The unit 10^{-28} m^2 is sometimes known as a barn. Nuclear quadrupole moments also possess signs, since there are two types of distribution of positive charge, shown in Fig. 5-7. The distribution must be consistent with the symmetry axis determined by nuclear spin, so either a cigar-shape (Fig. 5-7a) or a discus-shape (Fig. 5-7b) is possible. Such distributions may be regarded as the sum of positive monopoles (spherical distribution of positive charge) and quadrupoles of the form shown in Figs. 5-7c and 5-7d. The former is called a *prolate* quadrupole, and its moment is defined to be positive, whereas the latter is known as an *oblate* quadrupole, having a negative moment. Some typical values of Q are given in Table 5-2. The magnitudes cover a large range, which gives rise to very different patterns of behaviour in NMR. A more comprehensive list of data is contained in Appendix 2.

Table 5-2 Quadrupole moments, linewidth factors and receptivities of some spin $>\frac{1}{2}$ nuclides[(a)]

Nuclide	*Spin* I	*Quadrupole moment* $Q/10^{-28}$ m^2	*Linewidth factor* $10^{56}\,\ell/\text{m}^4$	*Relative receptivity* D^P	D^C
^{2}H	1	$2{\cdot}8\times10^{-3}$	$3{\cdot}9\times10^{-5}$	$1{\cdot}45\times10^{-6}$	$8{\cdot}21\times10^{-3}$
^{6}Li	1	-8×10^{-4}	$3{\cdot}2\times10^{-6}$	$6{\cdot}31\times10^{-4}$	3·58
^{7}Li	$\frac{3}{2}$	-4×10^{-2}	$2{\cdot}1\times10^{-3}$	0·272	$1{\cdot}54\times10^{3}$
^{10}B	3	8.5×10^{-2}	$1{\cdot}4\times10^{-3}$	$3{\cdot}93\times10^{-3}$	22·3
^{11}B	$\frac{3}{2}$	$4{\cdot}1\times10^{-2}$	$2{\cdot}2\times10^{-3}$	0·133	$7{\cdot}52\times10^{2}$
^{14}N	1	1×10^{-2}	5×10^{-4}	$1{\cdot}00\times10^{-3}$	5·69
^{17}O	$\frac{5}{2}$	$-2{\cdot}6\times10^{-2}$	$2{\cdot}2\times10^{-4}$	$1{\cdot}08\times10^{-5}$	$6{\cdot}11\times10^{-2}$
^{35}Cl	$\frac{3}{2}$	−0·10	$1{\cdot}3\times10^{-2}$	$3{\cdot}56\times10^{-3}$	20·2
^{37}Cl	$\frac{3}{2}$	$-7{\cdot}9\times10^{-2}$	$8{\cdot}3\times10^{-3}$	$6{\cdot}66\times10^{-4}$	3·78
^{59}Co	$\frac{7}{2}$	0·38	$2{\cdot}0\times10^{-2}$	0·277	$1{\cdot}57\times10^{3}$
^{75}As	$\frac{3}{2}$	0·29	0·11	$2{\cdot}53\times10^{-2}$	$1{\cdot}44\times10^{2}$
^{127}I	$\frac{5}{2}$	−0·79	0·20	$9{\cdot}50\times10^{-2}$	$5{\cdot}39\times10^{2}$
^{133}Cs	$\frac{7}{2}$	-3×10^{-3}	$1{\cdot}2\times10^{-6}$	$4{\cdot}82\times10^{-2}$	$2{\cdot}73\times10^{2}$
^{181}Ta	$\frac{7}{2}$	3	1·2	$3{\cdot}65\times10^{-2}$	$2{\cdot}07\times10^{2}$

[(a)] A more comprehensive list is contained in Appendix 2.

5-10 Nuclear quadrupole energy

An electric dipole moment, μ, at an angle ψ to an electric field, E, has an energy of interaction $\boldsymbol{\mu}\,.\,\mathbf{E} = \mu E \cos\psi$. For a simple electric quadrupole moment Q consisting of two point dipoles back-to-back a distance s apart in an electric field of gradient $\mathrm{d}E/\mathrm{d}z$ (perpendicular to $\mathbf{E}$), the classical energy will be (see Fig. 5-8)

$$\begin{aligned} U &= \mu E_1 \cos\psi + \mu E_2 \cos(180^\circ - \psi) \\ &= \mu \cos\psi\left[E_1 - \left(E_1 + \frac{\mathrm{d}E}{\mathrm{d}z}\, s \sin\psi\right)\right] \\ &= -\tfrac{1}{2}\mu s \sin 2\psi\ \mathrm{d}E/\mathrm{d}z \end{aligned} \tag{5-21}$$

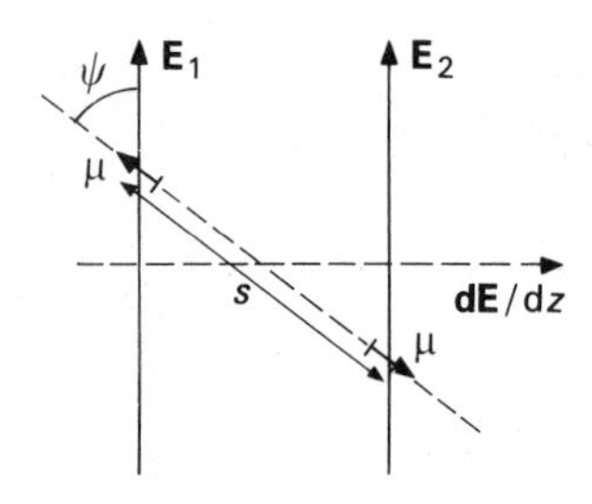

Fig. 5-8 The interactions between an electric quadrupole (viewed as two dipoles back-to-back) and an electric field gradient.

As mentioned in the previous section, this energy is zero if E is homogeneous. Figure 5-8 and Eqn 5-21 make it clear that the quadrupolar energy is invariant to the interchange of the two magnetic moments.

For a quadrupolar nucleus there will be an analogous interaction giving a quadrupolar Hamiltonian term, $\hat{\mathcal{H}}_Q$, which, for a situation characterized by a single electric field gradient $eq_{zz} = \mathrm{d}^2V/\mathrm{d}z^2$ (in the absence of a magnetic field) is

$$h^{-1}\hat{\mathcal{H}}_Q = \chi(3\hat{I}_z^2 - \hat{\mathbf{I}}^2)/[4I(2I-1)] \tag{5-22}$$

where χ is known as the *nuclear quadrupole coupling constant* and is given, in frequency units, by

$$\chi = e^2Qq_{zz}/h \tag{5-23}$$

The quadrupolar energy is therefore simply

$$h^{-1}U_Q = \chi[3m_I^2 - I(I+1)]/[4I(2I-1)] \tag{5-24}$$

where the spin quantization is in the axis system of the electric field gradient, which is generally fixed in the molecule. Clearly χ depends on the nucleus (through Q) and its local environment (which provides q_{zz}). The quadrupole energy (Eqn 5-24) does not depend on the *sign* of m_I, as is consistent with the invariance mentioned above.

Of course Eqn 5-24 describes a set of energy levels for any given quadrupolar nucleus such that transitions (subject to the selection rule $m_I = \pm 1$) may be caused by appropriate electromagnetic radiation. This gives rise to a form of spectroscopy, in the absence of a magnetic field, known as nuclear quadrupole resonance (NQR). Since quadrupole coupling constants vary widely (e.g. 3·161 MHz for ^{14}N in NH_3; 3008 MHz for ^{127}I in ICl), so do the transition frequencies, which are normally in the radiowave region of the spectrum. For reasons to be discussed later, NQR is usually only observable for solids. It yields information about electric field gradients, but the subject of pure NQR will not be considered any further here.

5-11 NMR of quadrupolar nuclei in solids

When a quadrupolar nucleus is placed in a magnetic field two energy terms need to be considered, viz. Zeeman and quadrupolar. Whether the resulting form of spectroscopy is essentially NMR or NQR depends on the relative magnitude of these terms. In this book only the 'high field' case will be considered, for which the Zeeman term predominates. This implies that the magnetic field rather than the electric field determines the direction of quantization. The energy levels are essentially those of NMR perturbed by quadrupolar effects, and for a single type of nucleus the appropriate first-order equation is

$$h^{-1}U = -\nu_0 m_I + \chi[3m_I^2 - I(I+1)](3\cos^2\theta - 1)/[8I(2I-1)] \tag{5-25}$$

where ν_0 is the Larmor frequency, $\gamma B_0(1-\sigma)/2\pi$, and θ is the angle between B_0 and q_{zz}. Since the selection rule is still $\Delta m_I = \pm 1$, the transition frequencies are given by

$$\Delta\nu = \nu_0 - \tfrac{3}{8}\chi \frac{(2m_I - 1)}{I(2I-1)}(3\cos^2\theta - 1) \tag{5-26}$$

where m_I ranges from I to $-I+1$. Therefore, for a single crystal containing nuclei with only one orientation, the spectrum will consist of $2I$ lines characterized by a splitting $\frac{3}{4}(3\cos^2\theta - 1)\chi/[I(2I-1)]$, and hence χ may be obtained if θ is known or is systematically varied. The spectrum is said to exhibit quadrupolar fine structure. If a *powder* is examined, all values of θ are possible (but not equally likely) and a *powder pattern* results, which arises from the superposition of many doublets. For a spin-1 nucleus, the powder pattern takes the same form as that arising from dipolar interactions for a heteronuclear system of two spin-$\frac{1}{2}$ nuclei, such as is illustrated in Fig. 6-1 (page 143) except that the dipolar interaction constant (R) is replaced by $\frac{3}{4}\chi$. Again, χ may be obtained. Of course, dipolar interactions lead to additional broadening for NMR resonances of quadrupolar nuclei in solids, but for many cases quadrupolar splitting is the dominant effect because of the large magnitude of χ. However, the majority of quadrupolar nuclei have a spin quantum number which is an odd multiple of $\frac{1}{2}$. Equation 5-26 shows that the transition $m_I = \frac{1}{2} \Rightarrow m_I = -\frac{1}{2}$ is independent of θ, so a strong absorption may be readily seen at ν_0, whereas the other transitions are more difficult to observe because of the broadening resulting from the effects of θ.

5-12 NMR of quadrupolar nuclei in solution

For the solution or liquid state, rapid isotropic molecular tumbling will tend to average the quadrupolar energy by randomly varying θ. Consequently only a single NMR absorption is usually seen (when scalar coupling to other spins is ignored). However, in many (indeed, most) cases, the averaging is not sufficiently efficient to give a sharp NMR line. The width of NMR absorptions from quadrupolar nuclei ranges from a few Hz (or even less for certain nuclei such as ^{2}H, ^{6}Li and ^{133}Cs) to tens of kHz or more.

Actually, instead of viewing the width of NMR resonances from quadrupolar nuclei as arising from inadequate averaging of quadrupolar interactions, it is usually more productive to consider the effect of such interactions as causing relaxation among the energy levels established by the Zeeman term. The quadrupolar interaction is, of course, affected by the direction of the electric field gradient. This gradient is fixed in the molecular framework so that as the molecule tumbles the quadrupolar energy is modulated. If this occurs at the appropriate rate (ca. the transition frequency), spin–lattice relaxation will be induced. As for other relaxation mechanisms, T_{1Q} and T_{2Q} are equal for mobile isotropic fluids, and it can be shown that

$$T_{2Q}^{-1} = T_{1Q}^{-1} = \tfrac{3}{10}\pi^2 \frac{2I+3}{I^2(2I-1)}\chi^2\tau_c \tag{5-27}$$

When the extreme narrowing condition, $\omega_0^2\tau_c^2 \ll 1$, is not fulfilled, τ_c must be replaced by $\tau_c/(1+\omega_0^2\tau_c^2)$. The correlation time, τ_c, is that for molecular tumbling, and is therefore the same as that used in expressions for dipolar relaxation (see Chapter 4). It increases as viscosity increases and therefore as temperature decreases. Note that Eqn 5-27 shows that the *slower* the molecular tumbling, the faster the spin relaxation.

The quadrupolar relaxation mechanism is the only one which depends on electrical interactions rather than magnetic ones. Generally, the quadrupolar mechanism dominates the relaxation of nuclei with spin $\geqslant 1$, and leads to linewidths substantially in excess of those caused by inhomogeneities in B_0. Therefore linewidths are given (see Section 3-4) by

$$\Delta\nu_{1/2} = \tfrac{3}{10}\pi \frac{2I+3}{I^2(2I-1)} \chi^2 \tau_c \qquad (5\text{-}28)$$

Relaxation times can be readily obtained by linewidth measurements, and there is often little point in measuring T_1 or in undertaking spin-echo experiments to obtain T_2. However, the large linewidths commonly encountered for quadrupolar nuclei do introduce problems for normal pulse Fourier transform NMR since they imply rapid decay of the FID. For linewidths in excess of ca. 10 kHz most of the signal will already have been lost during a dead-time (see Section 3-13) of ca. 30 μs, so that sensitivity is impaired. Moreover, phase-adjustment becomes difficult.

Apart from the influence of τ_c, linewidths are affected by the spin quantum number, I, by the quadrupole moment, Q, and by the field gradient, q_{zz}. Sharper lines are obtained from nuclei of high spin than those of low spin (other factors being equal). Both I and Q are characteristic of a given nuclide, so it is useful to define a linewidth factor:

$$\ell = Q^2(2I+3)/[I^2(2I-1)] \qquad (5\text{-}29)$$

This factor gives a measure of the suitability of a nuclide for NMR. Values of ℓ are given in Table 5-3 and Appendix 2. Where an element has more than one quadrupolar nuclide (e.g. $^{63}Cu/^{65}Cu$ or $^{95}Mo/^{97}Mo$) the proportionality of ℓ to Q^2 implies that ratios of quadrupole moments for the isotopes can be obtained from linewidth measurements on a given compound.

The other factor of importance in governing linewidths is the electric field gradient, which may be related to the local symmetry at the nucleus. Any sites of tetrahedral, octahedral, cubic or spherical symmetry will have, in principle, zero field gradient. Thus, sharp lines are obtained for suitable nuclei in simple ions such as the alkali or alkaline earth cations or the halide anions in aqueous solutions. Analogous situations occur for ^{14}N in $[NMe_4]^+$, ^{33}S in $[SO_4]^{2-}$ and ^{59}Co in

Table 5-3 NMR linewidths, quadrupole coupling constants and asymmetry parameters for ^{14}N in some organic compounds[a]

Molecule	$\Delta\nu_{1/2}$/Hz[b]	χ/MHz[c]	η/%[c]
$[NMe_4]^+$	<0·5	not available	not available
NMe_3	77	5·194	0
MeCN	78	3·738	0·46
MeNC	0·3	0·272[d]	0[d]
MeNCO	35	0·495[e]	0[e]
PhCN	290	3·885	10·7
pyrrole	172	2·060	26·9
carbazole	1500	not available	not available
quinoline	650	not available	not available
pyridine	170	4·584	39·6
$MeNO_2$	14	1·45[d]	0(assumed)
$PhNO_2$	53	1·76[d]	0(assumed)
aniline	1300	3·933	26·9

[a] Data taken from 'Nitrogen NMR', Eds. M. Witanowski & G. A. Webb, Plenum Press (1973)

[b] The results vary with solvent. For details the reader is referred to the original literature via the reference in footnote [a].

[c] From NQR measurements (solid state) except where otherwise noted.

[d] From NMR measurements on solutions in nematic solvents.

[e] From microwave spectra (gas phase)

$[Co(CN)_6]^{3-}$. Of course, even in such cases, long-range and/or temporary effects produce some quadrupolar relaxation, and therefore line broadening. In particular, ion-pairing will broaden lines and so linewidth measurements form a useful method of studying this phenomenon. Lower local symmetry has a profound effect, as can be seen from the linewidths in Table 5-3 for the ^{14}N nucleus, which is a reasonably favourable one. It is extremely difficult to observe resonances from some nuclei (e.g. ^{127}I in organic compounds). However, it is *electrical* symmetry that is important, and this fact sometimes leads to linewidths which are lower than expected for simple symmetry considerations, e.g. for ^{14}N in isocyanides (see Table 5-4). Figure 5-9 is of a ^{17}O spectrum, showing differences in linewidth between two types of oxygen in the same molecule. The sharpness of the lines due to the phosphoryl oxygen would not be intuitively predicted from simple considerations of the symmetry of the atomic environment.

The sensitivity of quadrupolar nuclei for NMR may be discussed in a similar way to that used for spin-$\frac{1}{2}$ nuclei in Section 3-6. However, a further spin-dependent factor (see Problem 5-7) appears because transition probabilities depend on I_+ or I_- and hence involve m_I. Therefore the equation for the receptivity of nucleus X must be modified to

$$D_X = |\gamma^3|\, CI_X(I_X+1) \tag{5-30}$$

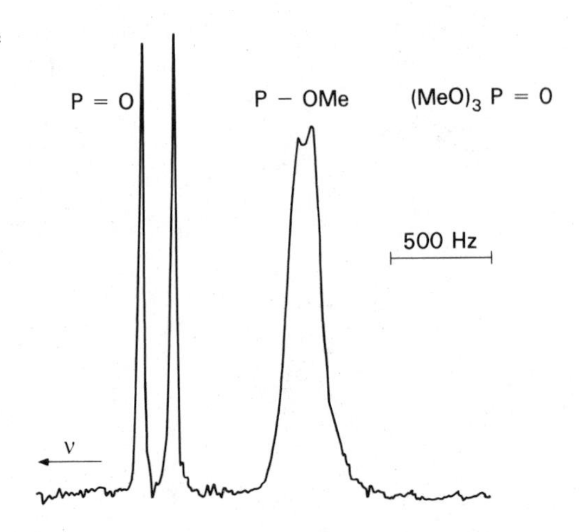

Fig. 5-9 Oxygen-17 FT NMR spectrum of trimethyl phosphate at 13·6 MHz and 80°C, illustrating linewidth variations due to oxygen atoms in the same molecule, i.e. arising primarily from electric field gradient differences. The data acquisition rate was 0·06 s, and the total experimental time 60 min. Splittings due to (P, O) coupling are visible for both bands ($^1J_{PO}$ = 150 Hz for P=O and ca. 90 Hz for P−OMe). The chemical shifts (with respect to the signal for H_2O) are 74.6 ppm (P=O) and 21·9 ppm (P−OMe). [Reproduced, with permission, from C. Rodger, Ph.D. Thesis, University of East Anglia (1976).]

Values of D_X^C and D_X^p for quadrupolar nuclei are given in Table 5-3 and Appendix 2.

The suitability of a quadrupolar nuclide for NMR is, therefore, governed by three considerations, viz. (i) receptivity, (ii) the linewidth factor, and (iii) the chemical likelihood of the existence of a relatively symmetrical local environment.

Table 5-4 Nuclear quadrupole coupling constants for ^{35}Cl

Molecule	Cl atom	Cl_2	KCl	HCl	FCl
State	gas	solid	gas	solid	gas
χ/MHz	−109·7	−109·0	<0·04	−53·4	−146·0

5-13 Quadrupole coupling constants and asymmetry factors

Values of χ can be obtained in principle from NMR linewidths or, alternatively, from NQR spectra (or from fine structure on pure rotational spectra). However, separation of the information to yield Q and q_{zz} independently is not easy. Atomic-beam techniques can yield values of Q directly. Both Q and q_{zz} are signed quantities, but since the NMR spectra described in preceding sections depend on χ^2 the signs are irrelevant. However, second-order effects are influenced by the sign of χ.

As discussed in the preceding section, differences in electric field gradients are a major influence on NMR linewidths for quadrupolar nuclei and, in fact, they are the only cause of variation in quadrupole coupling constants for a given nuclide. A detailed discussion of such field gradients is beyond the scope of this book, but some insight can be achieved by a very simple approach. Firstly, it is assumed that because electric influences are relatively short-range in effect discussion can be limited to the electrons of the nucleus in question. Now *s* electrons and filled shells are spherically symmetrical, and contributions from partly-filled *d* or *f* shells (if such exist) can be neglected because of the distance-dependence. To this approximation, therefore, q_{zz} (and hence χ) will depend only on the lack of spherical distribution of *p* electrons. Table 5-4 gives data for ^{35}Cl that can be used as a basis for discussion. The value for the Cl atom characterizes the situation for the lack of one *p* electron. A closely-similar value is found for the homonuclear covalent diatomic molecule Cl_2, as expected. Gaseous KCl shows a negligible value for χ, indicating the bonding is essentially pure ionic, giving a spherically symmetrical Cl^- ion. The value of χ for HCl indicates lack of ca. $\frac{1}{2}$ a *p* electron, i.e. ca. 50% covalent character, whereas the result for FCl shows lack of more than one *p* electron, providing evidence for a contribution from the structure F^-Cl^+. Quantitative use of such concepts is fraught with difficulties, and it should be noted that the information is, in any case, derived at the nucleus, whereas chemists usually seek data about the electronic situation in *the bond*, i.e. between nuclei.

So far it has been assumed that the nucleus under consideration lies at a position of axial electrical symmetry, since only a single field gradient, eq_{zz}, has been invoked. This is frequently a good approximation for monovalent atoms such as ^{35}Cl or ^{2}H, when the *z* direction is that of the relevant single bond. In other cases, such as ^{14}N in NMe_3, molecular symmetry ensures the appropriateness of the equations presented so far.[8] However, in many other cases, this is manifestly not so, and in general two field gradient parameters are needed. One might think that three (q_{xx}, q_{yy} and q_{zz}) are necessary, but Laplace's Equation, 5-31, applies, so these three parameters are not independent.

$$q_{xx}+q_{yy}+q_{zz}=0 \tag{5-31}$$

The two parameters used are q_{zz} (as heretofore) and the *asymmetry parameter*, η:

$$\eta=(q_{xx}-q_{yy})/q_{zz} \tag{5-32}$$

The axes are chosen such that $q_{zz}>q_{xx}>q_{yy}$, so that η lies between 0 and 1. It should be explained that the subscripts to *q* refer to molecule-fixed axes which are not directly compatible with the laboratory frame.

It is not possible in general to give a simple analytical expression for the quadrupolar energy to include the effects of field gradient asymmetry, but for the $I=1$ case Eqns 5-33 and 5-34 hold:

$$h^{-1}U=\tfrac{1}{4}\chi(1\pm\eta) \quad \text{for} \quad m_I=\pm1 \tag{5-33}$$

$$h^{-1}U=-\tfrac{1}{2}\chi \quad \text{for} \quad m_I=0 \tag{5-34}$$

It can be seen that η introduces an asymmetry into the (solid-state) NQR spectrum, so that χ and η can be separately determined.

For the interpretation of solution-state NMR of quadrupolar nuclei the existence of η hinders simple interpretation of bandshapes, and it is therefore often ignored! However, in principle an extra factor $(1+\frac{1}{3}\eta^2)$ is required in Eqn 5-27 for the quadrupolar relaxation rate. Table 5-3 gives some values of asymmetry parameters for ^{14}N.

5-14 The effect on spin-$\frac{1}{2}$ spectra of coupling to quadrupolar nuclei

Suppose a molecule contains a spin-$\frac{1}{2}$ nucleus such as ^1H spin-coupled to a quadrupolar ($I=1$) nucleus such as ^{14}N. The principles discussed in Chapter 1 would lead one to expect the ^1H NMR spectrum to be split into three lines, with the splitting equal to the scalar coupling constant J_{NH}. The intensities of the three lines should be equal because each spin state, $m_I(^{14}N)=+1, 0$, and -1, is equally likely. Such splittings are sometimes observed, but not commonly, because of the rapid relaxation of ^{14}N discussed above. There are several ways of understanding the bandshapes that are normally observed in this situation. Perhaps the most instructive is to consider the effect of ^{14}N relaxation to be equivalent to that of chemical exchange in the sense that if the transition $m_I(^{14}N)=-1 \Rightarrow m_I(^{14}N)=0$ occurs the proton resonance frequency will change from ν_H+J_{NH} (the high-frequency member of the triplet) to ν_H (the central member of the triplet). Increased rate of ^{14}N relaxation therefore exchanges the ^1H resonance frequency among the three members of the triplet, thus broadening the lines (ascribable to the Uncertainty Principle). When the exchange rate (i.e. T_{1N}^{-1}) becomes comparable to J_{NH} the triplet lines begin to merge, and a single line is seen when $T_{1N}^{-1} > J_{NH}$. The width of this line decreases as T_{1N}^{-1} increases still further. At this stage of the process the ^{14}N has become *self-decoupled* from the ^1H. In the intermediate situations the bandshape can be shown[9] to be given by

$$g(x) \propto \left(\frac{2w}{\pi J}\right) \frac{45+w^2(5x^2+1)}{225x^2+2w^2(17x^4-x^2+2)+w^4x^2(x^2-1)^2} \qquad (5\text{-}35)$$

where $w=10\pi T_{1N}J_{NH}$ and $x=(\nu-\nu_H)/J_{NH}$. Figure 5-10 shows some calculated shapes and Fig. 5-11 shows an observed spectrum. It may be observed that the outer lines are broadened more easily than the central line (see Problem 5-8). This is because relaxation between $m_I(^{14}N)=+1$ and $m_I(^{14}N)=-1$ is actually more efficient than for $\Delta m_I(^{14}N)=\pm 1$. Figure 5-12 illustrates the ^{14}N spectra for the same values of w.

For $[NH_4]^+$, because of the low value of q_{zz}, T_{1N} is rather long and the ^1H bandshape is near the 'slow-exchange' limit of a sharp triplet. For most other NH groups, such as those in amides, only broad single lines are observed. The situation is, however, complicated by the existence of real proton exchange for NH groups of all descriptions, which of course also causes decoupling. The bandshape for a given chemical system clearly depends (through T_{1N}) on q_{zz} and τ_c. The former reflects the effect of symmetry whereas the latter ensures that

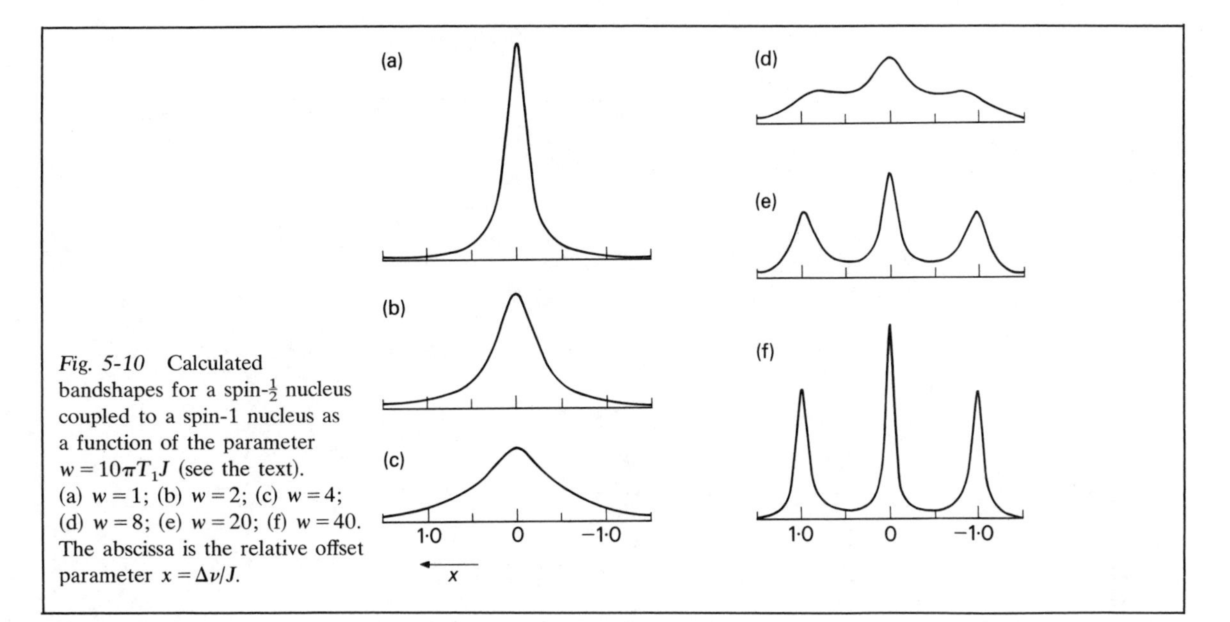

Fig. 5-10 Calculated bandshapes for a spin-$\frac{1}{2}$ nucleus coupled to a spin-1 nucleus as a function of the parameter $w = 10\pi T_1 J$ (see the text). (a) $w = 1$; (b) $w = 2$; (c) $w = 4$; (d) $w = 8$; (e) $w = 20$; (f) $w = 40$. The abscissa is the relative offset parameter $x = \Delta\nu/J$.

cooling a solution increases the likelihood of self-decoupling. The other factor involved is J_{NH}. It is clearly easier to self-decouple a small coupling constant than a large one, so that the chance of observing long-range (N, H) coupling is correspondingly reduced.

Similar effects are found for other quadrupolar nuclei and it is, indeed, rare to observe splitting due to spin–spin coupling with such nuclei as ^{35}Cl or ^{17}O. However, coupling of ^{13}C or ^{1}H to ^{2}H is frequently observable, giving rise, for instance, to a triplet in ^{13}C spectra from $CDCl_3$ solvent and a 1:2:3:2:1 quintet in ^{1}H spectra from residual CHD_2 groups in deuterated dimethylsulphoxide. Cases like the latter enable J_{DH} to be measured, and thus J_{HH} may be calculated for coupling between equivalent protons (see Section 8-20).

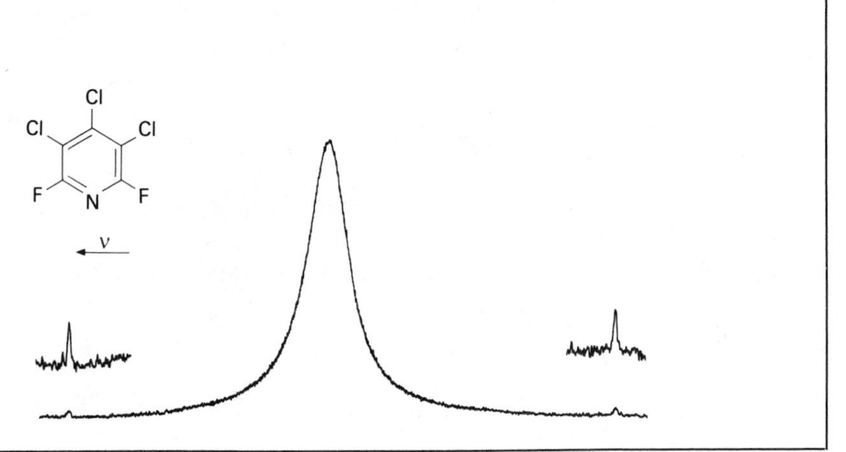

Fig. 5-11 94·1 MHz ^{19}F resonance of 2,6-difluoro-3,4,5-trichloropyridine at ambient probe temperature, showing the broadening effect of coupling to the ^{14}N nucleus. The linewidth arising from other relaxation processes is shown by the outer lines which are due to ^{15}N satellites. These give $|^2J(^{15}\text{N, F})| = 52.2$ Hz, which leads to $|^2J(^{14}\text{N, F})| = 37{\cdot}2$ Hz. The central band is Lorentzian to within experimental error. [See A. V. Cunliffe & R. K. Harris, *Mol. Phys.*, **15,** 413 (1968)].

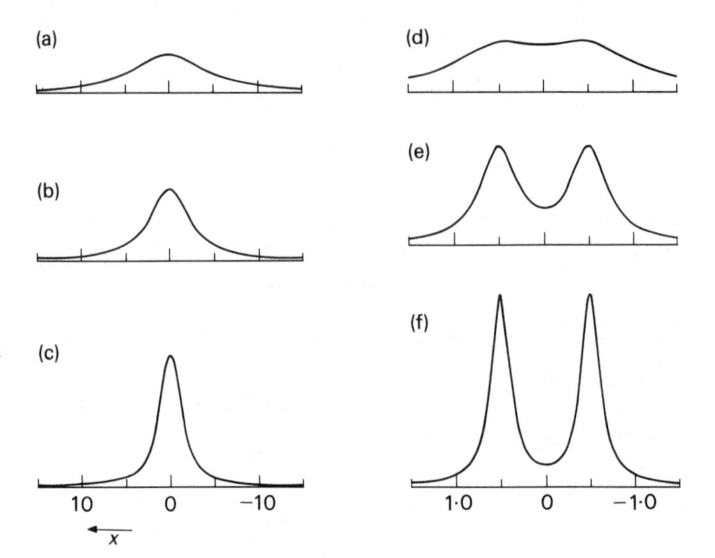

Fig. 5-12 Calculated bandshapes for a spin-1 nucleus coupled to a spin-$\frac{1}{2}$ nucleus as a function of the parameter w (see the text). The scale of the right-hand half of the figure and the values of w used are identical to those of Fig. 5-10. The left-hand half of the figure has an ordinate scale expansion by a factor of 4 and an abscissa scale contraction by a factor of 10. The calculated linewidths (ignoring coupling) are, in units of $\Delta\nu_{\frac{1}{2}}/J$, (a) 10; (b) 5; (c) 2·5; (d) 1·25; (e) 0·5; (f) 0·25.

When $T_{1N}^{-1} \gg |J_{NH}|$, Eqn 5-35 may be simplified (Problem 5-6) to the Lorentzian expression 3-19 with a relaxation time T_{2H} given by

$$T_{2H}^{-1} = 8\pi^2 J_{NH}^2 T_{1N}/3 \qquad (5\text{-}36)$$

At this extreme the situation may be viewed as one of scalar relaxation of the second kind causing line broadening. If the quadrupolar relaxation is much more rapid it can affect T_{1H} also. Generally this will only happen when the quadrupolar nucleus and the spin-$\frac{1}{2}$ nucleus resonate at similar frequencies, since this maximizes the correlation factor $\tau_{sc}/[1+(\omega_A-\omega_X)^2\tau_{sc}^2]$ (see Table 3-2)—the correlation time is equivalent to T_1 for the quadrupolar nucleus. Such behaviour has been observed for ^{13}C ($\Xi = 25{\cdot}145$ MHz) in the presence of ^{79}Br ($\Xi = 25{\cdot}054$ MHz)—for instance, for the quaternary carbon of the bromobenzene isotopomer containing ^{79}Br the spin–lattice relaxation time is[10] 3·6 s at 38 °C, much shorter than a typical value (~100 s) for a monosubstituted benzene).

It may be noted that the phenomenon of self-decoupling is not limited to quadrupolar nuclei. The ^{205}Tl nucleus has been observed to self-decouple from ^{1}H at high magnetic fields when relaxation due to the shielding anisotropy mechanism becomes efficient, and ^{31}P self-decouples from ^{1}H in, for example, dimethyl methyl phosphonate, $MePO(OMe)_2$, in the presence of paramagnetic hexaaquo cobalt(II)perchlorate at a concentration of ca. 0·05M or greater.

Notes and references

[1] In principle this may be regarded as averaging of the coupling between the value when the nuclei are in the same molecule and an infinitely large number of situations with zero coupling when the nuclei are in different molecules.

[2] H. M. McConnell, *J. Chem. Phys.*, **28,** 430 (1958).

[3] R. K. Harris, T. Pryce-Jones & F. J. Swinbourne, *J.C.S. Perkin II*, 476 (1980).

[4] D. A. Kleier & G. Binsch, *J. Magn. Reson.*, **3,** 146 (1970).

[5] NMR Program Library (see note 5 to Chapter 2).

[6] See, for example, G. H. Whitesides & H. L. Mitchell, *J. Amer. Chem. Soc.*, **91,** 5384 (1969).

[7] T. J. Swift & R. E. Connick, *J. Chem. Phys.*, **37,** 307 (1962).

[8] However, it should be noted that even for molecules such as NMe_3 crystal packing effects could lead to non-zero η for the solid.

[9] J. A. Pople, *Mol. Phys.*, **1,** 168 (1958).

[10] G. C. Levy, *J.C.S. Chem. Comm.*, 352 (1972).

Further reading

'Chemical rate processes and magnetic resonance', C. S. Johnson, *Adv. Magn. Reson.*, **1,** 33 (1965).

NMR of Chemically Exchanging Systems, J. L. Kaplan & G. Fraenkel, Academic Press (1980).

Dynamic NMR Spectroscopy, Eds. L. M. Jackman & F. A. Cotton, Academic Press (New York) (1975).

'NMR of the less common quadrupolar nuclei', F. W. Wehrli, *Ann. Repts. NMR Spectrosc.*, **9,** 126 (1979).

Dynamic NMR Spectroscopy, J. Sandström, Academic Press (New York) (1982).

Problems

5-1 (a) State which has the greater ^{35}Cl quadrupole coupling constant, $[ClO_4]^-$ or $[ClO_2]^-$. Give your reasoning.
(b) Which has the broader ^{35}Cl NMR resonance in solution, CCl_4 or NaCl?

5-2 How does the presence of the quadrupolar ^{11}B nucleus ($I = \frac{3}{2}$) affect the ^{19}F NMR spectrum of $^{11}BF_3$ in the solution state if molecular tumbling is (a) slow and (b) fast?

5-3 Figure 5-13 shows ^{19}F NMR spectra of PCl_2F_3 obtained at high and low temperatures. Explain the appearance of the spectra. What conclusions can you draw regarding the molecular structure?

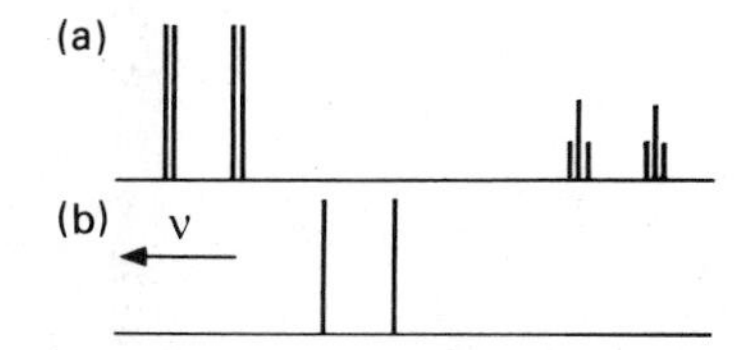

Fig. 5-13 Schematic fluorine-19 NMR spectra of PCl_2F_3 (a) at very low temperature, and (b) at somewhat higher temperature. The intensities in (b) are not on the same scale as those in (a).

5-4 Calculate the form of the quadrupole spectrum of the ^{35}Cl nucleus in an axially symmetric environment. Pure quadrupole transitions are observed at 25·8 MHz in K_2PtCl_6 and at 54·248 MHz in Cl_2. Calculate the field gradients and comment on the variation. [$Q(^{35}Cl) = -0{\cdot}10 \times 10^{-28}\ m^2$.]

5-5 Show, from the lineshape for exchange between two uncoupled, equally populated sites (Eqn 5-12), that the coalescence condition is given by Eqn 5-14.

5-6 Show that when $T_{1N}^{-1} \gg |J_{NH}|$ the lineshape 5-35 becomes Lorentzian with a width governed by the effective transverse relaxation time given in Eqn 5-36.

5-7 Using the concepts of Sections 2-4 and 2-6, show that the receptivity for a nucleus of spin I must contain a spin-dependent factor $I(I+1)$ as in Eqn 5-30.

5-8 In the laboratory frame of reference the quadrupolar Hamiltonian contains terms which are not included in Eqn 5-25 and which, for the solution state, are time-dependent due to molecular tumbling. These terms cause relaxation. They are proportional to:

$$2 \sin\theta \cos\theta[\hat{I}_z(\hat{I}_+ + \hat{I}_-) + (\hat{I}_+ + \hat{I}_-)\hat{I}_z] + \sin^2\theta(\hat{I}_+^2 + \hat{I}_-^2)$$

where θ is the angle between B_0 and q_{zz} (an axially symmetric electric field gradient is assumed).

Using the theory of Appendix 3, show the transition rates for relaxation for a spin-1 nucleus are in the ratio $W_1 : W_2 = 1:2$. Hence show that, for the resonance of a spin-$\frac{1}{2}$ nucleus coupled to a spin-1 nucleus, the ratio of linewidths for the observed triplet in the limit $w \gg 1$ (see Fig. 5-10) is 3:2:3. Check that this result is consistent with Eqn 5-35. The problem is discussed in reference [9] of Chapter 5.

6 NMR of the solid state

6-1 The two-spin system

In solids the molecules are generally held rather rigidly, so that dipolar interactions are not averaged to zero in the manner described for liquids in Section 4-2. If scalar coupling and shielding are ignored the nuclear spin Hamiltonian becomes the sum of Zeeman and dipolar terms, and may therefore, for a two-spin heteronuclear AX spin system, be written

$$h^{-1}\mathscr{H}_{\mathrm{NS}} = -(\nu_{\mathrm{A}}\hat{I}_{\mathrm{Az}} + \nu_{\mathrm{X}}\hat{I}_{\mathrm{Xz}}) + h^{-1}\hat{\mathscr{H}}_{\mathrm{dd}} \tag{6-1}$$

where ν_{A} and ν_{X} are the appropriate Larmor frequencies, and $\hat{\mathscr{H}}_{\mathrm{dd}}$ is given by Eqn 4-3. Since Zeeman energies are much larger than dipole energies, only term A in Eqn 4-3 is effective and the *truncated* Hamiltonian becomes

$$h^{-1}\hat{\mathscr{H}}_{\mathrm{NS}} = -(\nu_{\mathrm{A}}\hat{I}_{\mathrm{Az}} + \nu_{\mathrm{X}}\hat{I}_{\mathrm{Xz}}) - R\hat{I}_{\mathrm{Az}}I_{\mathrm{Xz}}(3\cos^2\theta - 1) \tag{6-2}$$

The energies of the spin states are immediately available as

$$h^{-1}U = -(\nu_{\mathrm{A}}m_{\mathrm{A}} + \nu_{\mathrm{X}}m_{\mathrm{X}}) - Rm_{\mathrm{A}}m_{\mathrm{X}}(3\cos^2\theta - 1) \tag{6-3}$$

The usual selection rules apply so that for a single crystal with only one orientation for $\mathbf{r}_{\mathrm{AX}}$ there are two A transitions at

$$\nu = \nu_{\mathrm{A}} \pm \tfrac{1}{2}R(3\cos^2\theta - 1) \tag{6-4}$$

and two corresponding X lines. The doublet splitting is $R(3\cos^2\theta - 1)$, and its variation with orientation in the magnetic field will yield both R and the relation of θ to the crystal setting.

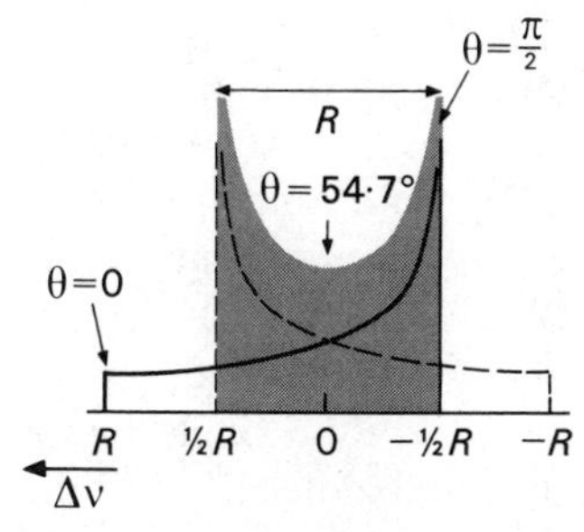

Fig. 6-1 A powder pattern (schematic) arising from dipolar coupling effects for, say, the A nucleus of a two-spin heteronuclear system (AX). The separate subspectra corresponding to A transitions for molecules with $m_{\mathrm{X}} = \frac{1}{2}$ and $m_{\mathrm{X}} = -\frac{1}{2}$ are indicated by full and dashed lines respectively. The total pattern is shown by the stippled area. Values of θ for the m_{X} subspectrum are given.

However, for a powder all values of θ exist at random, and the resulting A spectrum forms a *powder pattern* (Fig. 6-1), with a similar pattern in the X region. The figure indicates how the full pattern is constituted by the subspectra for the $m_{\mathrm{X}} = +\frac{1}{2}$ and $m_{\mathrm{X}} = \frac{1}{2}$ states. Such patterns yield a value for R and hence r_{AX}. Unfortunately, few materials contain isolated two-spin systems, and longer-distance interactions cause extensive further broadening, as will be discussed in Section 6-3, so that rather featureless spectra result for most powders.

For a *homo*nuclear pair of nuclei, assumed here to be equivalent, the Zeeman energies of the $\alpha\beta$ and $\beta\alpha$ states are equal so that term B of the dipolar Hamiltonian can no longer be ignored and the truncated

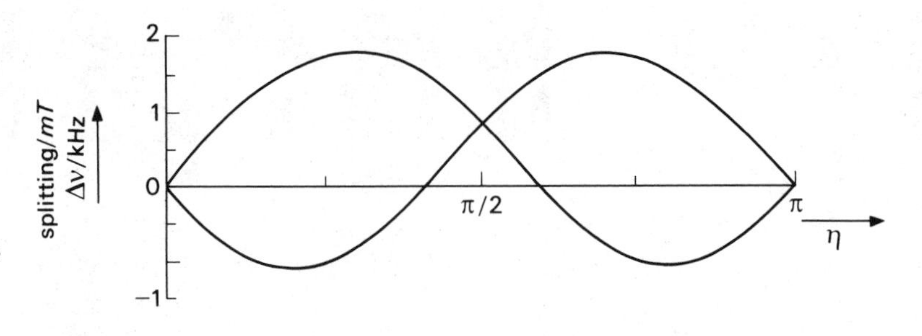

Fig. 6-2 The variation in dipolar doublet splittings for the protons in gypsum as a function of the angle between $\mathbf{B}_0$ and the x crystallographic axis, with $\mathbf{B}_0$ in the (001) plane. The maximum splitting is less than $3R$ because the H---H vectors are at an angle to the (001) plane. The splitting magnitudes in the figure are in magnetic field units. [Adapted from data given by G. E. Pake, *J. Chem. Phys.*, **16,** 327 (1948).]

expression becomes

$$h^{-1}\mathscr{H}_{NS} = -\nu_0(\hat{I}_{1z}+\hat{I}_{2z}) - R(3\cos^2\theta - 1)[\hat{I}_{1z}\hat{I}_{2z} - \tfrac{1}{4}(\hat{I}_{1+}\hat{I}_{2-} + \hat{I}_{1-}\hat{I}_{2+})] \tag{6-5}$$

The flip-flop term mixes the $\alpha\beta$ and $\beta\alpha$ states. The eigenstates are the symmetric and antisymmetric combinations $(\alpha\beta \pm \beta\alpha)/\sqrt{2}$. The antisymmetric level is not involved in transitions, so that only a pair of lines (Problem 6-1) is observed for a single crystal with one orientation for $\mathbf{r}_{12}$, at

$$\nu = \nu_0 \pm \tfrac{3}{4}R(3\cos^2\theta - 1) \tag{6-6}$$

that is, a doublet of spacing $\tfrac{3}{2}R(3\cos^2\theta - 1)$ is seen. Note the extra factor $\tfrac{3}{2}$ compared to the heteronuclear case (also found for dipolar relaxation in liquids—see Sections 3-16 and 4-3).

The H_2O protons of water of crystallization approximate to the isolated homonuclear spin-pair situation. The classic case is that of gypsum, $CaSO_4.2H_2O$. There are two crystallographic sites for water in the unit cell, with different orientations, so that in general *two* doublets are seen for single-crystal work. The doublet splittings are shown in Fig. 6-2, where η indicates the angle between the applied field B_0 and the x crystallographic axis. At $\eta = 0$ only a single line is seen, implying that $\theta = 54{\cdot}7°$ (54°44′) for the water molecules in both crystallographic sites. This gives information about the orientation of the molecules in the unit cells. The splitting constant $\tfrac{3}{2}R$ is 46·0 kHz, yielding $r = 0{\cdot}158$ nm. Such geometry measurements are very useful since X-ray diffraction studies only yield positions of hydrogen atoms with difficulty.

6-2 Magic-angle rotation

As shown in Section 6-1 the terms in the dipolar Hamiltonian which cause NMR line broadening for solids involve the geometric factor $(3\cos^2\theta - 1)$. It has already been indicated in Section 4-2 that rapid isotropic molecular tumbling, such as occurs in non-viscous solutions, averages this geometric factor to zero, thus explaining the very narrow NMR lines observed for such solutions. Therefore it is important to

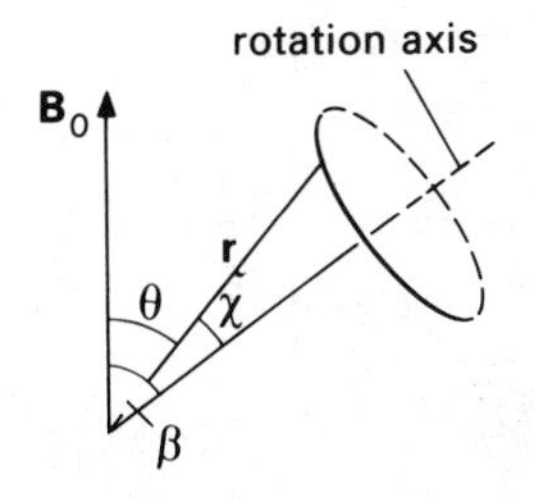

Fig. 6-3 Macroscopic sample rotation at an angle to the applied magnetic field $\mathbf{B}_0$, showing the geometric relationships involved.

ask if an experiment can be devised for solids which simulates such motion. Consider, then, coherent rotation of a solid sample about an axis at angle β to B_0 (see Fig. 6-3). It is necessary to find the average of $(3\cos^2\theta - 1)$ about the conical path indicated for the internuclear vector, $\mathbf{r}$. Appendix 4 shows that the result is

$$\langle 3\cos^2\theta - 1\rangle = \tfrac{1}{2}(3\cos^2\beta - 1)(3\cos^2\chi - 1) \tag{6-7}$$

The parameter χ is fixed for a rigid solid, though (like θ) it takes all possible values if the material is a powder. The term $\frac{1}{2}(3\cos^2\beta - 1)$ therefore acts as a scaling factor on a dipolar powder pattern. Fortunately the angle β is under the control of the experimentalist. If $\beta = 0$ (i.e. rotation about B_0), $\frac{1}{2}(3\cos^2\beta - 1) = 1$, so there is no net effect on the spectrum. If $\beta = \pi/2$ (i.e. rotation about an axis perpendicular to B_0), $\frac{1}{2}(3\cos^2 - 1) = \frac{1}{2}$, so the powder pattern is scaled down by a factor of 2 and reversed in direction (this reversal is not observable). The case of most interest is that when $\beta = 54{\cdot}7°$, since then $\cos\beta = 1/\sqrt{3}$ and $\frac{1}{2}(3\cos^2\beta - 1) = 0$, so that $\langle 3\cos^2\theta - 1\rangle = 0$ for all orientations (i.e. all values of χ). Thus, just as for isotropic tumbling, the dipolar interaction is averaged to zero, and so dipolar broadening is eliminated, giving much higher resolution. This situation is referred to as magic-angle rotation (MAR) and 54·7° is called the magic angle. The technique applies to both homonuclear and heteronuclear cases.

However, problems arise because the rate of rotation required to average the dipolar interactions properly has to be greater than the static bandwidth expressed in Hz. This can be several tens of kHz; such speeds cannot be achieved in practice, thus limiting the generality of the technique. However, many special cases have been exploited, and details will be found in reviews by Andrew.[1]

6-3 Second moments and the effects of molecular motion

As mentioned in Section 6-1, in practice powder spectra of rigid solids are relatively featureless. It is desirable to report some parameters which give information about the bandshape. The parameters most frequently chosen are known as the *moments* of the spectrum. The jth moment, M_j, is defined (in angular frequency terms) for a bandshape $f(\omega)$ as

$$M_j = \int_{-\infty}^{\infty} (\omega - \langle\omega\rangle)^j f(\omega)\,\mathrm{d}\omega \Big/ \int_{-\infty}^{\infty} f(\omega)\,\mathrm{d}\omega \tag{6-8}$$

where the denominator is a normalizing factor and $\langle\omega\rangle$ is the mean angular frequency of the band:

$$\langle\omega\rangle = \int_{-\infty}^{\infty} \omega f(\omega)\,\mathrm{d}\omega \Big/ \int_{-\infty}^{\infty} f(\omega)\,\mathrm{d}\omega \tag{6-9}$$

If $f(\omega)$ is an even function of ω, $M_j = 0$ for all odd j, as is the case for dipolar interactions in high magnetic fields. Usually only the second moment, M_2, is used. For the spectrum of a single crystal containing isolated heteronuclear $I=\frac{1}{2}$ spin-pairs (see Sections 4-1 and 6-1), it is readily seen that this is

$$M_2(\text{hetero}) = [\tfrac{1}{2}R(1-3\cos^2\theta)]^2 \tag{6-10}$$

whereas for homonuclear spin-pairs the additional factor $\frac{3}{2}$ again appears (see Section 6-1) so that

$$M_2(\text{homo}) = [\tfrac{3}{4}R(1-3\cos^2\theta)]^2 \tag{6-11}$$

which is $\sim(\gamma B_L)^2$. These equations are readily generalized for a more realistic situation when all interactions are taken into account:

$$M_2(\text{hetero}) = \left(\tfrac{1}{2}\gamma_A\gamma_X\hbar\frac{\mu_0}{4\pi}\right)^2 \sum_k [(1-3\cos^2\theta_{jk})^2/r_{jk}^6] \tag{6-12}$$

$$M_2(\text{homo}) = \left(\tfrac{3}{4}\gamma^2\hbar\frac{\mu_0}{4\pi}\right)^2 \sum_k [(1-3\cos^2\theta_{jk})^2/r_{jk}^6] \tag{6-13}$$

where the sum runs over all relevant nuclei, k, in relation to the considered nucleus j. If both homo- and hetero-nuclear interactions are present, the total second moment will be the sum of the two terms. The corresponding powder-pattern equations can be obtained using the expression

$$\langle(1-3\cos^2\theta)^2\rangle = \tfrac{4}{5} \tag{6-14}$$

In many situations the frequency-distribution, $f(\omega-\omega_0)$ is approximately Gaussian:

$$f(\omega-\omega_0) = (2\pi)^{-1/2}\sigma^{-1}\exp-[(\omega-\omega_0)^2/2\sigma^2] \tag{6-15}$$

where σ^2 is the variance of the distribution (i.e. the mean squared width). In such a case the second moment is given by

$$M_2 = \sigma^2 \tag{6-16}$$

If the definition of T_2 is the time taken for the FID to decay to $1/e$ of its initial value following a 90° pulse (which is consistent with the usage for a Lorentzian lineshape), then for a Gaussian situation $T_2=\sqrt{2}/\sigma$. The FID is, of course, the Fourier transform of the lineshape. A Gaussian shape transforms to a Gaussian decay, which, for y magnetization, may be written $M_y(t)=M_y(0)\exp(-t^2/T_2^2)$.

Of course, any rapid molecular motion will tend partially to average the dipolar interactions, so that the second moment will appear to be decreased[2] below its rigid lattice value, and T_2 will increase. This occurs when the relevant correlation time becomes comparable to $M_2^{-1/2}$. In particular, at the melting point, when rapid isotropic motion supervenes, the second moment drops to a negligible value (see Section 4-2). Consider a simple molecular motion such as rotation about an axis fixed in the molecule (for example the internal rotation of a methyl group). This situation is entirely analogous to that considered in Section 6-2, and Fig. 6-3 may be used. The difference in the

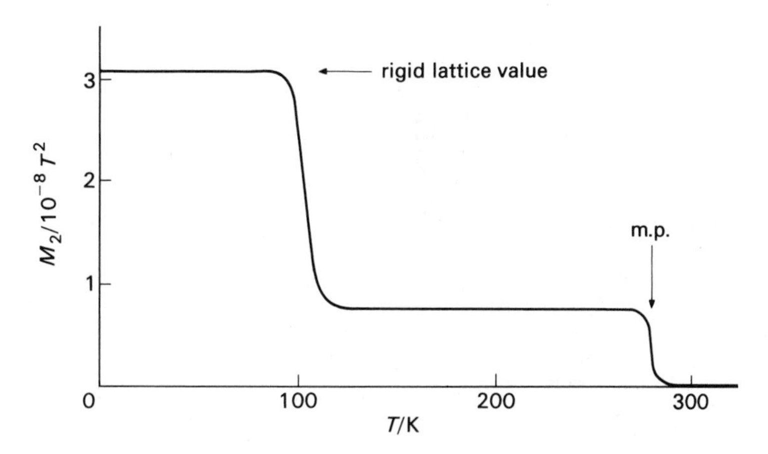

Fig. 6-4 Variation of the intramolecular homonuclear contribution to the second moment of the proton resonance of benzene as a function of temperature (diagrammatic). The second moment is given in terms of magnetic field units, rather than angular frequency units.

situation considered here is that χ now represents a constant angle for all molecules in a powdered sample, but β may take any value due to the random orientations of molecules. Once more the average value of $(1-3\cos^2\theta)$ determines the observed powder pattern, and Eqn 6-7 can again be invoked. In the present case, the powder pattern is scaled by the factor $\frac{1}{2}(3\cos^2\chi-1)$, and the second moment is scaled by $[\frac{1}{2}(3\cos^2\chi-1)]^2$. The classic case concerns the proton spectrum of powdered benzene. From studies of benzene itself and various deuterated isotopomers, Andrew and Eades[3] were able to plot the intramolecular (homonuclear) contribution to M_2 as a function of temperature (Fig. 6-4). Clearly some molecular motion at ca. 100 K is causing a reduction in the second moment by a factor of 4·03. Suppose the motion is of molecular rotation about the hexad axis of the molecule. Then the intramolecular interactions of relevance are in the plane of the ring, and therefore perpendicular to the rotation axis (i.e. $\chi=\pi/2$). Therefore $\frac{1}{2}(3\cos^2\chi-1)=\frac{1}{2}$ so M_2 should drop by $\frac{1}{4}$. This is sufficiently close to the observed value to indicate that the origin of the effect has been determined. For ^{13}C spectra of solids, for similar reasons, the rapid internal rotation of methyl groups should lead to a reduction factor of $\frac{1}{9}$ for M_2 if tetrahedral geometry is assumed.

The theory for the dependence of T_2 on the motional correlation time τ_c, discussed in Section 3-15 and in Chapter 4, is, of course, not valid near the rigid-lattice limit, i.e. when $\tau_c > T_2$, since the spins dephase due to differences in resonance frequencies before relaxation due to motion can occur. Figure 3-17 illustrates this fact since it shows T_2 reaching an asymptotic limit (the rigid-lattice value) rather than monotonically decreasing as $\tau_c \rightarrow \infty$. In this rigid-lattice limit T_2 will be governed entirely by the distribution of the local field, and dephasing will occur in a time of the order of $(\gamma B_L)^{-1}$.

6-4 Motional effects on relaxation times in solids

Of course large linewidths for rigid solids imply very short transverse relaxation times. Typical values for protons in organic solids are ca. 10 μs, but depend greatly on the (H, H) distances involved. How-

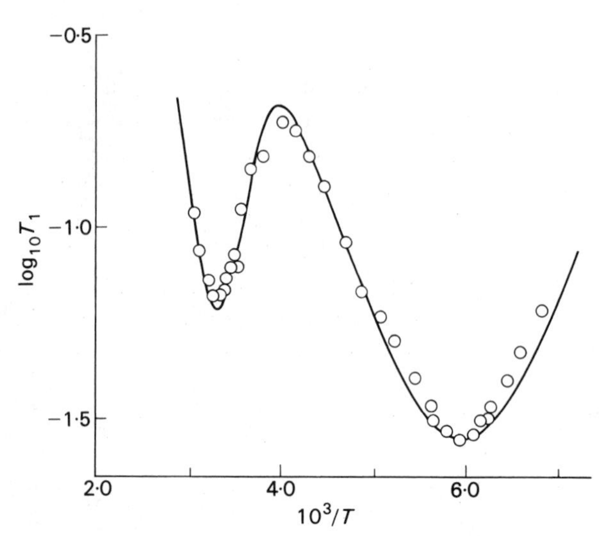

Fig. 6-5 Variation of the proton T_1 for solid trimethylphosphine sulphide, Me_3PS, as a function of temperature, showing minima due to two different motional processes. The solid line results from a theoretical interpretation. [Reproduced, with permission, from G. R. Hays, Ph.D. Thesis, University of East Anglia (1978).]

ever, T_1 and $T_{1\rho}$ for solids are long in the absence of any motion. The general effect of motion on relaxation times was considered in Chapter 3 from the viewpoint of mobile solutions. In the present section the reverse outlook is adopted, namely using the rigid solid situation as a basis. However, Fig. 3-18 is still relevant to the current discussion. Near the rigid lattice limit, correlation factors of the form $\tau_c/(1+\omega^2\tau_c^2)$ approximate to $(\omega^2\tau_c^2)^{-1}$. Consequently, as the temperature increases and begins to induce motion (τ_c decreasing), T_1^{-1} will increase. Eventually a maximum will be reached (a minimum in T_1) and T_1^{-1} will decrease again as the extreme narrowing situation begins to apply (see Fig. 3-17—a similar variation is obtained for dipolar relaxation). Of course motion in solids is usually not isotropic. In fact, frequently several independent motions, with different activation energies, will occur. Their major effect on T_1 will occur at different temperatures. Figure 6-5 illustrates this effect for the proton spin–lattice relaxation time of trimethylphosphine sulphide. Two motions dominate T_1 in the temperature range shown. The less energetic process is methyl group rotation about the C—P bond. This leads to the lower-temperature (169 K) T_1 minimum, whereas the minimum at 305 K may be attributed to overall rotation about the molecular symmetry axis. To interpret the temperature variation of T_1 properly requires equations of the Woessner type (see Section 4-13) but it proves feasible to determine the activation energies to the two processes, which are found to be 13·0 and 41 kJ mol^{-1}.

Similar considerations apply to spin–lattice relaxation in the rotating frame, but the minimum in $T_{1\rho}$ depends on the frequency ω_1 rather than ω_0. Therefore the $T_{1\rho}$ minimum is at a lower temperature than that for T_1. For example, the minimum in T_1 caused by overall rotation of Me_3PS is at 216 K. In effect $T_{1\rho}$ monitors motions in the tens of kHz range, whereas T_1 gives information about much more rapid motions (of the order of tens or hundreds of MHz).

It is clear from the foregoing that there are a number of distinct relaxation regimes in solids. For the homonuclear case involving a motion (assumed to be isotropic) with correlation time τ_c, the extreme narrowing situation, $w_0^2\tau_c^2 \ll 1$ (implying $\omega_1^2\tau_c^2 \lll 1$) yields

$$T_1^{-1} = T_{1\rho}^{-1} = T_2^{-1} = \tfrac{10}{3}M_2\tau_c \tag{6-17}$$

whereas when the system is 'rigid' as far as T_1 is concerned ($\omega_0^2\tau_c^2 \gg 1$) but is still in the extreme narrowing situation for $T_{1\rho}$ ($\omega_1^2\tau_c^2 \ll 1$):

$$T_{1\rho}^{-1} = M_2\tau_c = T_2^{-1} \gg T_1^{-1} \tag{6-18}$$

As the system approaches the rigid lattice limit, such that $\omega_1^2\tau_c^2 \gg 1$:

$$T_{1\rho}^{-1} = M_2/4\omega_1^2\tau_c^2 < T_2^{-1} \ggg T_1^{-1} \tag{6-19}$$

It should be noted that in the case of solids measurement of $T_{1\rho}$ requires high r.f. magnetic fields in order to spin-lock the whole bandshape. The condition for this is

$$B_1 > B_L \tag{6-20}$$

Values of $T_1(^1H)$ in homogeneous solids are usually strongly influenced by the phenomenon of *spin-diffusion.* Consider a moderately-sized organic molecule which is rigidly held in the solid with the exception of a methyl group which is mobile due to internal rotation. If the motion is of the appropriate rate T_1 for the methyl protons will in principle be substantially reduced. Although the dominant (H, H) dipolar interactions will be *within* the methyl group, dipolar coupling to other protons will probably be strong enough for flip-flop transitions to be efficient, thus transferring the relaxation to all the protons in the molecule. A single $T_1(^1H)$ will be observed, which will be longer than the theoretical methyl proton relaxation by a factor which is the ratio of the total number of protons in the molecule to the number of mobile relaxation-originating protons (three for a single methyl group).

6-5 Cross-polarization

For dilute spins such as ^{13}C, spectral broadening due to dipolar interactions with protons do not constitute the only difficulty in the way of getting satisfactory solid-state spectra. A second problem is the fact that spin–lattice relaxation times can be very long, so that multipulse methods are not very efficient. This difficulty has been largely overcome by an ingenious technique which derives the ^{13}C magnetization from the ^1H spins. The transfer, known as *cross-polarization,*[4] is made to occur in the rotating frame of reference by using the pulse sequence shown in Fig. 6-6. The first step is to apply a 90° pulse in the proton channel and to spin-lock the ^1H magnetization in the y direction of the rotating frame (see Section 3-17). At this point the r.f. in the ^{13}C channel is switched on, and the amplitude of the magnetic field

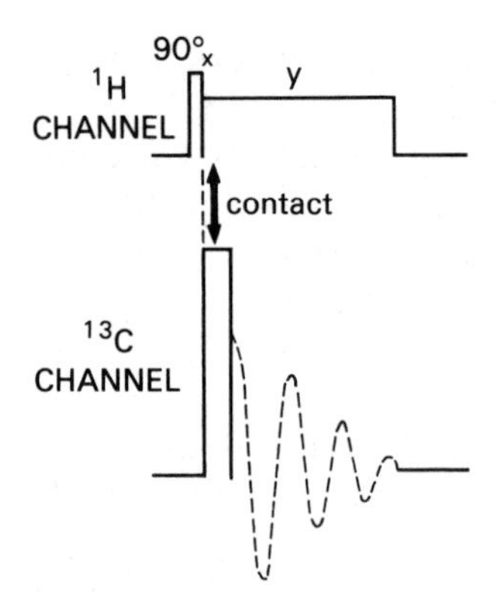

Fig. 6-6 The pulse sequence used to obtain cross-polarization from a proton spin reservoir to ^{13}C spins for solids.

B_{1C} adjusted so that the *Hartmann–Hahn matching condition*,[5] Eqn 6-21, is fulfilled:

$$\gamma_H B_{1H} = \gamma_C B_{1C} \tag{6-21}$$

This condition implies that in their respective rotating frames of reference the protons and carbons precess at equal rates and that the effective energies (also in the rotating frames) are comparable, thus allowing a rapid transfer of magnetization induced by the flip-flop term in the dipolar Hamiltonian. This transfer can be visualized thermodynamically as follows: The ^1H magnetization, produced in the laboratory frame by equilibration with the lattice, is given by the Curie Law (see Eqn 1-22) as

$$M_0(\mathrm{H}) = C_H B_0 / T_L \tag{6-22}$$

where $C_H = \frac{1}{4}\gamma_H^2 \hbar^2 N_H / k$ and T_L is the lattice temperature. This magnetization is transferred to the rotating frame by spin-locking, but it is, of course, no longer at equilibrium, since $B_1 \lll B_0$. In fact the situation can be expressed in terms of a spin temperature in the rotating frame, T_s. Thus

$$C_H B_0 / T_L = C_H B_{1H} / T_s$$

i.e.
$$T_s = (B_{1H}/B_0) T_L \tag{6-23}$$

In effect, the proton spins are at a very low temperature (~0·3 K), and their total spin energy is $-C_H B_0^2 / T_s$. However, at this point the magnetization of the carbon nuclei in their rotating frame is zero, equivalent to an infinite spin temperature. Naturally the spin energy will be repartitioned between the protons and the carbons to give a common spin temperature, T_s', as in Eqn 6-24:

$$C_H B_{1H}^2 / T_s = (C_H B_{1H}^2 + C_C B_{1C}^2) / T_s' \tag{6-24}$$

Since the proton spins are abundant whereas the ^{13}C are dilute, $C_C \ll C_H$ so $T_s' \simeq T_s$. In effect the protons form a high heat-capacity system, initially at a low temperature, whereas the carbons, initially at a high temperature, have a low heat-capacity. The protons lose only a very small part of their total magnetization, whereas the resulting magnetization of the carbons is

$$M(^{13}\mathrm{C}) = C_C B_{1C} / T_s' \simeq C_C B_{1C} / T_s \tag{6-25}$$

Substitution by Equations 6-21 and 6-23 gives

$$M(^{13}C) = C_C (\gamma_H / \gamma_C) B_0 / T_L \tag{6-26}$$

Equation 6-26 should be compared with the normal carbon magnetization at equilibrium in the laboratory frame, given by the equivalent of Eqn 6-22 as $M_0(^{13}\mathrm{C}) = C_C B_0 / T_L$. A gain of $\gamma_H / \gamma_C \sim 4$ is apparent (see Fig. 6-7). The enhanced magnetization in the carbon spins is detected by monitoring the FID following the cross-polarization.

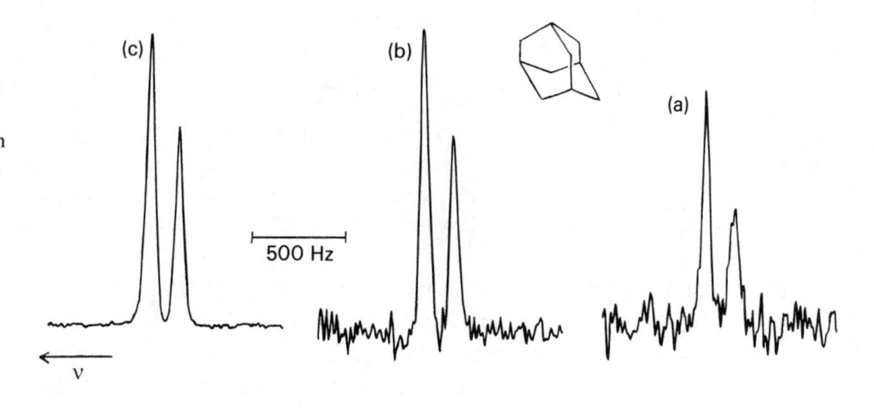

Fig. 6-7 Proton-decoupled ^{13}C NMR spectra at 22·6 MHz of static solid adamantane resulting from (a) a conventional FT experiment, (b) cross-polarization with one contact per pulse cycle, (c) multiple-contact cross-polarization (16 contacts per sequence). The acquisition time was 32 ms and the recycle time 3 s. For (b) and (c) the contact time was 5 ms. The spectra are scaled to give approximately equal peak heights. [Adapted, with permission, from R. K. Harris & K. J. Packer, *European Spectroscopy News*, **21**, 37–40 (1978).]

However, it is the next stage in the process which overcomes the problem of long ^{13}C relaxation times. Following the ^{13}C FID the carbon magnetization is again nearly zero, but the loss in proton magnetization is very small because the protons are still spin-locked. Renewed r.f. radiation in the ^{13}C channel with the Hartmann–Hahn condition will again result in a flow of magnetization to the carbons, giving a new FID which is co-added to the first, as usual. Note that no waiting time is required between the end of the FID and the re-institution of Hartmann–Hahn matching, i.e. $T_1(^{13}C)$ is irrelevant. In favourable cases this procedure can be carried out many times (*multiple-contact operation*—see Fig. 6-7(c)). Eventually, of course, the proton magnetization becomes substantially attenuated, for two reasons: (a) the transfer to the ^{13}C, though small as a proportion, is not zero, and (b) $M(^1H)$ will decay to the lattice in a time of the order of $T_{1\rho}(^1H)$. At this stage a waiting time of the order of $T_1(^1H)$ is necessary, before recommencing the whole sequence with a 90° pulse in the proton channel [note that $T_1(^{13}C)$ is still not involved and that $T_1(^1H)$ is generally less than $T_1(^{13}C)$]. The net result, after Fourier transforming the accumulated FID, is a spectrum which is considerably enhanced (see Fig. 6-7) over that which could be obtained from a simple pulse sequence such as is usually used for the solution state.

Consideration of the procedure indicates that the most favourable cases will be those with relatively long $T_{1\rho}(^1H)$ but relatively short $T_1(^1H)$, though the latter cannot, of course, be shorter than the former. Typical cross-polarization times are in the range 0·5–5 ms, so such times must be used for the Hartmann–Hahn contact. It is important to optimize the contact time, particularly for polymers and other materials which have sufficiently low values of $T_{1\rho}(^1H)$ that only single-contact operation is feasible. Such optimization will involve considerations of the *rate* of cross-polarization. Although this section has been written in terms of the common $^{13}C/^1H$ system, the technique can be applied for any dilute spin species (such as ^{15}N or ^{29}Si) in the presence of an abundant spin species. Moreover, 'dilution' of a spin species with 100% natural abundance may be achieved by chemical or physical means (incorporation on a large molecule in the former case; matrix isolation in the latter). Since the initial gain in sensitivity depends on a ratio of magnetogyric ratios, use of cross-polarization for spin species of low γ (e.g. ^{15}N) is particularly attractive.

6-6 Shielding anisotropy

So far it has been assumed that, provided dipolar interactions could be removed, solid-state NMR spectra would be entirely analogous in appearance with those of the solution state. This is not the case because of a facet of shielding not hitherto considered in any detail, namely that shielding constants depend on the orientation of the nuclear environment in the applied magnetic field. For instance the two orientations depicted for chloroform in Fig. 6-8, which have the molecular symmetry axis aligned parallel to and perpendicular to B_0, will have differing shielding constants for the ^{13}C, denoted $\sigma_{\parallel}$ and $\sigma_{\perp}$ respectively. Orientations between those shown will have intermediate values for the shieldings. A nuclear environment with less symmetry (and it should be noted that, strictly speaking, it is the crystallographic site symmetry that is important, not the molecular symmetry) will have its shielding characterized by *three* unique values. This is typical of a *tensor* property, for which the three values are referred to as the *principal components* and occur for orientations specified by the *principal axes* in a molecule-fixed system. Only in the case of a nuclear site of symmetry belonging to a cubic point group will the shielding constant be isotropic, i.e. independent of orientation in B_0. In the general case the observed shielding constant is denoted σ_{zz} and is a linear combination of the principal components, σ_{jj}:

$$\sigma_{zz} = \sum_{j=1}^{3} \sigma_{jj} \cos^2 \theta_j \qquad (6\text{-}27)$$

where the angles, θ_j, are those between the σ_{jj} and B_0. The convention $\sigma_{11} \leqslant \sigma_{22} \leqslant \sigma_{33}$ is usually adopted.

In solution, with molecules tumbling rapidly and isotropically, substantial averaging of $\boldsymbol{\sigma}$ occurs, so that only one-third of the trace, i.e. $\frac{1}{3}(\sigma_{11} + \sigma_{22} + \sigma_{33})$, is observed, and it is this quantity which was discussed in Chapter 1. In fact, Eqn 6-27 can be rewritten

$$\sigma_{zz} = \mathrm{Tr}\,\boldsymbol{\sigma} + \tfrac{1}{3} \sum_{j=1}^{3} (3\cos^2 \theta_j - 1)\sigma_{jj} \qquad (6\text{-}28)$$

where Tr stands for trace. In the case of axial symmetry, the corresponding equation is

$$\sigma_{zz} = \mathrm{Tr}\,\boldsymbol{\sigma} + \tfrac{1}{3}(3\cos^2 \theta_{\parallel} - 1)(\sigma_{\parallel} - \sigma_{\perp}) \qquad (6\text{-}29)$$

where Tr $\boldsymbol{\sigma}$ is now $(\sigma_{\parallel} + 2\sigma_{\perp})$, and $(\sigma_{\parallel} - \sigma_{\perp})$ is commonly referred to as the value of the shielding anisotropy.

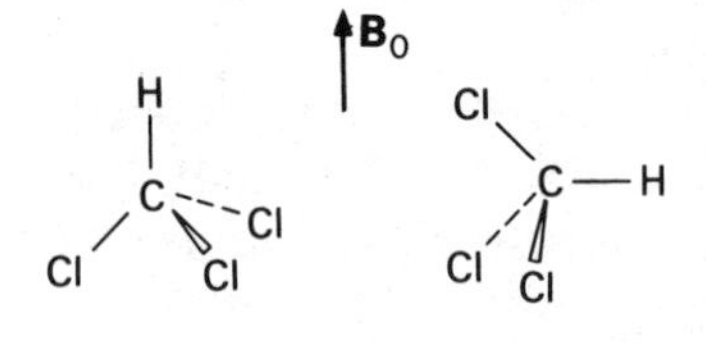

Fig. 6-8 Possible extreme orientations of a molecule of chloroform in a magnetic field, to illustrate the shielding constants $\sigma_{\parallel}$ and $\sigma_{\perp}$.

The consequence of these facts is that for a single crystal containing nuclei which are all in translationally equivalent positions, the NMR spectrum will consist of a single line whose frequency varies with the orientation of the crystal in $\boldsymbol{B}_0$. For a microcrystalline powder sample, on the other hand, the distribution of nuclear orientations results in absorption over a *range* of frequencies, giving a *powder pattern.* A schematic powder pattern for the case of axial symmetry is shown in Fig. 6-9(b), and a practical example is illustrated in Fig. 6-10(a). The extrema give immediate measures of $\sigma_{\parallel}$ and $\sigma_{\perp}$. The corresponding general case is depicted in Fig. 6-9(c). Thus values of the principal components of shielding tensors are available from solid-state NMR spectra provided (a) dipolar interactions can be removed, and (b) overlapping between powder patterns for non-equivalent sites is not serious. Carbon-13 shielding anisotropies typically range upwards to 438 ppm[6] (see Section 8-16).

6-7 High-resolution NMR of dilute spins

In order to achieve high-resolution spectra of nuclei like ^{13}C in solids, three problems have to be overcome, viz:

(a) Broadening due to heteronuclear dipolar interactions (typically (^{13}C, ^{1}H)).

(b) Low sensitivity due to long ^{13}C spin–lattice relaxation times.

(c) Broadening due to shielding anisotropy.

In the mid-1970s it was realized[7] that a combination of three previously-known techniques could overcome the difficulties.

The effects of heteronuclear dipolar interactions can be eliminated from the spectra by the double-resonance procedure of decoupling the protons using an appropriate second r.f. exactly as for the electron-coupled scalar interactions discussed in Sections 1-11, 1-18 and 4-8. However, because of the strength of dipolar coupling very high decoupling powers (ca. 100 watts) are required, and for a long time this gave technical problems. The technique is known as high-power or dipolar decoupling. The latter term is, however, somewhat misleading, since scalar decoupling occurs at the same time.

Section 6-5 showed that problem (b) can be overcome by the cross-polarization procedure. The pulse sequence shown in Fig. 6-6 incorporates high-power decoupling while the ^{13}C FID is recorded, thus overcoming problem (a) simultaneously.

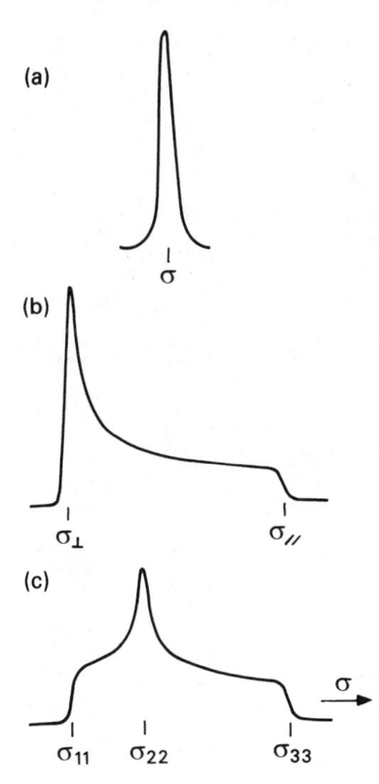

Fig. 6-9 Schematic powder patterns caused by shielding anisotropy for a site with (a) cubic symmetry, (b) axial symmetry, (c) lower symmetry.

The form of the second term in Eqns 6-28 and 6-29 recalls the discussion of Section 6-2 and suggests the solution to problem (c), viz. magic-angle rotation. As noted in Section 6-2 the achievable rates of MAR are inadequate to eliminate dipolar interactions in general, but broadening arising from ^{13}C shielding anisotropy is substantially lower than (^{13}C, ^{1}H) dipolar broadening, and so requires relatively modest MAR speeds (e.g. ca. 3 kHz if $B_0 = 2{\cdot}35$ T), which are achievable. If rates lower than the shielding anistropy (expressed in Hz) are used, then satellite resonances called *spinning sidebands* are seen. This implies that the difficulties in obtaining good high-resolution spectra

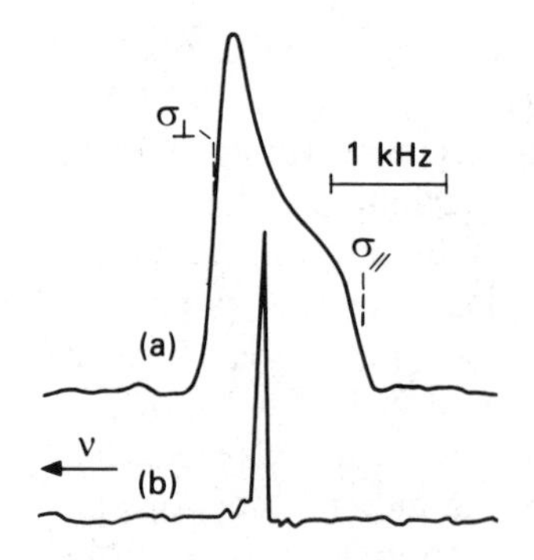

Fig. 6-10 Lead-207 spectra of solid lead nitrate, (a) static, and (b) with MAR. [Adapted, with permission, from D. J. Burton, R. K. Harris & L. H. Merwin, *J. Magn. Reson.*, **39**, 159–162 (1980), but with the direction of the frequency axis corrected.]

increase as the operating field B_0 increases. Happily MAR eliminates the small broadening due to dipolar interactions between the dilute spins themselves. Thus, in favourable cases, spectra with linewidths of only a few Hz may be obtained.

Figure 6-10 shows the improvement in resolution obtained by the final-stage MAR process in the case of lead nitrate, and Fig. 6-11 shows the effect of both high-power decoupling and MAR for calcium acetate. Although the aim of the techniques discussed in this section is to produce NMR spectra of solids which are comparable in resolution to those of liquids, it is not to be supposed that identical spectra will be obtained. Indeed the principal interest will often lie in the differences. Two such situations will be mentioned here. Firstly, crystallographic effects sometimes give rise to splittings of lines. In order for a particular type of carbon to yield a single line, all those carbons must be related by symmetry *in the crystal.* It is, however, not uncommon for such carbons to occupy crystallographically non-equivalent sites, in which case they will give rise to two (or more) resonances. Such non-equivalence may be intramolecular or intermolecular, and an example is given in Fig. 6-11(d). The second situation to be mentioned is, in a sense, similar: for non-crystalline (amorphous) solids, particularly many polymers, there is a *dispersion* of environments for a given carbon. This will lead to line broadening rather than line-splitting. Typical ^{13}C linewidths for glassy polymers are ca. 2 ppm from this cause. Figure 6-12 shows how widths vary for a cross-linked epoxide.

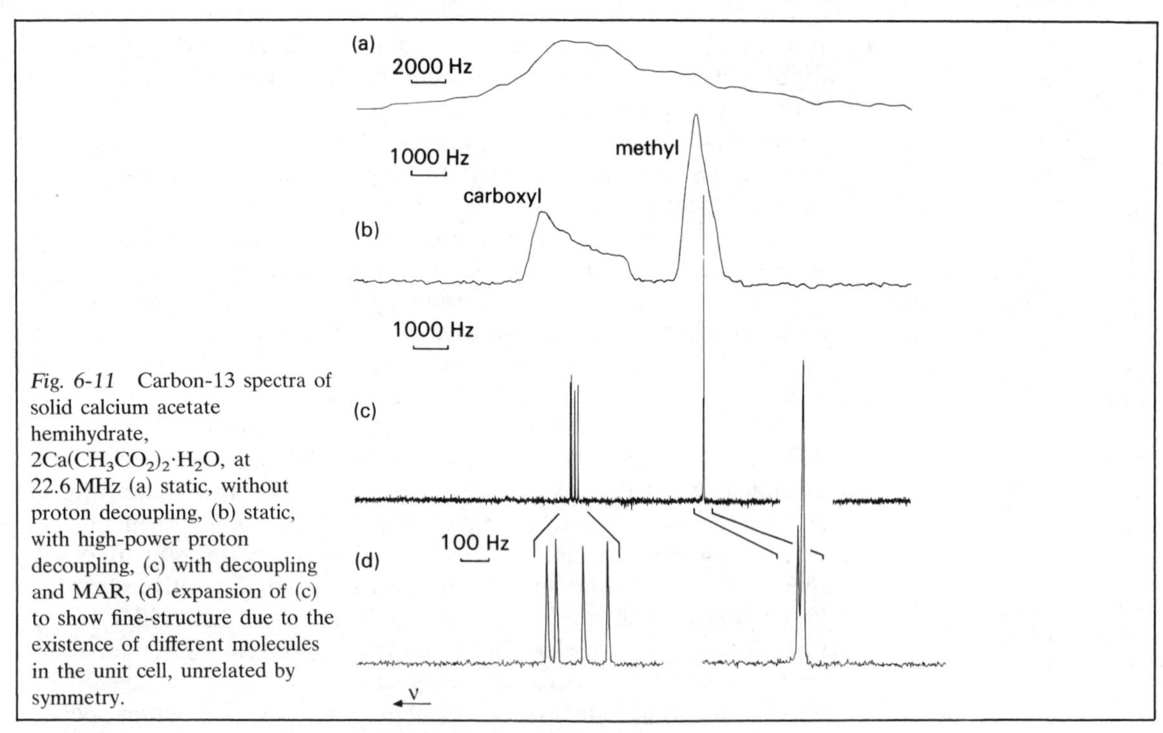

Fig. 6-11 Carbon-13 spectra of solid calcium acetate hemihydrate, $2Ca(CH_3CO_2)_2{\cdot}H_2O$, at 22.6 MHz (a) static, without proton decoupling, (b) static, with high-power proton decoupling, (c) with decoupling and MAR, (d) expansion of (c) to show fine-structure due to the existence of different molecules in the unit cell, unrelated by symmetry.

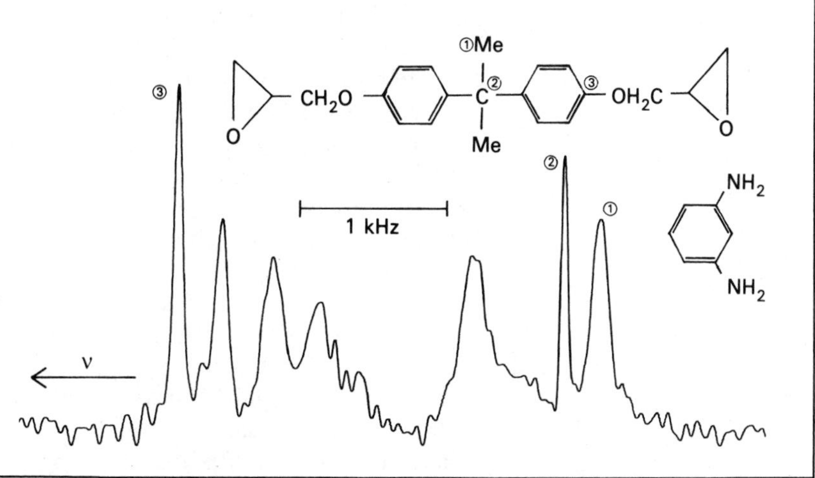

Fig. 6-12 Carbon-13 spectrum for a cross-linked epoxide at 22·6 MHz, obtained using high-power decoupling, cross-polarization and MAR. The epoxide is bis-phenol A diglycidyl ether (BADGE) and the cross-linking agent is *m*-phenylenediamine (both are shown on the figure). There are substantial variations in linewidths, due to dispersions in the environments of each type of carbon. The signal due to the central carbon of the BADGE moiety is narrow because it is sterically shielded from environmental variations.

Obviously chemical shifts for solids will differ to some extent from those for solutions of the same materials. This is clearly shown by cases with additional splittings (see Figs. 6-11, 6-13 and 6-16). Variations can sometimes be attributed to general 'medium' effects, i.e. changes in environment such as also occur for solutions. In other cases conformational changes may be involved, or tautomeric differences. One particularly important type of variation is provided by cases of

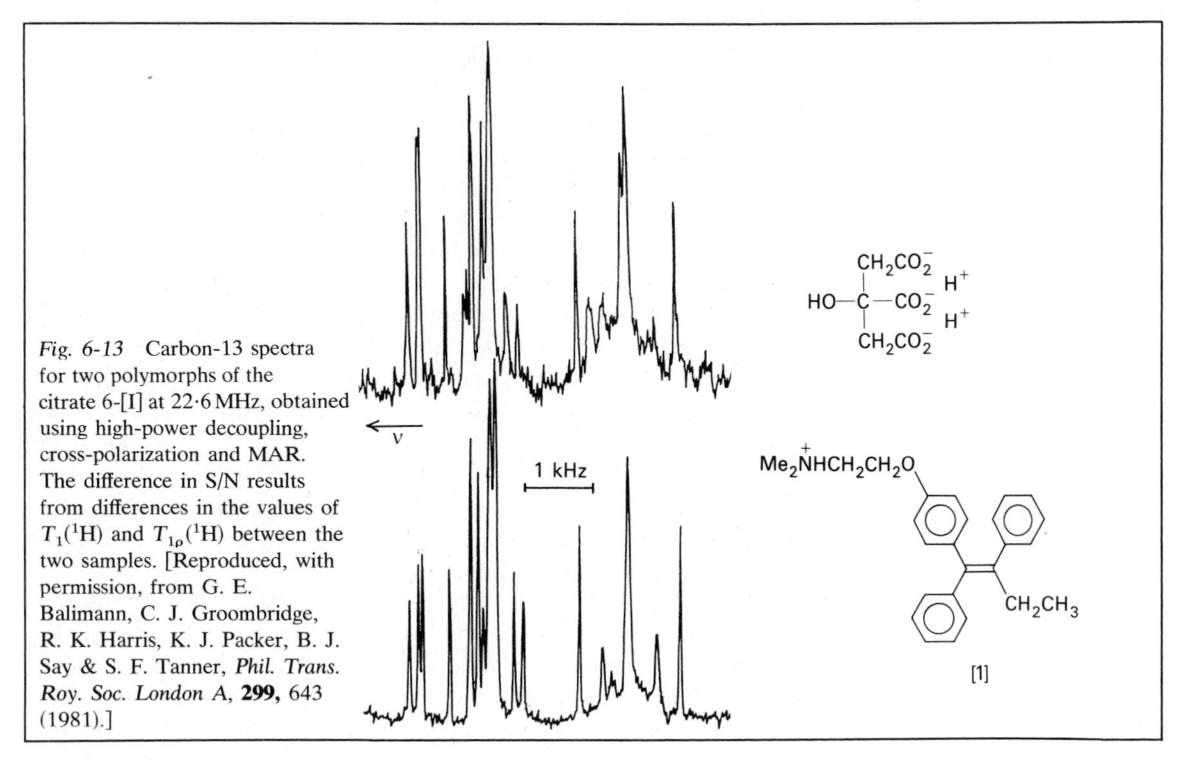

Fig. 6-13 Carbon-13 spectra for two polymorphs of the citrate 6-[I] at 22·6 MHz, obtained using high-power decoupling, cross-polarization and MAR. The difference in S/N results from differences in the values of $T_1(^1H)$ and $T_{1\rho}(^1H)$ between the two samples. [Reproduced, with permission, from G. E. Balimann, C. J. Groombridge, R. K. Harris, K. J. Packer, B. J. Say & S. F. Tanner, *Phil. Trans. Roy. Soc. London A*, **299**, 643 (1981).]

polymorphism, of importance in the pharmaceutical and polymer industries. Figure 6-13 illustrates a case where the solid-state ^{13}C spectra of two polymorphs are discernibly different. Such differences can be exploited for analytical purposes on the one hand and, if crystal structures are known, for the understanding of factors affecting chemical shifts on the other. The spectra of both polymorphs shown in Fig. 6-13 indicate the non-equivalence of all three of the citrate carboxyl carbons in the solid state; these carbons give rise to the three peaks at highest frequency in each case.

High-resolution carbon-13 NMR for solids is now well-established, and yields the same sort of detailed information on molecular structure and motion as already discussed for solutions. It is clearly a valuable tool for insoluble systems such as coal or cross-linked polymers. If other abundant spin-$\frac{1}{2}$ nuclei, such as ^{31}P or ^{205}Tl, are present, splittings due to 'scalar' coupling will be observed. Actually, such indirect coupling is, like the shielding interaction, a tensor property that is averaged to its isotropic value by molecular tumbling (in solution) or by magic-angle rotation.

The same three techniques which produce high-resolution spectra for ^{13}C in solids will also be applicable for any other dilute spin-$\frac{1}{2}$ nuclide such as ^{15}N or ^{29}Si. In fact, in practice ^{1}H and ^{19}F are normally the only spin-$\frac{1}{2}$ nuclei that are sufficiently 'abundant' to render other techniques necessary (see Section 6-9). Quadrupolar nuclei are, however, more of a problem.

The high-power-decoupling/CP/MAR suite of techniques is extremely powerful and can in principle be applied to any solid material, whether crystalline or non-crystalline, a powder or a glass, a pure system or heterogeneous material. However, certain considerations need to be borne in mind. Firstly, the optimum operating conditions (in terms of the duration of 'contact' and the recycle time) for, say, carbon spectra depend on the *proton* relaxation characteristics $T_1(^1H)$ and $T_{1\rho}(^1H)$. Therefore knowledge of these times is vital, and direct measurement is highly desirable. Moreover, for heterogeneous materials the different domains are likely to differ in $T_1(^1H)$ and/or $T_{1\rho}(^1H)$, so that knowledge of these parameters is essential to the interpretation of CP/MAR spectra in these cases. In fact, such spectra will be a superposition of subspectra from the different domains and will not necessarily quantitatively reflect their relative concentrations. Domains with values of $T_{1\rho}(^1H)$ less than the contact time will give a substantially-attenuated contribution to the CP spectrum. Isotactic polypropene provides an example. Plots of $T_{1\rho}(^1H)$ indicate three different domains (depending on the history of the sample). Values of $T_{1\rho}(^1H)$ as diverse as 0·83, 15 and 108 ms have been recorded for an annealed sample of the α-crystalline form.[8] The shortest value, presumably assignable to a relatively mobile domain, will not contribute to the observed CP/MAR spectrum under normal conditions, and thus information on it cannot be obtained in this way. However, spectra ob-

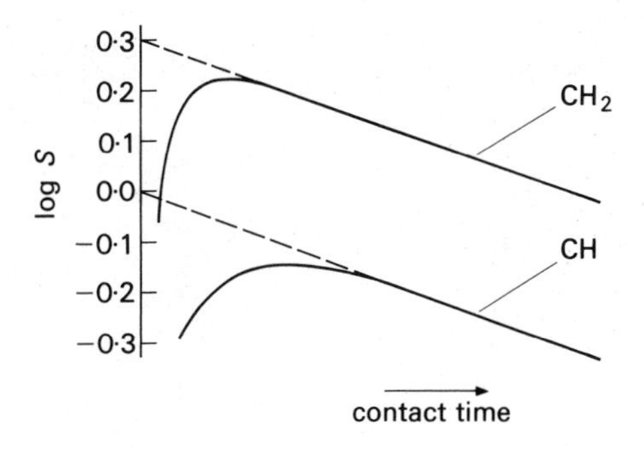

Fig. 6-14 Schematic plot to show the variation of ^{13}C signal intensity (arbitrary units) with contact time for an ideal homogeneous organic system containing two equivalent CH_2 carbons and one CH carbon. The rise time of the plots depends on the rate of cross-polarization, which is likely to be lower for the CH carbon than for the CH_2 carbons, whereas the decay is due to $T_{1\rho}(^1H)$, which will be the same for the protons relevant to all carbons in such an ideal case, due to spin diffusion.

tained using a simple $90°(^{13}C)$ pulse, as for solutions but with *high-power* decoupling, will discriminate in favour of domains with short $T_1(^{13}C)$ if the repetition rate of the pulses is reasonably high. Such domains are usually the more mobile ones, so spectra obtained by such *single pulse excitation* will complement those obtained by CP/MAR.

Clearly, quantitative interpretation of CP/MAR spectra is fraught with problems. Even for homogeneous systems the intensities of the various lines will depend on cross-polarization rates, which differ from carbon to carbon and are likely to be relatively long for quaternary carbons. Quantitative relative intensity measurement, therefore, is best obtained by varying the contact time to achieve a plot such as that of Fig. 6-14.

6-8 Non-quaternary suppression

The high-power decoupling discussed in the preceding section removes effects from scalar (C, H) coupling and so hinders the assignment of the resonances in solid-state spectra to specific carbons in an organic molecule. There is one simple pulse sequence,[9] shown in Fig. 6-15, which assists in assignment. This pulse sequence differs from that of Fig. 6-6 merely in the insertion of a 'window' in the proton decoupling

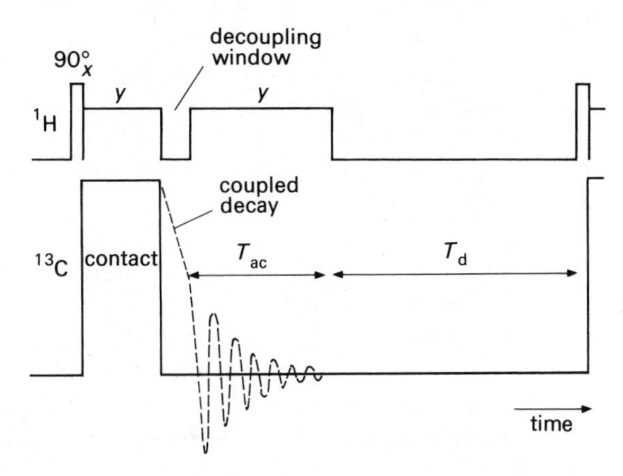

Fig. 6-15 The pulse sequence for non-quaternary suppression, showing the decoupling window.

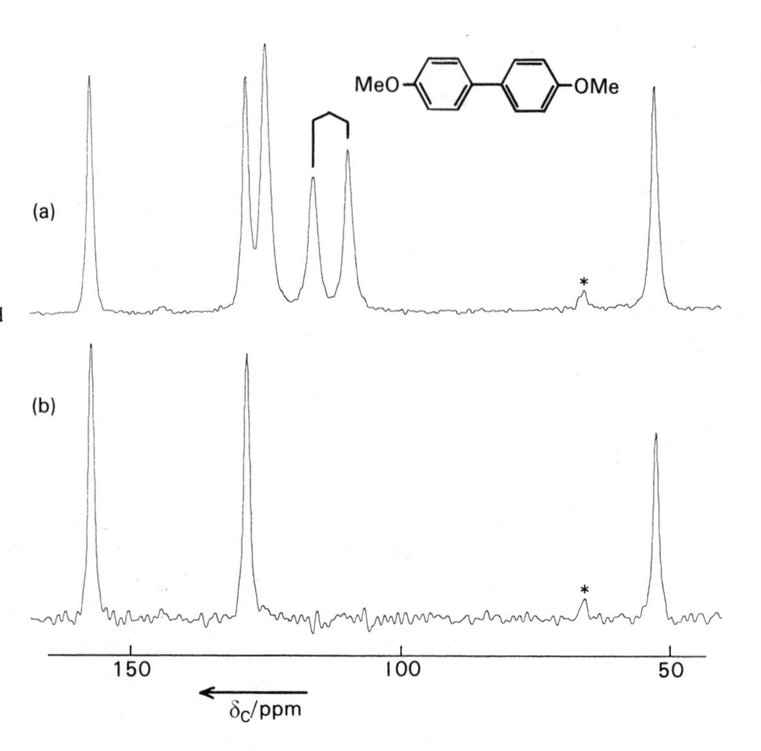

Fig. 6-16 An example of the use of the pulse sequence shown in Fig. 6-15, for the case of solid 4,4′-dimethoxybiphenyl: (a) normal CP spectrum, (b) spectrum with non-quaternary suppression obtained using a decoupling window of 40 μs. The peak marked with an asterisk is a spinning sideband. The linked pair of peaks is due to carbons *ortho* to the methoxy groups, rendered non-equivalent by the conformation of these groups. [R. S. Aujla, A. M. Chippendale, R. K. Harris, A. Mathias & K. J. Packer, unpublished work.]

before acquisition of the carbon FID commences. The duration of this window is typically ca. 40 μs. The result is to produce a spectrum, such as is illustrated in Fig. 6-16, which retains principally the quaternary carbons. This result can be understood if the form of the carbon spectrum for *individual* carbons during the decoupling window is considered. Protonated carbons have strong (C, H) interactions, thus giving broad bands, resulting in a rapid decay of transverse magnetization. Thus at the end of the window there is little signal left in the FID from such carbons. Quaternary carbons, on the other hand, give relatively narrow coupled spectra, so there is little decay of magnetization during the window, giving a substantial signal during the acquisition period. The technique thus achieves preferential suppression of the signal due to non-quaternary carbons. However, any other situation giving relatively narrow carbon lines also gives residual signals. Methyl carbons usually come into this category, as can be seen from Fig. 6-16, since rapid internal rotation causes line-narrowing, as discussed in Section 6-3.

6-9 Multiple-pulse techniques for abundant spins

When the dominant dipolar interactions are homonuclear, neither high-power decoupling nor cross-polarization is relevant. Moreover, at any rate for abundant-spin situations such as are commonly encountered for ^{1}H and ^{19}F, magic-angle rotation at the rates necessary to achieve line-narrowing (ca. tens of kHz) is not feasible. However, another method of line-narrowing is available, commonly (though not

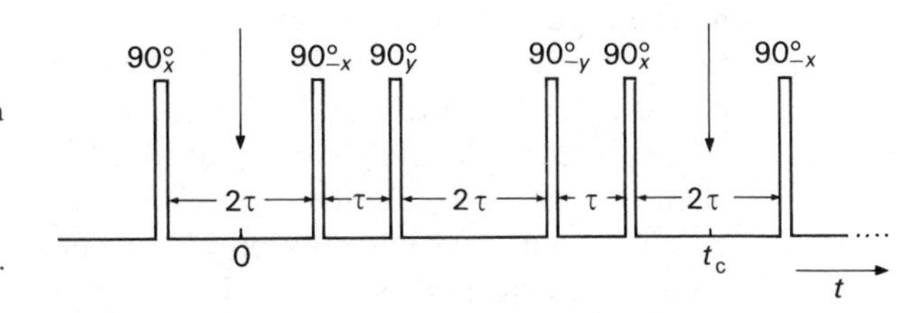

Fig. 6-17 The WAHUHA pulse sequence. The vertical arrows indicate sampling points in the observation windows. The first pulse is a preparation pulse akin to that used for observation of a simple FID. The next four form a pulse cycle, repeated as often as necessary to record the full FID in the observation windows. The cycle time, t_c, is 6τ.

very usefully) known as multiple-pulse operation. This method makes use of special pulse sequences involving changes of r.f. phase in such a way that *in certain windows* in the pulse sequence the effect of the dipolar Hamiltonian on the magnetization is zero. The pulse cycle is repeated with one detection point per cycle. Thus detection points and pulses are interspaced throughout the experiment and pulsing only lapses when the signal is fully decayed. The acquired signal is Fourier transformed as usual.

The simplest sequence consists of a cycle of four pulses known as the WAHUHA sequence after the names of its originators, Waugh, Huber and Haeberlen.[10] There are a number of equivalent sets of phase variations for the pulses, one of which is illustrated in Figure 6-17. It is not easy within the scope of the theory outlined in the present book to explain why the WAHUHA sequence results in an extension of the phase coherence of the magnetization. Suffice it to say that instead of operating on the geometrical factors of the dipolar Hamiltonian (as does MAR), WAHUHA operates on the spin factors, and the procedure is therefore known as averaging in spin space. The WAHUHA sequence requires pulses that are very accurate in their timing and phases. Moreover, there are stringent criteria to fulfil with respect to pulse power and sampling rates. The cycle time, t_c, must be short, so that in principle

$$t_c \ll \pi T_2 \qquad (6\text{-}30)$$

For a linewidth in the absence of narrowing of ca. 40 kHz this implies $t_c \ll 20\ \mu s$; however, suppression of dipolar interactions starts to become effective at somewhat longer cycle times, so that in practice typical values of t_c for the WAHUHA sequence are ca. 25 μs. The duration of the 90° pulses must be an order of magnitude shorter than t_c (e.g. ~1·5 μs) so very high B_1 powers are necessary. Pulse sequences more complicated than WAHUHA are required to compensate for pulse-timing and phase errors—cycles containing 8, 24, and 52 pulses have been proposed and used.

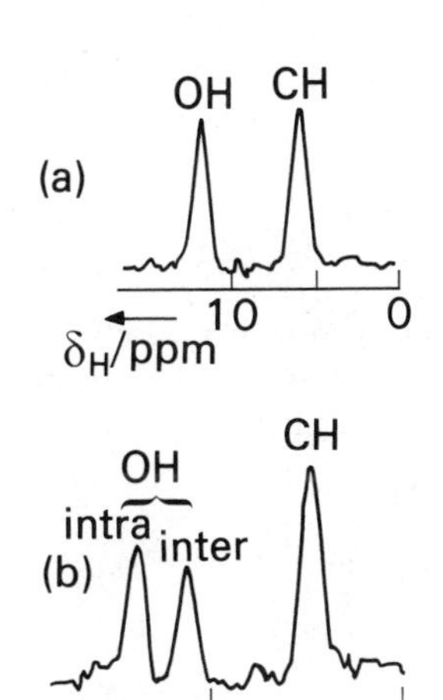

Fig. 6-18 Proton spectra of (a) solid fumaric acid, and (b) solid maleic acid, obtained with the WAHUHA pulse sequence and magic-angle rotation. [Reproduced, with permission, from H. Rosenberger, R. Sonnenberger, G. Scheler, U. Haubenreisser and T. Uhlich, Fourth Specialised Colloque Ampère, Leipzig (1979)—poster A5.34].

These sequences do not leave the other Hamiltonian terms unchanged. Although homonuclear scalar couplings are retained in full, chemical shifts (including their anisotropies) and heteronuclear scalar couplings are scaled by a factor which depends on the pulse sequence (it is $1/\sqrt{3}$ for WAHUHA).

Residual shielding anisotropies clearly mean that the phase-cycling multiple-pulse techniques alone do not produce spectra comparable to those of solutions. However, MAR may be employed to remove these effects, so that the combination of WAHUHA (or an analogous sequence) with MAR yields[11] proton linewidths of a few tens of Hz. Such resolution is still inadequate for some purposes, but is enough to give very valuable information in certain cases, particularly for problems involving hydrogen bonding. Figure 6-18, for example, shows spectra of solid fumaric and maleic acids. The hydrogen-bonding in the former is entirely intermolecular, giving rise to a single OH resonance, whereas for the latter two OH peaks are seen, corresponding to intra- and intermolecular hydrogen bonds[12]. The OH chemical shifts correlate with the $\dot{O}$—H · · · O distances in the hydrogen bonds, which are 0·2684 nm for fumaric acid, and 0·2643 nm (inter) and 0·2502 nm (intra) for the two types of hydrogen bond in maleic acid.

Notes and references

[1] E. R. Andrew, *Prog. NMR Spectrosc.*, **8,** 1 (1971);
E. R. Andrew, *Internat. Rev. Phys. Chem.* **1,** 195 (1981).

[2] The phrase 'appear to be decreased' is used because in principle sidebands appear and the true second moment remains constant. However the second moment of the centreband only is the quantity usually discussed.

[3] E. R. Andrew & R. G. Eades, *Proc. Roy. Soc. A*, **218,** 537 (1953).

[4] A. Pines, M. G. Gibby & J. S. Waugh, *J. Chem. Phys.*, **59,** 569 (1973).

[5] S. R. Hartmann & E. L. Hahn, *Phys. Rev.*, **128,** 2042 (1962).

[6] The value 438 ppm is for CS_2 (H. W. Spiess, D. Schweitzer, U. Haeberlen & K. Hausser, *J. Magn. Reson.*, **5,** 101 (1971)).

[7] J. Schaefer & E. O. Stejskal, *J. Amer. Chem. Soc.*, **98,** 1031 (1976).

[8] A. Bunn, M. E. A. Cudby, R. K. Harris, K. J. Packer & B. J. Say, *Polymer*, **23,** 694 (1982).

[9] S. J. Opella & M. H. Frey, *J. Amer. Chem. Soc.*, **101,** 5854 (1979).

[10] J. S. Waugh, L. M. Huber & U. Haeberlen, *Phys. Rev. Lett.*, **20,** 180 (1968).

[11] L. M. Ryan, R. E. Taylor, A. J. Paff & B. C. Gerstein, *J. Chem. Phys.*, **72,** 508 (1980).

[12] G. Scheler, U. Haubenreisser & H. Rosenberger, *J. Magn. Reson.*, **44,** 134 (1981).

Further reading

'High-resolution ^{13}C NMR of solid polymers', J. Schaefer & E. O. Stejskal, *Top. ^{13}C NMR Spectrosc.*, **3,** 283 (1979).

'High-resolution ^{13}C NMR studies of bulk polymers', J. R. Lyerla, *Contemp. Top. Polym. Sci.*, **3,** 143 (1979).

NMR Spectroscopy in Solids (Proceedings of a Symposium), The Royal Society (London) (1981).

'Pulsed NMR in solids', P. Mansfield, *Prog. NMR Spectrosc.*, **8,** 41 (1972).

'High-resolution NMR in solids: Selective averaging', U. Haeberlen, *Adv. Magn. Reson.*, Suppl. 1 (1976).

'High-resolution NMR spectroscopy in solids', M. Mehring, *NMR Basic Principles & Prog.*, **11** (1976).

Problems

6-1 Consider a nuclear spin system in a static solid for which there are isolated equivalent spin-pairs of $I=\frac{1}{2}$ nuclei.

(a) Prove that there is no mixing between the symmetrized wave functions $(\alpha\beta+\beta\alpha)/\sqrt{2}$ and $(\alpha\beta-\beta\alpha)/\sqrt{2}$.

(b) Show that $\langle(\alpha\beta+\beta\alpha)/\sqrt{2}|\,\mathscr{H}\,|(\alpha\beta+\beta\alpha)/\sqrt{2}\rangle=-Rh(1-3\cos^2\theta)/2$ where $R=(\mu_0/4\pi)r^{-3}\gamma^2\hbar/2\pi$.

(c) Thence deduce that the spectrum consists of a pair of lines separated by $\frac{3}{2}R(3\cos^2\theta-1)$

6-2 Show that the width at half-height of a Gaussian lineshape is $2{\cdot}36\sigma$ in angular frequency units.

7 Special pulse sequences and two-dimensional NMR

7-1 Introduction

Much of the power of modern FT NMR results from the ability to manipulate the behaviour of nuclear spins by using sequences of pulses of varying duration, phase, frequency and amplitude. Special pulse sequences have been derived to achieve a number of specific aims. Descriptions have already been given of several such sequences, e.g. for measuring T_1 (Section 3-12) or T_2 (Section 3-13), or for eliminating the NOE (Section 4-12). This chapter surveys a selection of other special pulse sequences for use with solutions. Inevitably, the choice of sequences to be included is somewhat arbitrary.

7-2 Spin-echoes in coupled systems

Many special pulse sequences involve the spin-echo phenomenon described earlier (Section 3-13). It must be recalled that such a sequence refocuses inhomogeneities in the static magnetic field B_0, and also refocuses chemical shift differences. It is of interest to ask how spin–spin coupling affects a spin-echo experiment. The answer depends on how the pulses are applied to the different spins. Consider a system containing two coupled non-equivalent spins under first-order conditions, i.e. a *hetero*nuclear AX system. The net magnetization of A can be considered as composed of two separate magnetizations, $M_A^{X\alpha}$ and $M_A^{X\beta}$, for A nuclei with their X neighbours in the α and β states respectively. Following a 90°(A) pulse these magnetizations will precess about B_0 independently at different rates. If J_{AX} is positive, $M_A^{X\beta}$ will precess faster (at $\nu_A + \frac{1}{2}J_{AX}$) than $M_A^{X\alpha}$ (which precesses at $\nu_A - \frac{1}{2}J_{AX}$), so that after a short time τ, if inhomogeneities in B_0 are ignored, the situation will be as in Fig. 7-1(b). If at this time a 180°_y pulse is applied, the magnetizations become as in Fig. 7-1(c), but since $M_A^{X\beta}$ is still precessing faster than $M_A^{X\alpha}$, the two will come together a further time τ later, and an echo will appear. The coupling thus has the same effect as a chemical shift or inhomogeneities in B_0, i.e. refocusing occurs. However, we have ignored any direct effect of the pulses on the X spins, since this will be absent for a heteronuclear AX system.

If, on the other hand, a *homo*nuclear AX system is considered, with the pulses affecting A and X equally, a different situation is obtained. The 180° pulse on the X spins inverts the populations of the X-spin levels, and thus has the effect of changing the labels of the $X\alpha$ and $X\beta$

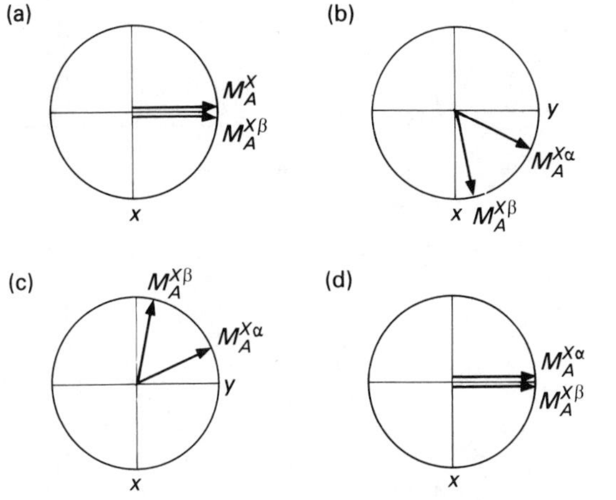

Figure 7-1 The evolution in the x/y plane of the A magnetization vectors of a *heteronuclear* AX spin system during a spin-echo pulse sequence:
(a) immediately after the $90^\circ_x(A)$ pulse; (b) a time τ later;
(c) immediately after the $180^\circ_y(A)$ pulse; (d) a further time τ later (showing refocusing).

states. Therefore $M_A^{X\alpha}$ and $M_A^{X\beta}$ are interchanged, and this would normally occur at the same time as the 180° rotation of these magnetizations about the y direction, so Figure 7-2(c) applies. The faster-precessing component is now still ahead of the slower one, so that refocusing does not occur, but the situation a time τ later is as in Fig. 7-2(d). It is easy to calculate the phase angle ϕ generated at time 2τ since the difference in angular frequency between the two components is $2\pi J_{AX}$. Thus

$$\phi = 4\pi J_{AX}\tau \tag{7-1}$$

If the FID of the second half of the echo is recorded and Fourier-transformed to obtain a spectrum, the two A lines will have phases $\pm\phi/2$, which are time-dependent, as is shown in Fig. 7-3(a), which also shows what happens to the A lines of an AX_2 spin system under the same conditions.

Suppose, however, that the signal is simply sampled at the echo maximum, and the net magnetization detected as an absorption mode

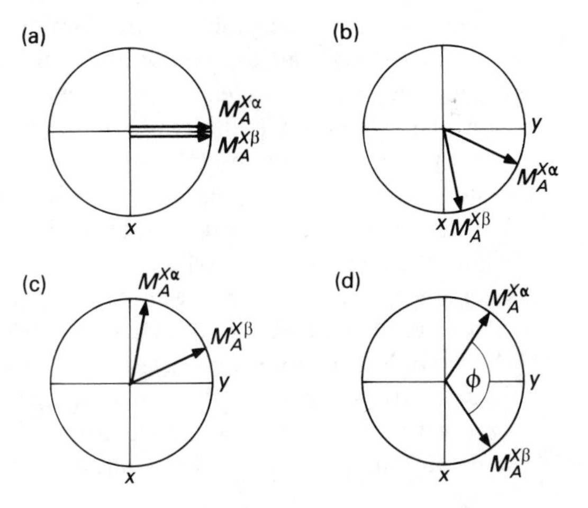

Fig. 7-2 The evolution in the x/y plane of the A magnetization vectors of a *homonuclear* AX spin system during a spin-echo experiment:
(a) immediately after the $90^\circ_x(A)$ pulse; (b) a time τ later;
(c) immediately after the $180^\circ_y(A)/180^\circ(X)$ pulse pair;
(d) a further time τ later (showing the phase angle developed).

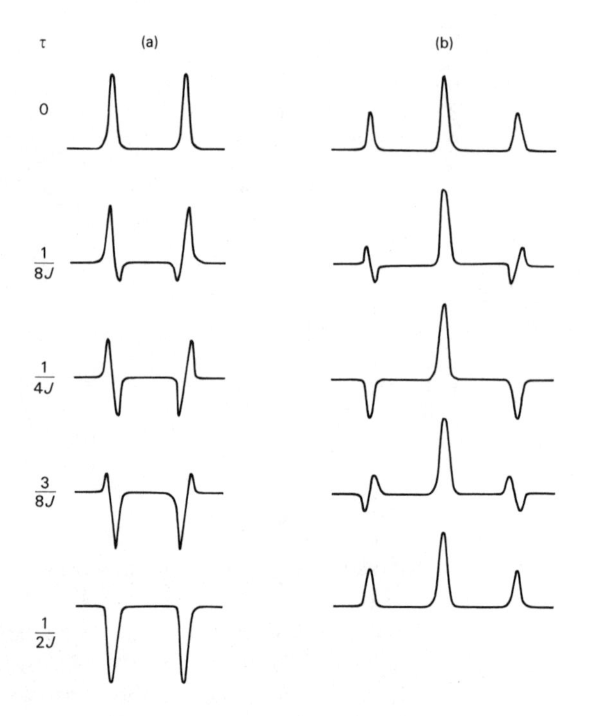

Fig. 7-3 The effects of echo modulation on the spectrum of the A nucleus for (a) an AX, and (b) an AX_2 homonuclear spin system following a spin-echo pulse sequence, $90^\circ_x - \tau - 180^\circ_y - \tau -$ echo.

signal, viz. $(M_A^{X\alpha} + M_A^{X\beta})\cos \phi/2$. Since $M_A^{X\alpha} = M_A^{X\beta} = \frac{1}{2}M_A^0$, this becomes

$$M_A(2\tau) = M_A^0 \cos 2\pi J_{AX}\tau \tag{7-2}$$

Clearly this implies the echo height varies cosinusoidally with τ, i.e. the echo is *modulated* as a function of τ. A series of measurements of echo heights (e.g. in a CPMG experiment—see Section 3-13) yields data that contains modulation from all the relevant couplings, so that Fourier transformation yields a spectrum (i.e. a function of frequency). However, since chemical shifts are refocused at the echo maxima, only coupling constants affect the result of the above procedure, which is therefore termed *J-spectroscopy*.[1] One advantage of this procedure is that inhomogeneity effects are eliminated, so that linewidths reflect true T_2 values, and extremely high resolution (~mHz) is obtained in favourable cases. An example is given in Fig. 7-4. Simple application of the method is, however, limited to first-order spectra.

Fortunately, echo modulation is not limited to truly homonuclear systems. Heteronuclear cases (e.g. AX) can be converted to effectively homonuclear ones by the simple device of implementing 180° pulses simultaneously for both types of spin (i.e. employing a form of double resonance). In such a case one would normally employ *selective detection* (e.g. observe A only,), and it would readily allow for recording of the complete echo FID *while decoupling the X spins* (see Fig. 7.5). Fourier transformation then gives a spectrum which appears normal except for the phases of the signals, which will depend on the time τ and the magnitude of the coupling constants involved. Important

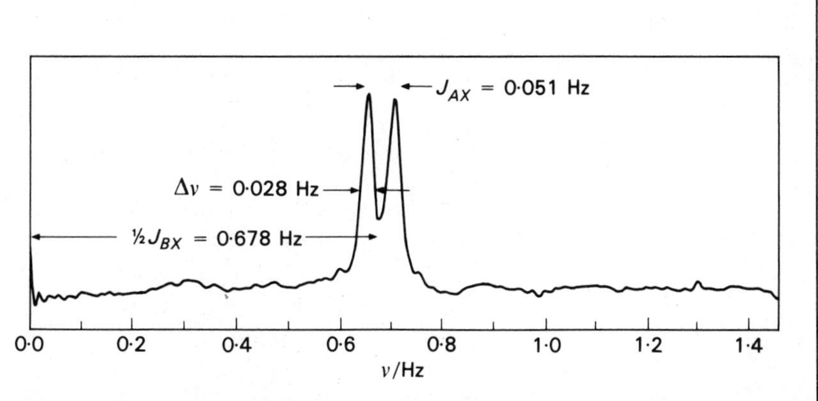

Fig. 7-4 Partial J spectrum of 3-bromothiophene-2-aldehyde. Carr–Purcell spin-echo responses of the aldehyde proton were detected selectively at a repetition rate of 2·5 Hz. 150 samples were taken, one at the peak of each echo, and subjected to Fourier transformation in a digital computer. Resonance responses appear at $(J_{BX}-J_{AX})/2$ and $(J_{BX}+J_{AX})/2$ with widths that correspond to a spin–spin relaxation time of 11·2±0·5 s. [Reproduced, with permission, from R. Freeman & H. D. W. Hill, *J. Chem. Phys.*, **54,** 301 (1971).]

applications of such an experiment are discussed in the next two sections. It may be noted that decoupling causes magnetizations $M_A^{X\alpha}$ and $M_A^{X\beta}$ to precess at the same rate. This implies relative phases are fixed at the value determined by the sequence of events prior to the decoupling.

The echo sequence $90°(A)-\tau-180°(A)/180°(X)-\tau-$ can thus be used in general for magnetization splitting, or, of course, for magnetization reforming (see Section 7-4(d)). The echo effect provides a way of refocusing chemical shifts but not coupling constants. To achieve the reciprocal result, i.e. to refocus coupling constants but not chemical shifts, is very simple for a heteronuclear system. All that is necessary for, say, proton shifts in a CH system is to use the sequence

$$90°(^1\text{H})-\tau-180°(^{13}\text{C})-\tau-\text{detect} \tag{7-3}$$

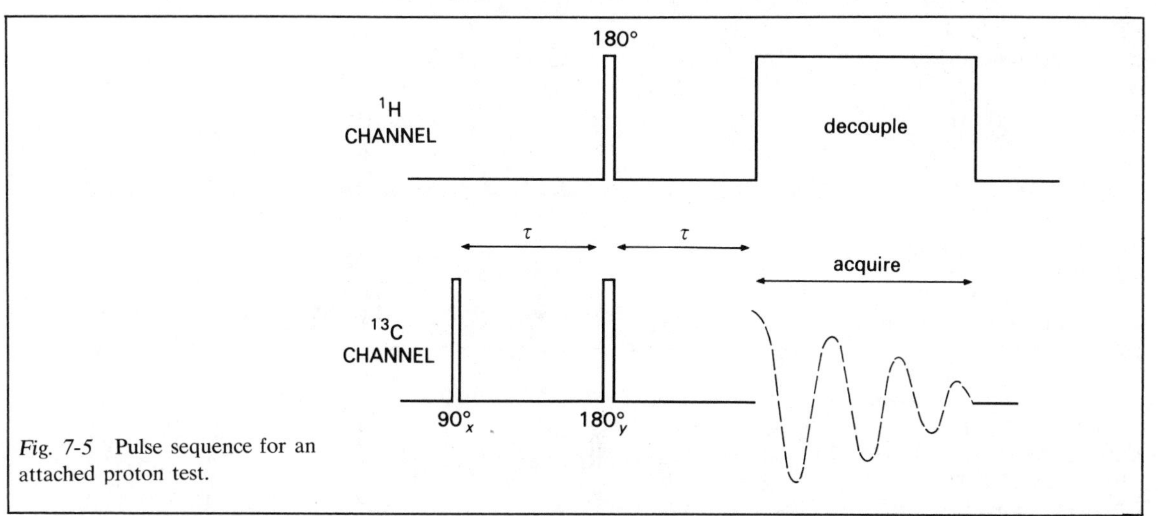

Fig. 7-5 Pulse sequence for an attached proton test.

The 180°(^{13}C) pulse switches labels on $M_H^{C\alpha}$ and $M_H^{C\beta}$ so that the phase difference between the two signals is eliminated at time 2τ, whereas the influence of chemical shifts on phase is unaffected.

7-3 Attached proton tests

Carbon spectra with noise decoupling of protons are relatively simple in appearance, but the information regarding the multiplicity of bands is lost. Proton-coupled carbon spectra, on the other hand, may be very complicated and difficult to interpret. Thus it may not be easy to assign the resonances in the decoupled spectra to methyl, methylene, methine or quaternary carbons. There are several relevant ways of deducing such assignments. It must first be recognized that values of $^1J_{CH}$ are much larger than any other (C,H) coupling constants and generally lie in the range 120–150 Hz. Suppose, therefore, the pulse sequence of Fig. 7-5 is applied.[2] Then, as described in the preceding section, a spin-echo is formed in which chemical shifts are refocused at time 2τ, but for which the effect of coupling depends on τ. If $\tau = \frac{1}{2}[^1J_{CH}(\text{ave})]^{-1}$, then the components of ^{13}C magnetization for a CH group will have a phase difference at time 2τ of $\pm180°$ with respect to the Larmor frequency. Decoupling at time 2τ will therefore give a full *negative* signal. The three components of a CH_2 triplet, however, will have phase differences of $+360°$, $0°$ and $-360°$ at 2τ, giving a full *positive* signal. Methyl signals behave rather like those for CH groups, and quaternary carbons, being unaffected by the ^{1}H channel, behave in a similar fashion to CH_2 carbons (long-range coupling may be ignored). The result is a clear distinction between CH_3 or CH carbons on the

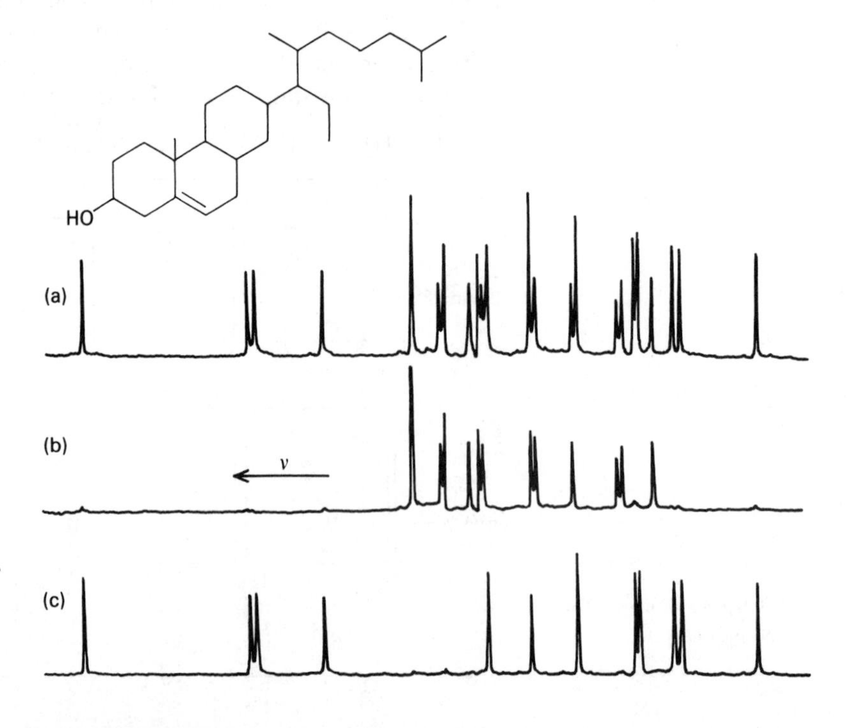

Fig. 7-6 Selection of ^{13}C sub-spectra for cholesterol using the pulse sequence described in the text. (a) the normal decoupled spectrum, (b) the sub-spectrum for the CH_2 and quaternary carbons (from the addition of the two types of FID), (c) the sub-spectrum for the CH_3 and CH carbons (from the subtraction of the two types of FID). In each case the olefinic region of the spectrum is omitted. [Adapted, with permission, from M. R. Bendall, D. M. Doddrell & D. T. Pegg, *J. Amer. Chem. Soc.*, **103,** 4603 (1981), copyright 1981, American Chemical Society.]

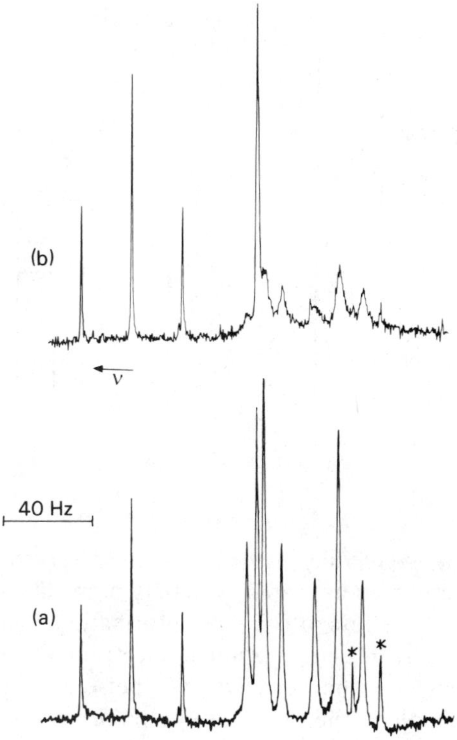

Fig. 7-7 Selection of quaternary carbons by low-power noise decoupling. The figure shows $^{13}C-\{^{1}H\}$ spectra of tetra-t-butyl diphosphane, $Bu^t_4P_4$, at −60 °C and 25 MHz, with (a) high-power decoupling, and (b) low-power decoupling. Figure (b) shows clearly a triplet and a singlet due to quaternary carbons in two different environments. The asterisks indicate lines due to impurities. [See S. Aime, R. K. Harris, E. M. McVicker & M. Fild, *J.C.S. Chem. Comm.*, 426 (1974).]

one hand and CH_2 or quaternary carbons on the other. Sub-spectra from the two types of carbon may be displayed separately if the pulse sequence of Fig. 7-5 is alternated with one in which the 180° proton pulse is omitted.[3] Addition of the free induction decays from the two sequences gives a spectrum of the CH_2 and quaternary carbons only, whereas subtraction of the free induction decays gives a CH_3 and CH sub-spectrum, as is illustrated in Fig. 7-6.

Quaternary carbons can be readily distinguished from all others by low-power noise decoupling. This is carried out at an amplitude sufficient to remove splittings due to long-range interactions but inadequate for decoupling directly-bonded protons. Resonances from CH_3, CH_2 and CH carbons give broad complicated spectra whereas quaternary carbons give sharp single peaks (Fig. 7-7). Composite particle principles (see Section 2-11) show that CH_2 carbons with magnetically-equivalent attached protons give sharp single lines from $\frac{1}{4}$ of their total intensity.

7-4 The effects of selective pulses

Selective pulses were mentioned in Section 4-5 under the heading of selective population transfer (SPT). A number of valuable experiments using selective pulses (many of which involve a special case of SPT, namely selective population inversion (SPI)) will be described in the present section.

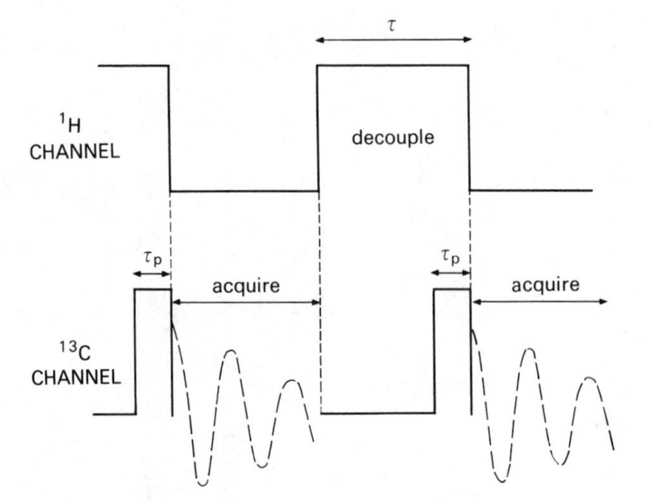

Fig. 7-8 Two cycles of a pulse sequence for the selective observation of coupled carbon-13 spectra. The pulse τ_p selectively excites one line during the decoupling regime.

(*a*) *Selective coupled carbon*-13 *spectra* Coupled ^{13}C spectra can be rather complicated, whereas noise-decoupled spectra, though simple, involve loss of coupling information. The pulse sequence shown in Fig. 7-8 results in a coupled spectrum of *only one type of* ^{13}C. Note that because the selective ^{13}C pulse is applied while the system is decoupled, the selectivity can be made very high. Freeman and coworkers give[4] a clear illustration of this type of experiment. The time τ need only be as long as the ^{13}C pulse duration, but if the NOE is desired, it must be at least of the order of $T_1(^{13}C)$.

(*b*) *Study of chemical exchange* Consider an uncoupled AX spin system undergoing mutual exchange at a rate that is slow on the NMR timescale (see Chapter 5) so that the spectrum consists of two single lines. A selective 180° pulse applied to, say, the A line will, of course, invert it, while leaving the X line unaffected. However, with passage of time the A magnetization will undergo change for *two* reasons, viz. spin–lattice relaxation and exchange with the X magnetization. Moreover, the X magnetization will be affected by exchange. Consequently, if the system is monitored by a 90° pulse as a function of the time after the 180° pulse the spectrum will show an evolution which depends on T_{1A}, T_{1X} and the exchange lifetime ($\tau_A = \tau_X = k_r^{-1}$), thus allowing the determination of τ_A provided this is comparable to or shorter than T_{1A} and T_{1X}. Suitable equations were developed by Forsen and Hoffman[5] for the related case of saturated transfer. If $T_{1A} = T_{1X}$ ($= T_1$ say) they take a simple form. The time-dependence of the A magnetization may be written

$$dM_{zA}/dt = -(T_1^{-1} + \tau_A^{-1})(M_{zA} - M_0) + \tau_A^{-1}(M_{zX} - M_0) \qquad (7\text{-}4)$$

and a similar expression holds for the X magnetization. If it is assumed that initially $M_{zX} = M_0$ but that the A spins are perturbed in some way, then the coupled pair of equations for the two magnetizations has the general solution:

$$M_{zA}(t)-M_0=\tfrac{1}{2}[M_{zA}(0)-M_0][\exp(-T_1^{-1}t)+\exp\{-(T_1^{-1}+2\tau_A^{-1})t\}] \quad (7\text{-}5)$$

$$M_{zX}(t)-M_0=\tfrac{1}{2}[M_{zA}(0)-M_0][\exp(-T_1^{-1}t)-\exp\{-(T_1^{-1}+2\tau_A^{-1})t\}] \quad (7\text{-}6)$$

In the case of an initial inversion of the A magnetization $M_{zA}(0)=-M_0$, and the equations may be re-arranged to yield

$$\frac{M_0-M_{zA}(t)}{M_0}+\frac{M_0-M_{zX}(t)}{M_0}=2\exp(-T_1^{-1}t) \quad (7\text{-}7)$$

$$\frac{M_0-M_{zA}(t)}{M_0}-\frac{M_0-M_{zX}(t)}{M_0}=2\exp\{-(T_1^{-1}+2\tau_A^{-1})t\} \quad (7\text{-}8)$$

Thus both T_1 and τ_A may be obtained from the sums and differences of the A and X intensities. The exchange lifetime is, in fact, given directly by

$$\frac{(M_0-M_{zA}(t))-(M_0-M_{zX}(t))}{(M_0-M_{zA}(t))+(M_0-M_{zX}(t))}=\exp(-2\tau_A^{-1}t) \quad (7\text{-}9)$$

Figure 7-9 gives an example. In favourable cases this type of experiment yields results for exchange rates substantially slower than those obtained by bandshape analysis (Section 5-4), thus usefully extending the temperature range accessible for kinetic studies by NMR (particularly important for accurate determination of $\Delta H^{\ddagger}$ and $\Delta S^{\ddagger}$).

(c) ***Peak suppression*** If the peaks of interest in a spectrum are weak with respect to an extraneous strong signal, difficulties arise in a FT

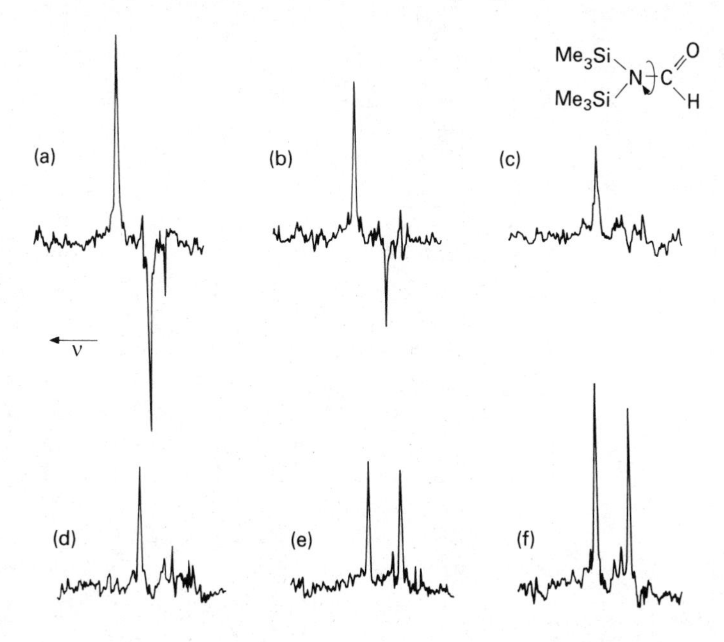

Fig. 7-9 Selective pulse experiments on bis(trimethylsilyl) formamide. The figure shows 19·87 MHz $^{29}Si-\{^1H\}$ spectra obtained at −66 °C using a pulse sequence of the type [180° (selective) − τ − 90° (nonselective) − $T_{ac}-T_d]_n$ with acquisition time $T_{ac}=4$ s, delay time $T_d=150$ s, and $n=26$. The values of τ are (a) 0·05 s, (b) 1·2 s, (c) 2·0 s, (d) 5 s, (e) 10 s, (f) 20 s. The data may be evaluated to give the exchange lifetime for internal rotation about the N—C bond, $\tau_A=1{\cdot}7$ s, and $T_1=12{\cdot}0$ s.

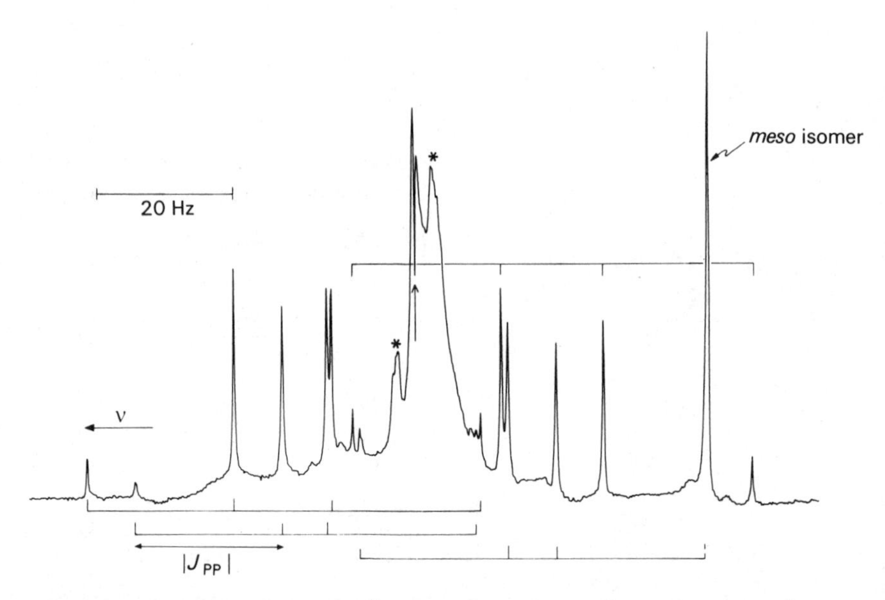

Fig. 7-10 Peak suppression by a selective pulse technique. The figure shows the 40·5 MHz ^{31}P-{^{1}H} spectrum of *rac*-[MePhP(:S)]$_2$ with suppression of the major peak due to the per-^{12}C-isotopomer (indicated by the vertical arrow), revealing the ^{13}C satellites (for the carbons directly bonded to the phosphorus) which form *ab* sub-spectra as indicated. A minor peak due to the *meso* isomer of the compound is visible. The pulse sequence used is [180°(selective) – τ – 90° (nonselective) – $T_{ac}]_n$, with τ adjusted to the null condition for the peak to be suppressed. [Reproduced, with permission, from R. K. Harris, R. H. Newman & A. Okruszek, *Org. Magn. Reson.*, **9**, 58 (1980).]

spectrum, because of the requirement for a large *dynamic range* of the computer and other spectrometer components. There are two common situations where this arises: firstly, when a solvent gives an intense signal, and, secondly, when it is desired to observe 'satellite spectra' (see Section 2-9(c)) for isotopomers containing rare nuclei in the presence of those with the abundant isotope. This situation can be alleviated by means of the pulse sequence:

$$180° \text{ (selective)} - \tau - 90° \text{ (non-selective)} - \text{FID} \qquad (7\text{-}10)$$

where the 180° pulse is only applied to the strong peak and the time τ is adjusted so that the null condition (Eqn 3-31) is obeyed for the strong peak. Figure 7-10 illustrates the method for a ^{31}P spectrum in which the interest lay in satellite peaks due to molecules containing ^{13}C.

(d) Cross-polarization One of the most pressing problems in NMR is that of sensitivity, particularly for nuclides of low natural abundance and small magnetic moment. Part of the difficulty arises because of unfavourable Boltzmann factors. The proton has the highest receptivity of any nuclide, so that sensitivity for other species may be dramatically improved if suitable population-transfer experiments can be implemented. In fact, for a heteronuclear AX spin system a simple selective pulse in the X region will achieve such a transfer. Consider the appropriate energy level diagram (Figure 7-11). In (a) is shown the Boltzmann situation. Suppose we apply a selective 180° pulse to the $\alpha\alpha \rightarrow \alpha\beta$ X transition (i.e. X_2). This inverts the populations of these levels, giving the situation in Fig. 7-11(b). Now if the A spectrum is 7-3(a) for $\tau = 3/4J$. Decoupling during the ^{13}C FID would result in no

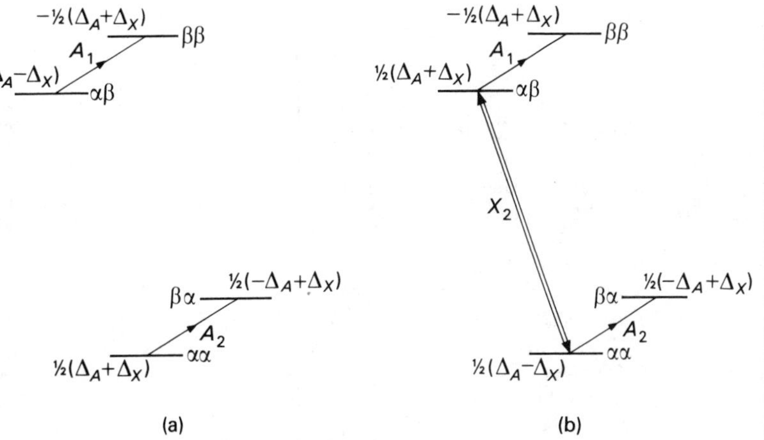

Fig. 7-11 Energy level diagram for an AX spin system showing the excess populations (a) for Boltzmann equilibrium, and (b) following a selective 180° pulse at the frequency of transition X_2. The symbols Δ_A and Δ_X refer to the normal Boltzmann population differences across the A and X resonances respectively. They are proportional to γ_A and γ_X respectively.

monitored by a 90°_x pulse, the two A transitions will have intensities governed by the relevant population differences. For the situation of Fig. 7-11(a) these will both be Δ_A. However, following the selective X pulse, the population difference is increased across the A transition which is progressive with the selectively-irradiated X line (i.e. A_1) and becomes $\Delta_A + \Delta_X$. The population difference for the regressively-connected A line (i.e. A_2) is reduced to $\Delta_A - \Delta_X$. Since Δ_A and Δ_X are proportional to γ_A and γ_X respectively, the progressive A line is enhanced by a factor $1+(\gamma_X/\gamma_A)$ over the Boltzmann situation whereas the regressive A line is affected by a factor $1-(\gamma_X/\gamma_A)$. Usually, the experiment will be used for cases where $|\gamma_X|>|\gamma_A|$ so that the regressive line will actually be negative. For the ^{13}C-$\{^1H\}$ case the two ^{13}C lines will show changes in intensity by factors ca. 5 and −3 over the normal spectrum, that is, they will show a phase difference of 180°, as illustrated by Fig. 7-12. If a 90°_y pulse is used to monitor the magnetization, instead of a 90°_x pulse, the spectrum would appear as in Fig.

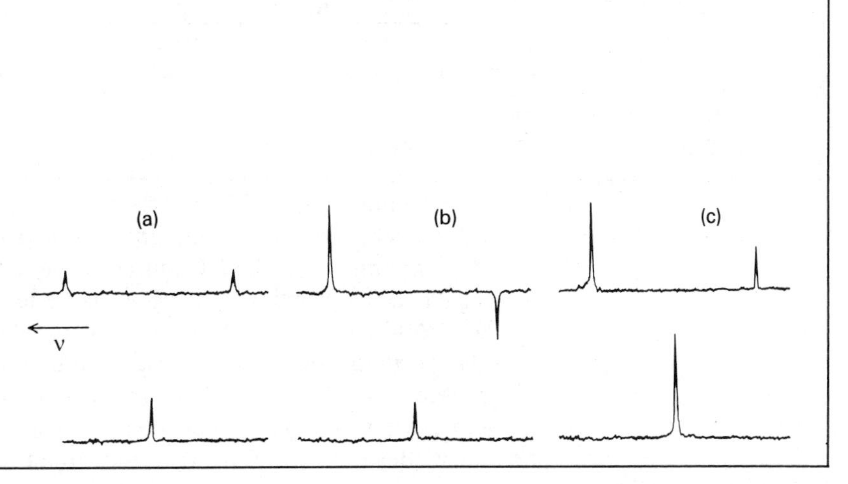

Fig. 7-12 $^{13}C-\{^1H\}$ cross-polarization for liquid chloroform from the use of selective pulses as described in the text. The upper spectra are coupled and the lower ones decoupled. (a) normal spectra, with gated decoupling so as to eliminate the NOE; (b) spectra obtained following a selective pulse on transition X_2; (c) spectra obtained following successive selective pulses on transitions X_2 and A_2 (see [6]). [Reproduced, with permission, from R. A. Craig, R. K. Harris & R. J. Morrow, *Org. Magn. Reson.*, **13**, 229 (1980).]

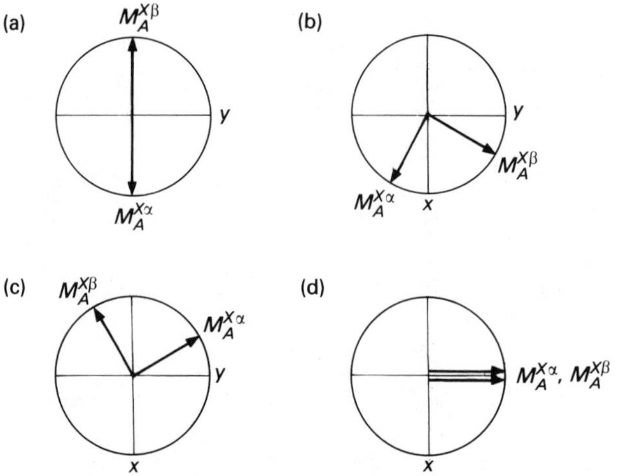

Fig. 7-13 Refocusing effects for the A magnetization vectors of an AX spin system following a selective 180° pulse on the X spins (see the text). The magnetization vectors in the x/y plane are shown: (a) immediately after the 90°_y (A) pulse; (b) after a time $\tau = 1/4J_{AX}$; (c) after simultaneous 180° (X) and 180°_y (A) pulses (see Fig. 7-14); (d) after a further time $\tau = 1/4J_{AX}$.

net gain in sensitivity. However, if acquisition is delayed without decoupling for a time $1/2J_{CH}$ following a 90° ^{13}C pulse, then (as should be clear from Section 7-2), the phase difference will be eliminated[6] (Fig. 7-13) and decoupling will give a signal enhanced by $\gamma_H/\gamma_C(\sim 4)$ over normal (see Fig. 7-12). The pulse sequence is shown in Fig. 7-14.

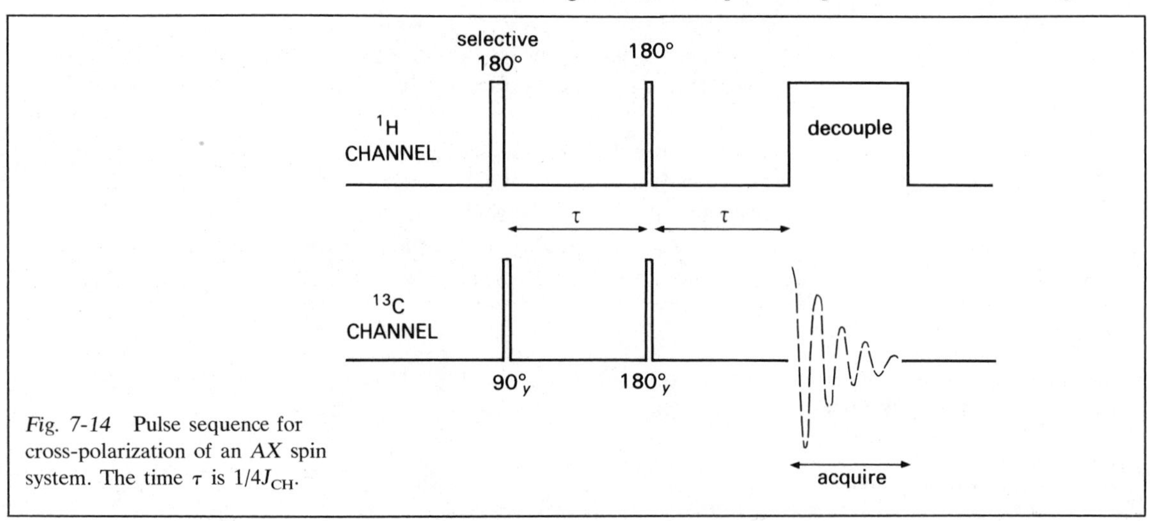

Fig. 7-14 Pulse sequence for cross-polarization of an AX spin system. The time τ is $1/4J_{CH}$.

The simultaneous 180° pulses in the ^{1}H and ^{13}C channels are to refocus chemical shifts but not coupling constants (see Section 7-2). Note the phases of the ^{13}C pulses, designed to give an absorption mode signal. Enhancements will be even better for some other nuclei (e.g. for ^{15}N the gain is ~10). Moreover no NOE is involved (since it is the ^{1}H magnetization that is used), thus side-stepping the null signal problem (see Section 4-10) when observing nuclei with negative magnetogyric ratios. A further advantage is that repetition of the experiment depends on $T_1(^1H)$ rather than the relaxation time of the observed nucleus—the former may be substantially less than the latter. The optimum delay time between the 90° pulse and acquisition will be different for spin systems AX_n with $n>1$ (Problem 7-4).

7-5 The INEPT experiment[7]

Consider a two-spin AX system subjected to the pulse sequence of Fig. 7-15 with $\tau = \frac{1}{4}J_{AX}^{-1}$. For the X spins, the two components of magnetization (differing by the spin-state of the associated A nucleus) behave prior to the second 90°_x pulse as described in Section 7-2, developing a phase difference of 180°, as indicated in Fig. 7-16. The 90°_y pulse in the X channel then results in the situation of Fig. 7-16(e): $M_X^{A\beta}$ is unchanged from Fig. 7-16(a) but $M_X^{A\alpha}$ is inverted. The simple interpretation of Fig. 7-16(e) is that the X populations have been inverted for molecules with the A spins α only. This is precisely the same as the situation achieved by the selective population inversion experiment described in the preceding section and illustrated in Fig. 7-11. It implies that for the A spins, $M_A^{X\beta}$ is enhanced and $M_A^{X\alpha}$ is also enhanced in magnitude but is inverted (see Fig. 7-16(f)). Consequently the $90^\circ_x(A)$ pulse of Fig. 7-15 will give the enhanced positive and negative peaks as shown in Fig. 7-12. Moreover, if a $90^\circ_y(A)$ pulse is used instead of $90^\circ_x(A)$, and the pulse sequence of Fig. 7-15 is extended (see Fig. 7-17) to allow for refocusing and decoupling (as discussed in Section 7-4(d)), there will be full cross-polarization, and therefore enhanced spectra for nuclei of low magnetogyric ratio will be obtained. This method is known as INEPT from the words *I*nsensitive *N*uclei *E*nhanced by *P*olarization *T*ransfer. It may be noted that the term magnetization may be preferred to polarization, but this does not yield such a euphonious acronym!

The advantage of the INEPT sequence over the selective experiment described in the previous section is that it is not necessary in the INEPT work to know the precise frequencies of resonance. Moreover, *all* A nuclei in the sample which have similar J_{AX} values will be

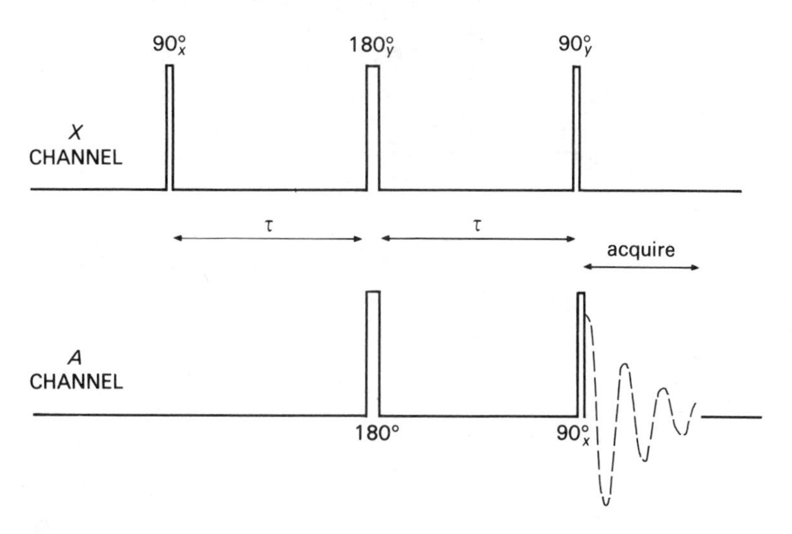

Fig. 7-15 The INEPT pulse sequence (without decoupling).

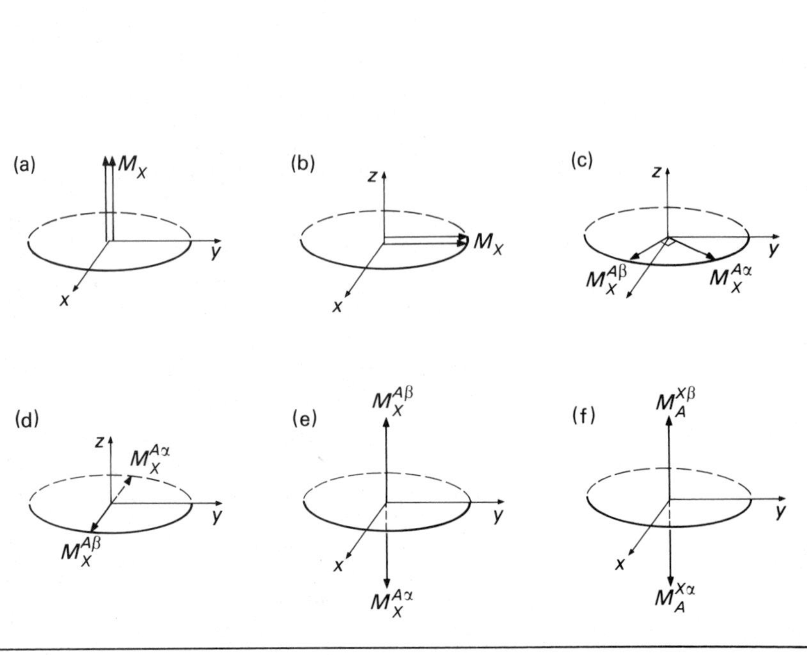

Fig. 7-16 Magnetization vectors for an AX spin system during an INEPT pulse sequence (see Fig. 7-15). (a) to (e) X magnetization vectors at the following times: (a) initially; (b) immediately after the $90^\circ_x(X)$ pulse; (c) a time $\tau = 1/4J_{AX}$ later, with a phase difference of 90° between the two vectors; (d) at time 2τ, with a phase difference of 180° between the two vectors (see Fig. 7-2); (e) immediately following the $90^\circ_y(X)$ pulse. (f) shows the A magnetization vectors at the same time as for (e) (but prior to the $90^\circ_y(A)$ pulse)—the relationship between (e) and (f) is explained in Fig. 7-11. The pulse sequence of Fig. 7-15 would result in $M_A^{X\alpha}$ and $M_A^{X\beta}$ differing in magnitude, but they are equalized if the $90^\circ_y(X)$ pulse is replaced by $90^\circ_{-y}(X)$ in alternate cycles of the sequence, and the two free induction decays from the alternate cycles are subtracted.

enhanced in a single experiment. This is a big advantage for ^{13}C work, as most CH_3, CH_2 and CH groups fall into this category.

At the intermediate stage given by the pulse sequence of Fig. 7-15, which may be referred to as *coupled* INEPT, use of a $90^\circ_{-y}(X)$ pulse instead of $90^\circ_y(X)$ will result in inverting $M_A^{X\beta}$ and enhancing $M_A^{X\alpha}$ (being equivalent, for a CH group, to selective inversion of line X_1). If pulse sequences of the type 7-15 with $90^\circ_y(X)$ and $90^\circ_{-y}(X)$ pulses are alternated, and the two types of FID subtracted, all influence of the Boltzmann populations of the A spins will be removed, and a $1:-1$ doublet will be observed, with intensities enhanced by γ_X/γ_A. For CH_2 groups the corresponding observation (see Problem 7-3) is of a

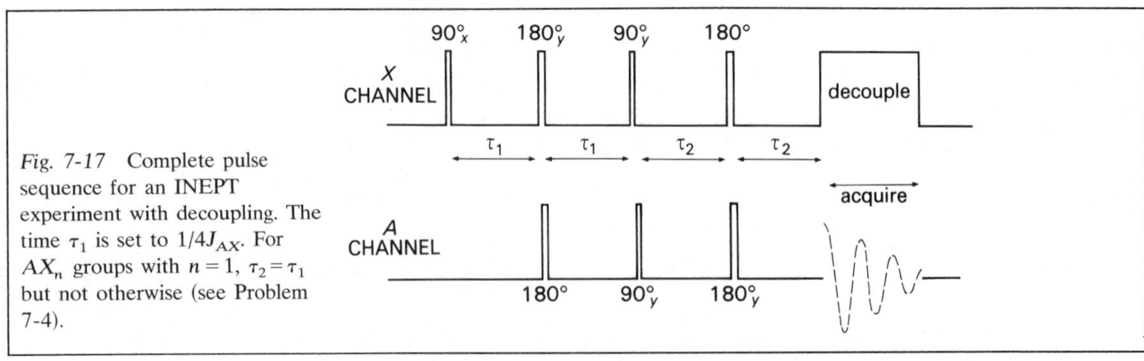

Fig. 7-17 Complete pulse sequence for an INEPT experiment with decoupling. The time τ_1 is set to $1/4J_{AX}$. For AX_n groups with $n = 1$, $\tau_2 = \tau_1$ but not otherwise (see Problem 7-4).

1:0:−1 triplet, with the intensities of the outer lines enhanced by $2\gamma_X/\gamma_A$. The result for CH_3 groups is a 1:1:−1:−1 quartet, with the intensities of the outer lines enhanced by $3\gamma_X/\gamma_A$. When the full INEPT sequence of Fig. 7-17 is applied to CH_2 and CH_3 groups, however, optimum refocusing is obtained when $\tau_2 \neq 1/4J_{CH}$, as mentioned in Section 7-4(a) (see also Problem 7-4). Consequently, for a general organic system a compromise value of τ_2 must be chosen, not only because values of $^1J_{CH}$ vary, but also because CH, CH_2 and CH_3 groups are all likely to be present. Intensities will no longer be simply related to carbon concentrations. Quaternary carbons will not be visible in the spectra at all when the $90^\circ_{\pm y}(X)$ pulse alternation is used. It may be noted that INEPT experiments can also use long-range coupling, as for ^{29}Si-{^1H} studies of compounds containing $\equiv SiCH_3$ groups.

Coupled INEPT spectra, as noted above, show relative intensities of multiplet components which differ from those of spectra obtained using simple single-resonance multipulse operation. Doddrell *et al.*[8] have introduced a pulse sequence which overcomes this problem and is therefore referred to as DEPT (*D*istortionless *E*nhancement by *P*olarization *T*ransfer). The DEPT sequence is as follows:

$$90^\circ_x(X) - \tau - 180^\circ_x(X)/90^\circ_x(A) - \tau - \theta_y(X)/180^\circ(A) - \tau - \mathrm{FID}(A)$$

where τ is set to $1/2J_{AX}$ and the value of θ may be varied. Unwanted natural A polarization may be removed by phase-alternation (±y) of the $\theta_y(X)$ pulse, followed by appropriate addition/subtraction of the FIDs. For an AX_n system with $A \equiv {}^{13}C$, $X \equiv {}^1H$, $J_{AX} \equiv {}^1J_{CH}$(ave), choice of $\theta = 90^\circ$ gives a ^{13}C subspectrum due to CH groups only (compare Section 7-3). Moreover, if FIDs are acquired for $\theta = 45^\circ$, 90° and 135°, using twice as many transients for $\theta = 90^\circ$ as for the other two cases, CH_2 subspectra are obtained from subtracting the accumulated FID for $\theta = 135^\circ$ from that for $\theta = 45^\circ$, and CH_3 subspectra are produced from FID(45°)+FID(135°)−0·707 FID(90°). Obviously, since multiplet components are not distorted by the DEPT pulse sequence, decoupling may be carried out during FID acquisition, so the organic chemist may be presented with separate decoupled spectra for CH, CH_2 and CH_3 groups with correctly-related intensities, providing a very powerful technique for structure determination. Unfortunately, events during the second τ period of the sequence cannot be satisfactorily described using the simple spin-vector model used in this chapter, but a detailed discussion is available[8].

7-6 Two-dimensional NMR—basic principles

As discussed in Chapter 2, Fourier transform NMR spectroscopy involves the conversion of a time-domain signal (the FID) into a frequency presentation (the spectrum). However, for many pulse sequences, such as those described earlier in this chapter, the FID depends on a second time variable. For the spin-echo sequence, for instance, this second time is the interval between the original 90° pulse and the echo. Suppose such a time variable is designated t_1, and the elapsed time of recording the FID is defined as t_2. Then the FID may be formally considered as a function of both times, i.e. as $S(t_1, t_2)$. If experiments are carried out for a number of times t_1, then normal Fourier transformation of the free induction decays will produce a series of spectra, these being a function of both frequency (f_2 say) and the time t_1. Now consider the array of intensities in these spectra at a given frequency f_2 for a range of times t_1. This array may itself be Fourier transformed to give a spectrum. Indeed, if this is done for a suitable number of frequencies f_2 the result will be a series of spectra which may be displayed as a function of two frequencies, f_1 and f_2. Thus the double transformation may formally be considered as

$$S(t_1, t_2) \xrightarrow{\text{FT over } t_2} S(t_1, f_2) \xrightarrow{\text{FT over } t_1} S(f_1, f_2) \tag{7-11}$$

The precise meaning to be attached to f_1 and f_2 depends on the particular experiment considered. There are, in fact, a number of different types, three of which will be considered in the following sections. The time involving the variable t_2 will be termed the *detection period* whereas that relating to the variable t_1 will be called the *evolution period.* The idea of two-dimensional NMR was first discussed by Jeener[9] at a conference in 1971, but the particular experiment he suggested will not be discussed here since the principles involved are more complex than those of several other 2-D experiments.

One may distinguish two general classes of 2-D NMR spectra: (a) *resolved* spectra, which spread the signal peaks of a single spectrum into two dimensions characterized by different NMR parameters; and (b) *correlated* spectra, which are essentially correlation diagrams between two spectra. Examples of both types will be discussed here.

There are various modes of presentation of 2-D spectra. One of these is to trace a stacked series of spectra for various values of f_2, similar in appearance to those already encountered for measurement of T_1 (see Fig. 3-13). Such a presentation has a familiar look about it. Alternatively, contour plots may be displayed, which are easier to interpret in complicated cases.

7-7 Heteronuclear 2-D *J*-resolved spectroscopy

The principles of *J*-spectroscopy have already been discussed in Section 7-2. The pulse sequence of Fig. 7-5 may be used as a basis for a 2-D version of *J*-spectroscopy, with the evolution time t_1 considered as 2τ (i.e. the time between the original 90° pulse and the echo). It is, perhaps, clear that the frequency dimension f_2 will yield ^{13}C chemical shift information only, since the FID is recorded during proton noise decoupling. However, the time interval t_1 causes modulation in the FID according to the value of the (C, H) coupling constants. This means that the frequency dimension f_1 contains a record of coupling constants only, the splitting pattern for each type of carbon being centred at zero. Thus, the two dimensional array of spectra represent a plot of ^{13}C chemical shifts vs. (C,H) coupling constants. Such a plot

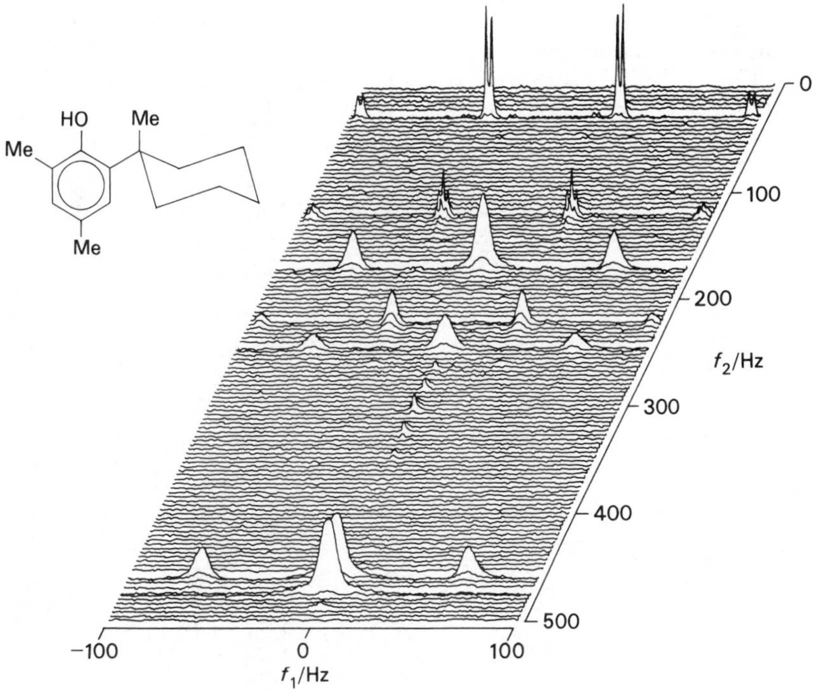

Fig. 7-18 The heteronuclear ^{13}C 2-D J spectrum of 2(methylcyclohexyl)-4,6-dimethyl phenol (aliphatic region only). The f_2 axis represents ^{13}C chemical shifts (relative to the signal for tetramethylsilane). The f_1 axis gives the (C, H) coupling constants. The signals due to the seven different aliphatic carbon types (3 methyl, 3 methylene and a quaternary) can be clearly seen. [Reproduced, with permission, from M. H. Levitt & R. Freeman, *J. Magn. Reson.*, **34,** 675 (1979).]

therefore overcomes the problems caused by the complexity of normal coupled ^{13}C spectra. The projection of the spectra onto the f_2 axis gives the equivalent of the noise-decoupled carbon spectrum. An example of heteronuclear 2-D J-spectroscopy is given in Fig. 7-18.

It is important to ask how many values of t_1 need to be used to obtain satisfactory results, since this will determine the total time taken by the experiment. In fact the usual rule of Fourier transformation regarding spectral width (see Eqn 3-30) applies, viz:

$$F = N/2T \tag{7-12}$$

The interpretation to be placed on this equation for the t_1 dimension is as follows: the time t_1 must take N values between 0 and a maximum T, which means, in effect, that the total time taken by the experiment is N times that required for accumulation for a single value of t_1. Fortunately, for ^{13}C-{^{1}H} work the necessary value of F is only that needed to accommodate a methyl quartet, so for typical values of $^1J_{CH}$ F should be set to ca. 400 Hz. If a resolution of 2 Hz is required, then T needs to be 0·5 s and N must be 400.

The method of obtaining heteronuclear 2-D J-spectra described above may be referred to as the spin-flip mode. An alternative[2] is to use proton decoupling during the second half of t_1 in place of the 180°(^{1}H) pulse—known as the gated decoupler mode. The former suffers from the disadvantage that second-order features, which arise from strong coupling in the ^{1}H region, are not properly reproduced. If a homonuclear coupled system (typically protons) is subjected to the

7-8 Homonuclear 2-D *J*-resolved spectroscopy

spin-echo pulse sequence, the echo is automatically modulated by the homonuclear couplings. The two-dimensional procedure described in the preceding section may therefore be implemented. The result will differ from that for the heteronuclear case in one important respect, because decoupling during the recording of the FID is not feasible. This means that the frequency dimension f_2 contains information about both proton chemical shifts and (H,H) coupling constants. Since the dimension f_1 depends on coupling constants alone, the result is that

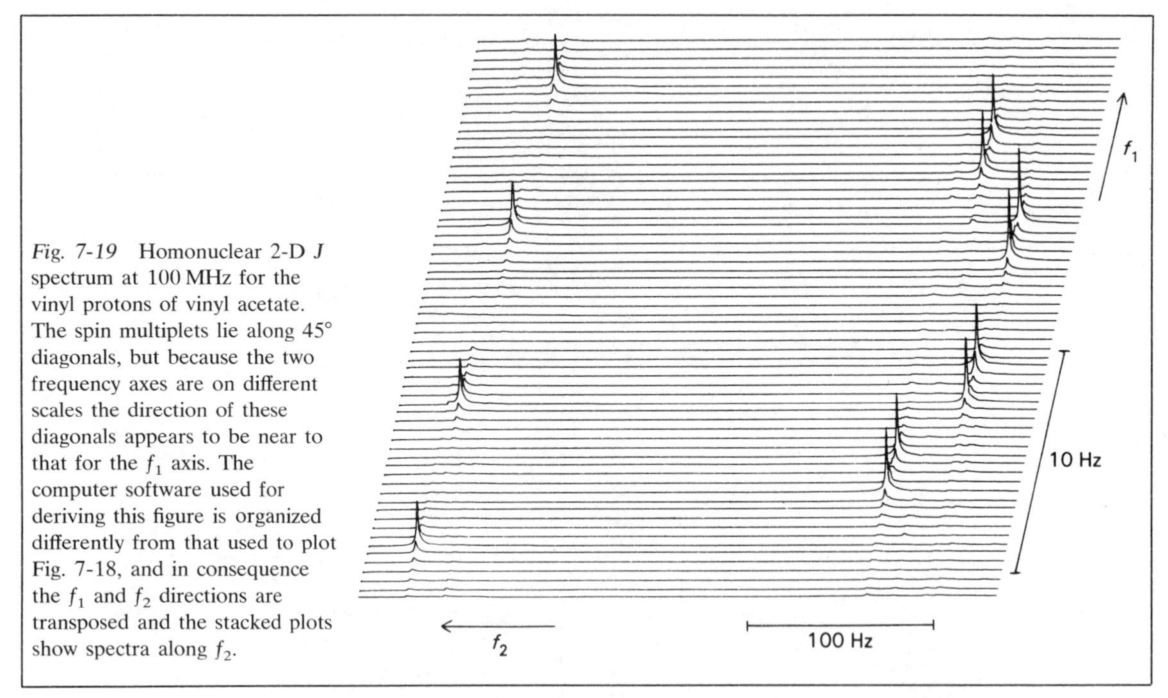

Fig. 7-19 Homonuclear 2-D *J* spectrum at 100 MHz for the vinyl protons of vinyl acetate. The spin multiplets lie along 45° diagonals, but because the two frequency axes are on different scales the direction of these diagonals appears to be near to that for the f_1 axis. The computer software used for deriving this figure is organized differently from that used to plot Fig. 7-18, and in consequence the f_1 and f_2 directions are transposed and the stacked plots show spectra along f_2.

spin multiplets lie along 45° diagonals, as is shown in Fig. 7-19. However, this can be rectified by suitable computation, resulting in a plot of ^1H chemical shifts against (H,H) coupling constants. A projection along the 45° diagonals prior to rectification results in a fully

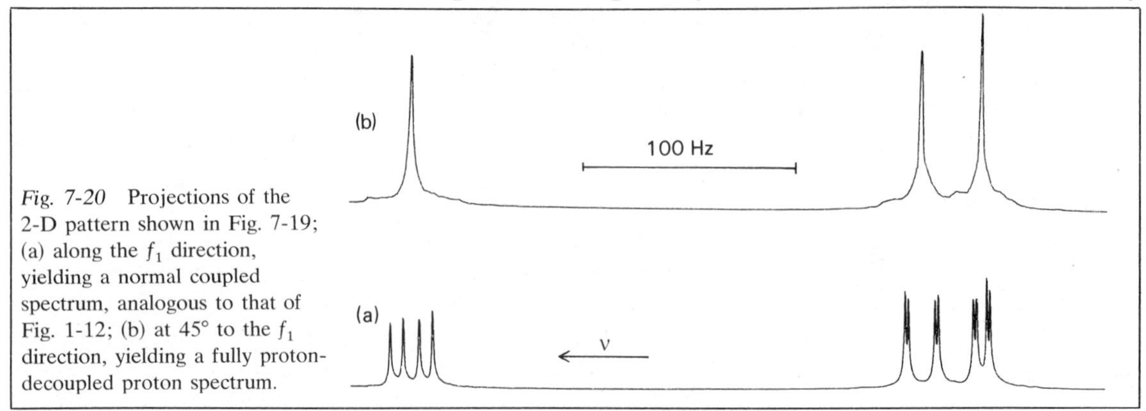

Fig. 7-20 Projections of the 2-D pattern shown in Fig. 7-19; (a) along the f_1 direction, yielding a normal coupled spectrum, analogous to that of Fig. 1-12; (b) at 45° to the f_1 direction, yielding a fully proton-decoupled proton spectrum.

proton-decoupled proton spectrum, which can render proton spectra as simple to interpret (one peak for each type of proton, in the absence of ^{19}F, ^{31}P etc.) as noise-decoupled carbon spectra. This is illustrated in Fig. 7-20. A skew projection results in scaled-down splittings due to coupling and can usefully simplify complex spectra while retaining information on multiplicities. Simple homonuclear 2-D J-resolved spectra are only obtained under first-order conditions of coupling.

7-9 Proton/carbon chemical shift correlation

In order to interpret NMR spectra of organic molecules fully it is frequently desirable to relate the peaks in the carbon and hydrogen regions, e.g. for a directly bonded ^{13}C–H pair of nuclei. This may be done by selective decoupling experiments, but must be repeated for each band in the proton spectrum, and the whole procedure is tedious and time-consuming for a moderately complex organic molecule. This section describes a two-dimensional method of achieving full correlation of all peaks.

Consideration needs to be given first to the effects of a $90°(^1H) - t_1 - 90°(^1H)$ pulse sequence on a coupled pair of ^{13}C and ^1H nuclei with proton shift (offset from the carrier frequency) $\Delta\nu_H$. The two magnetization vectors $M_H^{C\alpha}$ and $M_H^{C\beta}$ need to be considered separately. If $t_1(\Delta\nu_H + \frac{1}{2}J_{CH}) = n$, where n is an integer, then $M_H^{C\beta}$ will precess by an integral number of revolutions (in the rotating frame of reference) between the pulses. The net effect, therefore, will be equivalent to that of a 180° pulse, i.e. $M_H^{C\beta}$ will be inverted. Similarly, $M_H^{C\alpha}$ will be inverted for $t_1(\Delta\nu_H - \frac{1}{2}J_{CH}) = n$. Due to the difference, J_{CH}, these special times act to give selective magnetization inversion for one or other of the ^1H doublet lines. As discussed in Sections 7-4(d) and 7-5, this results in cross-polarization (otherwise known as magnetization transfer) and the intensities of the doublet lines in the ^{13}C spectrum, as would be monitored by a 90°(C) pulse, are affected. This provides a link between the ^1H and ^{13}C spectra, with variation of t_1 modulating the appearance of the ^{13}C spectrum. A 2-D experiment of this type would give a plot of the coupled ^1H spectrum vs. the coupled ^{13}C spectrum, but with phases and intensities of multiplet lines governed by the considerations discussed in Section 7-4(d). Clearly, multiplet splittings cannot be eliminated by decoupling during acquisition, since mutual cancellation of component signals (which are in antiphase) would occur.

A way forward is provided by the sequence of Fig. 7-21. The first 180° pulse in the ^{13}C channel interchanges the spin labels of $M_H^{C\alpha}$ and $M_H^{C\beta}$ so that they are refocused over the interval t_1. The time τ_1 is set to $[2J_{CH}(\text{ave})]^{-1}$ so that $M_H^{C\alpha}$ and $M_H^{C\beta}$ are 180° out of phase at time $t_1 + \tau_1$. The second $90°(^1H)$ pulse will then cause cross-polarization, as discussed above, for appropriate values of t_1 only, such that $(t_1 + \tau_1)$ $\Delta\nu_H = n$. The $90°(^{13}C)$ pulse will produce antiphase signals for $M_C^{H\alpha}$ and $M_C^{H\beta}$ (for these special times t_1 only), and the second interval $\tau_2 = [2J_{CH}(\text{ave})]^{-1}$ brings these magnetization components in phase

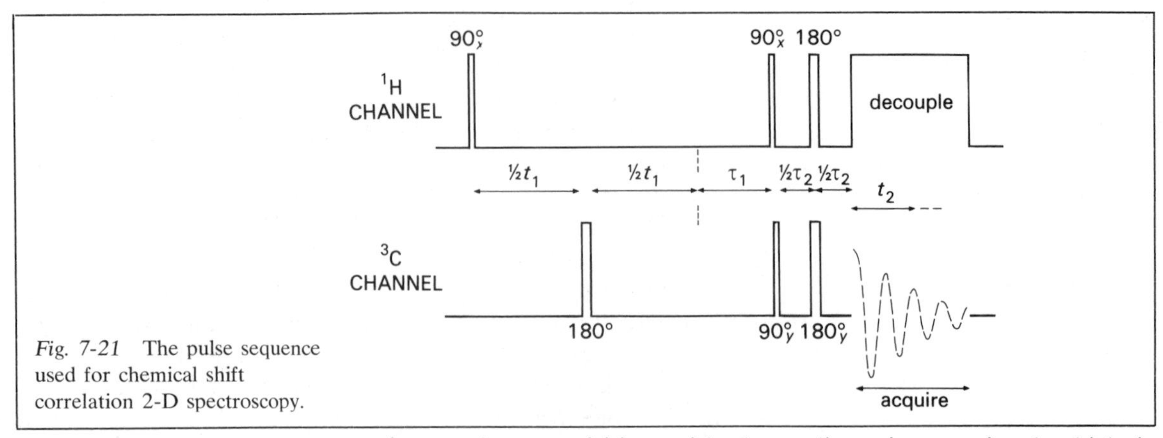

Fig. 7-21 The pulse sequence used for chemical shift correlation 2-D spectroscopy.

again, so that acquisition with decoupling gives a signal which is enhanced by γ_H/γ_C. Usually it will be required to cross-polarize CH_3, CH_2 and CH groups simultaneously. Since the ^{13}C component magnetization vectors dephase at different rates for these groups, a compromise for times τ_1 and τ_2 is required. The $180°(^1H)/180°(^{13}C)$ pair of pulses in the middle of the interval τ_2 serves to refocus ^{13}C chemical shifts. Since the FID depends on t_1 as well as t_2, double Fourier transformation yields a plot which is the decoupled ^1H spectrum (since $M_H^{C\alpha}$ and $M_H^{C\beta}$ are refocused over the time t_1 by the $180°(^{13}C)$ pulse) vs. the decoupled ^{13}C spectrum, thus correlating δ_C and δ_H. If tetramethylsilane is added to the solution it forms a

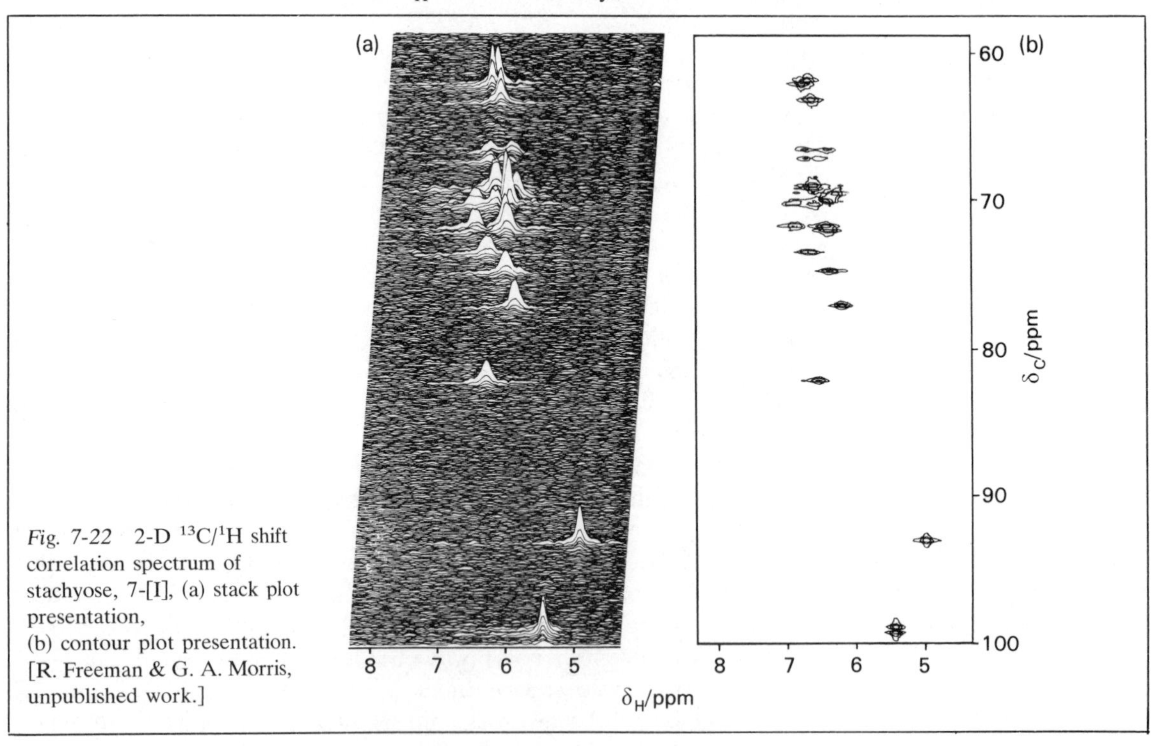

Fig. 7-22 2-D $^{13}C/^1H$ shift correlation spectrum of stachyose, 7-[I], (a) stack plot presentation, (b) contour plot presentation. [R. Freeman & G. A. Morris, unpublished work.]

convenient reference in both frequency dimensions. Figure 7-22 gives an example of this type of 2-D NMR, for the molecule stachyose, 7-[I]. Long-range (C,H) couplings may be used as a basis for the magnetization transfer if τ_1 and τ_2 are suitably adjusted.

7-[I]

Notes and references

[1] R. Freeman and H. D. W. Hill, *J. Chem. Phys.*, **54,** 301 (1971).

[2] An analogous effect is obtained if the time intervals are doubled and the 180° pulse in the ^{1}H channel is replaced by decoupling for the second time interval as well as for the acquisition of the ^{13}C signal.

[3] M. R. Bendall, D. M. Doddrell & D. T. Pegg, *J. Amer. Chem. Soc.*, **103,** 4603 (1981).

[4] G. A. Morris & R. Freeman, *J. Magn. Reson.*, **29,** 433 (1978).

[5] S. Forsèn & R. A. Hoffman, *J. Chem. Phys.*, **39,** 2892 (1963); **40,** 1189 (1964).

[6] This can also be achieved by a selective *carbon* pulse following the selective proton pulse [R. A. Craig, R. K. Harris & R. J. Morrow, *Org. Magn. Reson.*, **13,** 229 (1980)]—the result is identical for an AX_n system with $n = 1$, though not for $n > 1$.

[7] G. A. Morris and R. Freeman, *J. Amer. Chem. Soc.*, **101,** 760 (1979);
G. A. Morris, *J. Amer. Chem. Soc.*, **102,** 428 (1980);
D. P. Burum and R. R. Ernst, *J. Magn. Reson.*, **39,** 163 (1980).

[8] D. M. Doddrell, D. T. Pegg & M. R. Bendall, *J. Magn. Reson.* **48,** 323 (1982).

[9] J. Jeener, Ampère International Summer School, Basko Polje, Yugoslavia (1971), unpublished.

Further reading

'Two-dimensional Fourier transformation in NMR', R. Freeman & G. A. Morris, *Bull. Magn. Reson.*, **1,** 5 (1979).

'Spin-echo Fourier transform NMR spectroscopy', D. L. Rabenstein & T. T. Nakashima, *Anal. Chem.* **51,** 1465A (1979).

Two-dimensional NMR in Liquids, A. Bax (D. Reidel Publishing Co., Dordrecht, Netherlands) (1982).

Problems

7-1 Draw diagrams analogous to those of Fig. 7-2 to explain the phase modulation for an AX_2 spin system, as illustrated in Fig. 7-3.

7-2 Obtain a diagram analogous to those of Fig. 7-3 but for an AX_3 spin system. Hence, plot the intensity observed for the carbon of a methyl group subject to the pulse sequence of Fig. 7-5 as a function of τ.

7-3 By considerations analogous to those of Section 7-4(d) for an AX spin system, use appropriate energy level diagrams to show that the INEPT pulse sequence (Fig. 7-15) acting on AX_2 and AX_3 spin systems gives 1:0:−1 and 1:1:−1:−1 multiplets, respectively, with enhancements $2\gamma_X/\gamma_A$ and $3\gamma_X/\gamma_A$, respectively, when A magnetization is suppressed by alternating the sign of the phase of the $90^\circ_y(X)$ pulse.

7-4 Obtain a general expression for the A multiplet intensities of an AX_n spin system for a coupled INEPT experiment (assuming suppression of the A magnetization). Hence derive an equation for the signal intensity for such a system in a decoupled INEPT experiment (Fig. 7-17) as a function of the time interval τ_2. A discussion relating to this question is to be found in D. M. Doddrell, D. T. Pegg, W. Brooks & M. R. Bendall, *J. Amer. Chem. Soc.*, **103,** 727 (1981).

8 Chemical shifts and coupling constants

8-1 Introduction

The values of chemical shifts and coupling constants are often used in an entirely empirical way for structure determination. The organic chemist makes frequent use of such facts as (a) methyl protons of CH_3—CH_2—C groups give rise to resonances near δ 1·0, or (b) proton–proton coupling constants for *vicinal* protons *trans* across C=C are greater than those for corresponding *cis* protons. However, this chapter will not be concerned with such applications, but rather will discuss the theoretical and physical attempts to understand the nature of the nuclear spin parameters. It will be assumed that the student will become familiar with the correlative use of proton and carbon chemical shifts and coupling constants through courses and practical work in organic chemistry; the book by Jackman and Sternhell given in the reading list at the end of this chapter deals with this aspect of NMR.

The energies involved in NMR are extremely small, but the spin parameters may be measured quite accurately, especially for the solution state. The nuclei may be regarded as probes in order to obtain detailed information about molecules. Both chemical shifts and coupling constants depend on the distribution of electron density near the nuclei and consequently they provide data on bond hybridization, bond dipole moments and susceptibilities, and even on electronic energies. The way in which electron density is affected by substituents and by interactions with solvent can also be investigated. A considerable mass of information has been accumulated in this way, and, whereas it is not yet possible to calculate chemical shifts and coupling constants *ab initio* with accuracy for any but the simplest molecules, many of the factors affecting these parameters are well known and predictions for new systems are often possible. Much of the discussion of this chapter will be about proton data, primarily because there have been more experimental NMR studies for protons than for any other nucleus. In general, only the parameters as averaged over molecular tumbling in the liquid or solution phase will be considered. This is the most common experimental situation, but it is important to remember that both chemical shifts and coupling constants are in principle anisotropic quantities, providing yet more detailed information about the molecules. Theoretically, the averaging process is incorporated by calculating σ or J for three Cartesian directions, adding these three contributions together, and dividing by 3, so that factors of $\frac{1}{3}$ appear in many equations, e.g. Eqns 8-17 and 8-33. Moreover, certain terms have zero contribution *on average*, which greatly simplifies the situation (see Section 4-2). In special situations, such as when liquid crystal solvents are used, partial orientation of the molecules occurs and the averaging process is not complete. It is usually only in such cases that the absolute signs of coupling constants can be determined; only the relative signs can be determined by measurement of absorption frequencies and intensities when there is no preferred orientation of the molecules.

PART A CHEMICAL SHIFTS

8-2 Atomic shielding

As discussed in Section 1-8, an external magnetic field induces motion of electrons such that a secondary magnetic field is produced which opposes the primary field at the centre of the motion (a diamagnetic effect) and therefore *lowers* the frequency needed for resonance. Such motion may be described as an effective rotation of the whole electronic cloud about the direction of $\mathbf{B}_0$. The rotation occurs at an angular velocity $\boldsymbol{\omega}_i$ given by

$$\boldsymbol{\omega}_i = \frac{e}{2m_e}\mathbf{B}_0 \qquad (8\text{-}1)$$

Now such an electronic motion is related to a current density $\mathbf{j}$ (see Fig. 8-1):

$$\mathbf{j} = -e(\boldsymbol{\omega}_i \times \mathbf{r})\rho_e = -\frac{e^2}{2m_e}(\mathbf{B}_0 \times \mathbf{r})\rho_e \qquad (8\text{-}2)$$

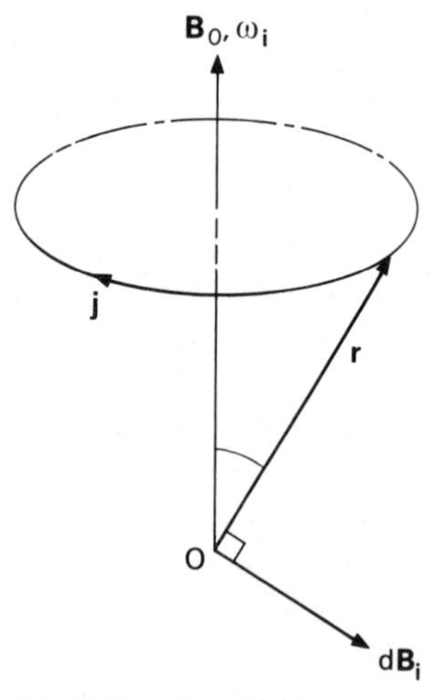

Fig. 8-1 Diamagnetic circulation of the electronic charge cloud under the influence of a magnetic field B_0.

In this equation $\boldsymbol{\omega}_i \times \mathbf{r}$ is the velocity of the electrons, $\mathbf{r}$ is the distance from the nuclear origin to the element of current density under consideration, and ρ_e is the charge density in units of the electronic charge. Such a current density produces an element $\mathrm{d}\mathbf{B}_i$ of magnetic field at the origin, O, which is given by Biot and Savart's Law (note: we have defined $\mathbf{r}$ in the opposite direction to that normally used in Biot–Savart calculations) as

$$\mathrm{d}\mathbf{B}_i = -\frac{\mu_0}{4\pi r^3}(\mathbf{j} \times \mathbf{r})\,\mathrm{d}V \qquad (8\text{-}3)$$

where μ_0 is the permeability constant, and $\mathrm{d}V$ is the element of volume with current density $\mathbf{j}$. Substitution for $\mathbf{j}$ gives

$$\mathrm{d}\mathbf{B}_i = \frac{\mu_0 e^2}{8\pi m_e r^3}(\mathbf{B}_0 \times \mathbf{r}) \times \mathbf{r}\rho_e\,\mathrm{d}V \qquad (8\text{-}4)$$

If $\mathbf{r}$ is at an angle θ to $\mathbf{B}_0$ (Fig. 8-1) this reduces to

$$\mathrm{d}B_i = \frac{\mu_0 e^2}{8\pi m_e r} B_0 \rho_e \sin\theta\,\mathrm{d}V \qquad (8\text{-}5)$$

This field has a component along $\mathbf{B}_0$ of $-\mathrm{d}B_i \sin\theta$ (components perpendicular to B_0 are zero on average). Integration over all space gives the total component of the induced field

$$B_{iz} = -\frac{\mu_0 e^2 B_0}{8\pi m_e} \int \left(\frac{\rho_e}{r}\right) \sin^2\theta \, \mathrm{d}V \tag{8-6}$$

The diamagnetic contribution, σ_d, to the shielding constant is therefore (see Eqn 1-27)

$$\sigma_d = \frac{\mu_0 e^2}{8\pi m_e} \int \left(\frac{\rho_e}{r}\right) \sin^2\theta \, \mathrm{d}V \tag{8-7}$$

Transformation to polar coordinates, followed by integration (Problem 8-1) under the assumption of a spherical charge distribution (valid for atoms) gives

$$\sigma_d = \frac{\mu_0 e^2}{3m_e} \int_0^\infty r\rho_e \, \mathrm{d}r \tag{8-8}$$

This result is known as Lamb's formula; an equivalent expression may be derived quantum mechanically. It may be evaluated if the electron distribution is known. For atoms it gives the *total* shielding constant. Some values are given in Table 8-1. For the hydrogen atom σ is small, but the value increases rapidly and monotonically with the nuclear atomic number (since the number of electrons increases). However, in NMR experiments we are only concerned with *differences* in screening constants between atoms of the same species in different molecules. Such differences may be expected to be very small fractions of the total diamagnetic effect.

Table 8-1 Free atom values of shielding[(a)]

Atom	H	Li	C	N	O	F	Si	P
σ_d/ppm	17·8	101·5	260·7	325·5	395·1	470·7	874·1	961·1
Atom	Mn	Co	Se	Mo	Xe	Pt	Tl	Pb
σ_d/ppm	1942·1	2166·4	2998·4	4000·6	5642·3	9395·6	9894·2	10 060·9

[(a)] Mostly taken from G. Malli & C. Froese, *Int. J. Quant. Chem.*, **1S** 95 (1967).

8-3 Molecular shielding: the paramagnetic term

In molecules the presence of the other nuclei hinders rotation of the electron cloud about the nucleus whose shielding is being considered (indeed, most electrons will probably be primarily associated with other nuclei and bonds). This statement, though oversimplified, indicates that in molecular systems there exists, in addition to the Lamb term, another contribution to shielding which normally opposes the diamagnetic effect and is therefore known as the paramagnetic[1] term.

The paramagnetic contribution to shielding may be estimated by finding the way in which the electronic wave functions are modified by the magnetic field. When a field is applied, new energy terms appear, and the electronic wave functions calculated in the absence of the field are no longer eigenfunctions of the electronic Hamiltonian. The new electronic wave functions may be obtained as linear combinations of the eigenfunctions for the field-absent case. Thus the field can be considered to mix 'excited' electronic states into the ground state wave function. The principal extra energy term which appears when the field is applied is the Zeeman energy of interaction of the electron orbital magnetic moment, $\boldsymbol{\mu}$, with the field, and it is given classically by Eqn 1-13. This energy is very much smaller than the total electronic energy and so the way in which it modifies the wave functions may be derived using perturbation theory. The resulting expression for the paramagnetic contribution to shielding (σ_p) involves a summation over 'excited' electronic states, with the energies of excitation in the denominator. The full equation, due to Ramsey, is rather complex and will not be given here; there is an analogous term in the Van Vleck formula for magnetic susceptibility.

The shielding constant for molecules may be expressed as the sum of diamagnetic and paramagnetic parts (the former analogous to the Lamb term for atoms).

$$\sigma = \sigma_d + \sigma_p \tag{8-9}$$

Evaluation of σ_p requires detailed knowledge of excited state wave functions, which is often not available. Nonetheless, accurate calculations have been made for a few simple molecules, such as H_2 and LiH. Moreover, some general comments are useful. The paramagnetic term vanishes for *s* electrons, since these are spherically symmetrical with zero orbital angular momentum, and they therefore have no magnetic moment to contribute as in Eqn 1–13. On the other hand, σ_p may be large (a) for an asymmetric distribution of *p* and *d* electrons near the nucleus in question, and (b) if there are low-lying excited singlet states of the correct symmetry. Clearly, both σ_p and σ_d will differ for different nuclei in the same molecule, owing to the uneven distribution of electron density. They also depend on the orientation of the molecule relative to the magnetic field, i.e. σ is an anisotropic property, but we will only consider the average obtained due to random molecular tumbling.

8-4 Shielding constants for large molecules

For large molecules σ_d and σ_p tend to become very large but of opposite sign, so that if they are calculated as indicated in Sections 8-2 and 8-3 there is a high uncertainty in the net σ. Exact calculations of σ are no longer feasible. However it proves possible to make use of alternative methods and/or semi-empirical calculations. The total screening constant may be written as a sum of local effects and long-range or molecular effects. These are further broken down into separate terms as follows

$$\sigma = \sigma_d(\text{local}) + \sigma_p(\text{local}) + \sigma_m + \sigma_r + \sigma_e + \sigma_s \tag{8-10}$$

where σ_m = neighbour anisotropy effect
σ_r = effects of ring currents
σ_e = electric field effects
σ_s = solvent (or medium) effects

This scheme is arbitrary but useful, since it does not suffer from drawbacks inherent in the Ramsey approach. The effects of σ_d (local) and σ_p (local) are similar in concept to the diamagnetic and paramagnetic terms already discussed, except that only electrons closely associated with the nucleus whose shielding constant is being estimated are considered. However, direct comparisons between the terms of the Ramsey expression 8-9 and those of Eqn 8-10 are to be avoided. Examples of the variations in σ_d(local) and σ_p(local) will now be considered and the remaining contributions to σ will be discussed later. It is pertinent to note at this point that σ_m is a *direct* effect on the magnetic field, whereas σ_e is an *indirect* effect (in that it acts to alter the local terms). The term σ_s contains in principle contributions of the same type as σ_m, σ_r, and σ_e but they are of *inter*- rather than *intra*-molecular origin.

The contributions σ_m and σ_r (plus analogous effects subsumed into σ_s) are in principle independent of the nucleus being shielded (though not, of course, of its position in the molecule). Therefore although they may be proportionately important for light atoms (especially hydrogen), where the total shielding is small, they are generally insignificant for heavier nuclei. Table 8-2 shows this situation for the relatively light nuclei C, N and O in some compounds involving multiply-bonded atoms. Non-local contributions are even less important for single-bonded atoms and for atoms outside the first row of the Periodic Table.

The shielding range for nuclei varies greatly with their position in the Periodic Table. The normal range for ^{1}H is only ca. 10 ppm, although for relatively unusual chemical situations this may extend up

Table 8-2 Some contributions to shielding (in ppm) for carbon, nitrogen and oxygen[a]

Molecule	*Nucleus*	σ_d(loc)	σ_d(non-loc)	σ_p(loc)	σ_p(non-loc)
CO	C	259·36	0·03	−206·24	−4·92
	O	395·39	0·64	−367·39	−7·25
H_2CO	C	257·77	−0·10	−208·24	5·61
	O	397·80	0·10	−651·47	1·85
HCN	C	259·08	0·11	−157·53	−6·10
	N	326·70	0·06	−301·30	−5·04

[a] Taken from G. A. Webb, Table 3-1 of *NMR and the Periodic Table*, Eds. R. K. Harris & B. E. Mann, Academic Press (1978).

Table 8-3 Chemical shift ranges for diamagnetic compounds

Element	boron	carbon	nitrogen	oxygen	fluorine
Range/ppm[a]		655	930		1313
	(200)	(400)		(700)	(800)

Element	aluminium	silicon	phosphorus	sulphur	chlorine
Range/ppm[a]	270	416	706	600	820
		(150)			

Element	vanadium	cobalt	rhodium	platinum	mercury
Range/ppm[a]	2400	18 350	8760	13 217	3587
				(6000)	

[a] Numbers in brackets refer to typical ranges, whereas numbers without brackets give the extreme ranges.

to ca. 70 ppm. Ranges for other nuclei are given in Table 8-3. Even for first-row elements the ranges are over an order of magnitude greater than for ^{1}H. For certain metals, e.g. ^{59}Co and ^{205}Tl, the range reaches several %. Of course, chemical functionality plays an important role in determining shielding ranges. Elements such as Na and K, which form mainly ionic compounds, tend to have small ranges. Covalent compounds of monovalent elements usually have smaller shielding ranges than for multivalent elements. As might be anticipated, shielding depends heavily on such chemical factors as tendency to form multiple bonds, participation of *d* electrons and ability to exist in several oxidation states. The latter factor is illustrated for ^{129}Xe in Table 8-4. The large shift range for some elements sometimes gives rise to problems of pulse power and/or digitization in the operation of FT NMR spectrometers.

For many elements, the extremes of shielding are given by chemically unusual molecules (such as those with low-lying paramagnetic electronic states). There is also a well-recognised 'heavy atom' effect, resulting in exceptionally high shielding for ^{13}C in CI_4 ($\delta_C = -293$ ppm) and for ^{29}Si in SiI_4 ($\delta_{Si} = -352$ ppm). This effect has been shown[2] to involve electron spin–orbital interactions arising on the heavy atom.

Small influences on the molecular energy states or their occupancy can influence shielding. For example, all shielding constants have an intrinsic temperature-dependence, quite apart from the effects on conformational populations or intermolecular interactions. Moreover, isotopic substitution causes changes in shielding. For instance, ^{19}F in

Table 8-4 Xenon-129 chemical shifts

Compound	Xe	XeF_2	XeF_4	$XeOF_4$	$[XeO_6]^{4-}$
Oxidation number	0	2	4	6	8
δ_{Xe}/ppm	−5331	−1750	+253	0	+2077

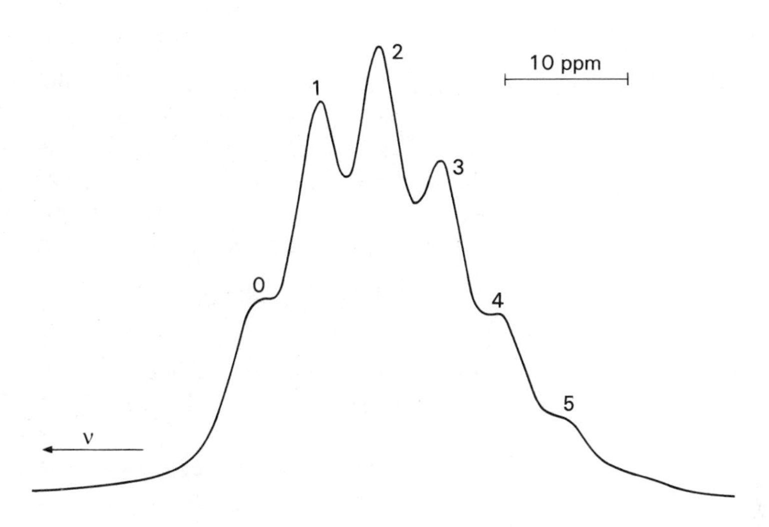

Fig. 8-2 Cobalt-59 NMR spectrum at 23·7 MHz of $[Co(en)_3]Cl_3$ 4·5 h after dissolution at pH 3·03 in a mixture of H_2O and D_2O in the molar ratio 3·03. The different peaks are due to the variety of isotopomers present, and the numbers indicate how many deuterons are in the relevant isotopomer. The peak intensities change as a function of time, allowing the kinetics of the $^1H/^2H$ exchange to be measured. The figure illustrates clearly the isotope shift arising from the $^1H \rightarrow {}^2H$ replacement. The individual lines are ca. 110 Hz in width, giving information about the rate of quadrupolar relaxation.

CF_3I is shielded by 0·149 ppm more in the $^{13}CF_3I$ isotopomer than in $^{12}CF_3I$. Both the intrinsic temperature-dependence of shielding and the isotopic shift are related to the nature of the vibrational energy levels or their occupancy. Both effects are likely to be greatest for elements with large shielding ranges. Thus for ^{59}Co the variation of shielding with temperature (which may, however, include an intermolecular contribution) is typically $d\sigma/dT \sim -1{\cdot}5$ ppm K^{-1}. The isotopic effect of the replacement $^1H \rightarrow {}^2H$ on ^{59}Co shielding in aqueous $[Co(en)_3]Cl_3$ is as high as 5·0 ppm der deuterium (see Fig. 8-2), and is reasonably additive, so that the shift difference between the fully protonated and fully deuterated species is 60·2 ppm. Note that this is for atomic replacements *two* bonds removed from the observed nucleus. Of course, the isotopic shift is greatest when the relative mass change of the replacement is the highest, i.e. for $^1H \rightarrow {}^2H$. It may be noted that the isotopic effects mentioned above are *secondary* in character. *Primary* isotopic shift effects, which would cause differences between the shielding of, say, ^{14}N and ^{15}N in corresponding isotopomers, are generally ignored. However, the ranges of shifts for two isotopes, which are the same in ppm, may be very different in Hz since the operating frequencies are proportional to γ at constant B_0. Thus for $B_0 = 2{\cdot}35$ T a range of 10 ppm for hydrogen represents 1000 Hz for ^1H at 100 MHz but only 153·5 Hz for ^2H at 15·35 MHz.

8-5 The local diamagnetic term

Clearly, if the electron density around the magnetic nucleus is reduced, the shielding due to the local diamagnetic term is decreased. This is important for ^1H resonance, as is shown by the following δ values for methyl protons:

$Si(CH_3)_4$	CH_3CH_3	MeI	MeCl	Me_2O
0·00	0·88	2·16	3·05	3·24

It has, in fact, been proposed that the chemical shift difference between CH_3 and CH_2 proton resonances in compounds of the type CH_3CH_2X be used to predict a substituent electronegativity, E_X, for X. The inductive effect of X affects the CH_3 resonance less than the CH_2 resonance because of the attenuation due to the extra intervening carbon atom. The relationship 8-11 has been suggested.

$$E_X = 0{\cdot}684(\delta_{CH_2} - \delta_{CH_3}) + 1{\cdot}78 \qquad (8\text{-}11)$$

However, such equations need to be used with caution in view of the other contributions to σ listed in Eqn. 8-10.

For nuclei other than ^1H, although σ_d(loc) is large (Table 8-2) its variation with chemical environment is small,[3] so that chemical shifts are usually discussed in relation to σ_p(loc) alone—see below.

8-6 The local paramagnetic term

The local paramagnetic effect depends on the excitation energies of electrons primarily associated with the nucleus under consideration. It produces an induced magnetic moment which reinforces the applied field and therefore leads to *deshielding*, i.e. a high frequency shift in NMR. The chemical shifts of nuclei other than ^1H are usually dominated by σ_p(local). For instance, in the vapour phase the ^{19}F resonance of F_2 occurs 643 ppm to high frequency of that for HF. The difference in σ_d is estimated to be about 50 ppm and therefore cannot account for the observed values. However, σ_p is zero for a spherically symmetric species such as an atom or an ion, e.g. F^-, whereas it is appreciable for an asymmetric distribution of *p* electrons as for covalently-bonded fluorine. There is therefore a strong dependence of σ_p on the ionic character of the bond to fluorine. In the case of HF the paramagnetic term is small (~ -60 ppm), while for F_2 it has been estimated at ~ -700 ppm; the calculated effect is adequate to account for the observed shift between the two species. The importance of variation in σ_p(local) is further illustrated in Table 8-3.

Expressions for σ_p (local) are usually developed from a molecular orbital (MO) approach using a linear combination of atomic orbitals (LCAO). If it is assumed that only *s* and *p* electrons are important it

Table 8-5 Carbon-13 shielding constants with respect to that for benzene[(a)]

Molecule[(b)]	CH_3^*CN	$^*CH_3CN$	HC≡CH	CH_3^*CHO	$^*CH_3CHO$
Calculated[(c)]	89·91	132·64	116·82	38·71	134·77
Experimental[(c)]	76 ± 10	193 ± 14	120 ± 10	−6 ± 10	162 ± 10

(a) K. A. K. Ebraheem & G. A. Webb, *Org. Magn. Reson.*, **9**, 241 (1977).
(b) The asterisk indicates the relevant carbon atom
(c) Given as $\sigma_{mol.} - \sigma_{benzene}$ in ppm.

can be shown that for atom A bonded to a number of other atoms B

$$\sigma_p(\text{local}, A) = -\frac{\mu_0 e^2 \hbar^2}{6\pi m^2} \langle r^{-3}\rangle_{np} \sum_j \sum_k (U_k - U_j)^{-1} \sum_{\ell,m} (c_{\ell Aj} c_{mAk} - c_{mAj} c_{\ell Ak}) \times \sum_B (c_{\ell Bj} c_{mBk} - c_{mBj} c_{\ell Bk}) \quad (8\text{-}12)$$

where $\langle r^{-3}\rangle_{np}$ is the average inverse cube distance of the valence p electron from nucleus A,

$l, m \equiv x$, y and z taken in pairs in cyclic order,
j, k refer to occupied and unoccupied molecular orbitals respectively,
$c_{\ell Aj}$ is the LCAO coefficient of the p_ℓ orbital on A in MO j.

and the summation over B includes A. It may be noted that s electrons do not contribute to the paramagnetic shielding term, and neither do $\pi \rightarrow \pi^*$ transitions. The value of $\langle r^{-3}\rangle_{np}$ is related to the atomic number and to the position of the element in the Periodic Table. It is a maximum for the noble gases and drops to a minimum for each alkali metal. Its variation with atomic number accounts for the fact that shielding ranges for corresponding elements in the first two rows of the Periodic Table are comparable but there are sharp increases for the third and fourth rows. Calculations based on Eqn 8-12 (plus σ_d(local)) give a reasonably satisfactory description of the chemical shifts of the first-row elements. Table 8-5 gives some data for ^{13}C.

The paramagnetic contribution to shielding is particularly marked for transition metal complexes because the summation 8-12 is dominated by the lowest electronic excitation energy, provided the upper state has the correct symmetry properties. Ligand field theory shows this energy difference is often rather small. Thus Table 8-6 lists the

Table 8-6 Calculated and observed ^{59}Co chemical shifts[(a)] for a series of cobalt complexes

Compound[(b)]	*Observed shift, $\Delta\sigma$/ppm*	*Calculated shift, $\Delta\sigma$/ppm*	$h^{-1}c^{-1}\,\Delta U/\text{cm}^{-1}$
$Co(NH_3)_3(NO_2)_3$	−6940	−6000	23 210
$[Co(en)_3]Cl_3$	−7010	−7700	21 400
$Na_3[Co(NO_2)_6]$	−7350	−8500	20 670
$[Co(NH_3)_5NO_2]Cl_2$	−7460	−7300	21 840
$[Co(NH_3)_6]Cl_3$	−8080	−8200	21 000
$[Co(NH_3)_5Cl]Cl_2$	−9070	−11 100	18 720
$Co(acac)_3$	−12 300	−14 000	16 900

(a) Negative shifts indicate decreased shielding with respect to the signal of aqueous $K_3[Co(CN)_6]$.
(b) en = ethylenediamine; acac = acetylacetonate ion.

^{59}Co chemical shifts for a series of cobalt complexes; in general, the shifts correlate with the energy separation, ΔU, between the ground and lowest excited states of the d^6 configuration. Thus the shielding decreases as the ligand-field strength of the ligands decreases, i.e. $|\sigma_p|$ is in the order acac>Cl^->NH_3>en>NO_2^->CN. Values of ΔU and calculated chemical shifts are given in Table 8.6.

A further comment is in order at this stage as to why ^1H shifts have a much smaller range (ca. 20 ppm, with a few exceptions) than for other nuclei (e.g. ca. 800 ppm for ^{19}F). There is a two-fold answer: in the first place, as already mentioned, the total number of electrons in the vicinity of ^1H is smaller than for other nuclei so the magnitude of σ is smaller and consequently *changes* in σ from one molecule to another are also small. Secondly, the lowest excitation energy involved in σ_p for ^1H is of the type $1s \rightarrow 2p$ (which implies a large value of $U_k - U_j$ and a small lower-state LCAO p-orbital coefficient), whereas for ^{19}F there are lower-lying orbitals and the ground-state bonding orbital involves p-electrons; σ_p(local) is thus very small for ^1H but large for ^{19}F. Clearly, σ_p(local) is not the dominant effect for ^1H chemical shifts.

8-7 The average energy approximation

Even in cases where the lowest excitation energy does not necessarily dominate Eqn 8-12 it is common practice to simplify the theory by assuming the existence of a somewhat ill-defined average excitation energy. A further simplification is the restriction of the coefficients in Eqn 8-12 to valence orbitals centred on the atom in question. These approximations are rather crude but they can lead to an understanding of the trends of chemical shifts. The expression for the paramagnetic term becomes

$$\sigma_p(A) = -\frac{\mu_0 e^2 \hbar^2}{6\pi m^2 \Delta U} \langle r_{np}^{-3} \rangle P_u \qquad (8\text{-}13)$$

where P_u is referred[4] to as the p-electron 'imbalance', and is given by:

$$P_u = (p_{xx} + p_{yy} + p_{zz}) - \tfrac{1}{2}(p_{xx}p_{zz} + p_{yy}p_{xx} + p_{yy}p_{zz}) - \tfrac{1}{2}(p_{xy}p_{yx} + p_{xz}p_{zx} + p_{zy}p_{yz}) \qquad (8\text{–}14)$$

where the individual terms are related to charge densities and bond orders, and are given by:

$$p_{\mu\nu} = 2 \sum c_{\mu j} c_{\nu j} \qquad (8\text{-}15)$$

and refer to the valence p-electrons only (a term similar to 8-14 may be added for the d-electrons if desired). The maximum value for P_u is 2, which occurs when two p orbitals are filled ($p_{\mu\mu} = 2$) and one is empty ($p_{\nu\nu} = 0$) or vice versa. The minimum value is, of course, zero.

Expression 8-13 suffices to explain the non-additivity of substituent effects in cases of multiple substitution for a number of nuclei (^{29}Si, ^{31}P, ^{119}Sn).[5] This non-additivity results in what is called a

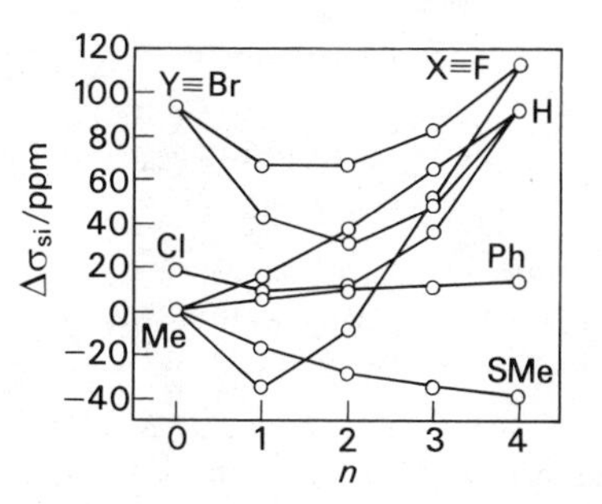

Fig. 8-3 The 'sagging pattern' for ^{29}Si shielding for several series of compounds of general formula SiX_nY_{4-n} [Adapted, with permission, from *NMR and the Periodic Table*, Eds. R. K. Harris and B. E. Mann, Academic Press (1978).]

'sagging' pattern for shielding, which is illustrated for ^{29}Si in Fig. 8-3. The intermediate compound SiX_2Y_2 in the series from SiX_4 to SiY_4 tends to have the largest p-electron imbalance, and therefore has a shielding below that anticipated by additivity. It is logical, on this basis, that the gross 'sagging' occurs when there are large differences in the nature (e.g. electronegativity) of X and Y. The shielding difference between SiX_4 and SiY_4, on the other hand, has to be attributed to variations in ΔU and/or $\langle r_{3p}^{-3}\rangle$ and/or σ_d(local).

The theory developed here helps to explain why when chemical shifts for different elements of the same column of the Periodic Table for corresponding chemical compounds are plotted against one another a straight line results. This has been shown to be true, for instance, for ^{77}Se and ^{125}Te in organoselenium and -tellurium compounds.[6] The linearity may be ascribed to the similarity in the variations of the relevant molecular orbitals. The slope is, however, not unity, but ca. 1·8 in the Se/Te example, which may be explained in part by differences in $\langle r^{-3}\rangle_{np}$ and ΔU for the selenium and tellurium cases.

8-8 Magnetic anisotropy of neighbouring bonds

The magnetic field B_0 will induce magnetic moments in all the electron orbitals of the molecule. Apart from those which are delocalized (Section 8-9), each electron will be associated with an atom or a bond. It is of importance to ask whether induced moments in a group G of bonding or atomic electrons will affect a nucleus X at a distance. Clearly there is such an effect in any given molecular orientation (Fig. 8-4). However, NMR is usually concerned with molecules that are rapidly tumbling, and it can be shown that provided the induced moment in G is constant (isotropic) the effect at X averages to zero. Induced moments are proportional to magnetic susceptibilities; thus, when the magnetic field intensity is along a principal axis (ℓ) of magnetic susceptibility for G, the induced magnetic moment is

$$\mu_{i\ell}(G)=\chi_\ell(G)H_\ell \tag{8-16}$$

where χ_ℓ is the susceptibility (per molecule of G in direction ℓ), and H_ℓ is the field intensity. It is clear that only *anisotropy* in bond susceptibilities causes shielding effects at a distance. In fact, it can be shown that if the induced magnetic moments are considered to act as point

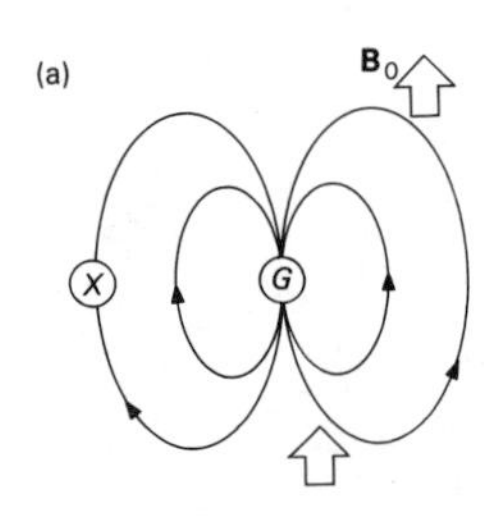

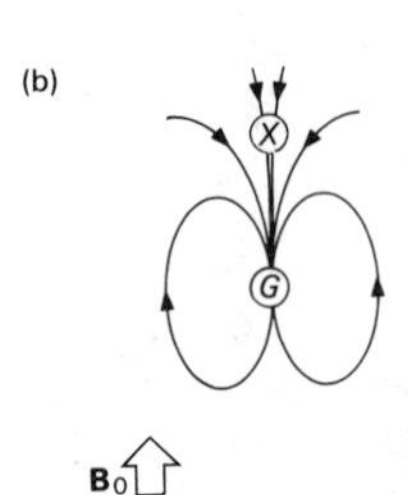

Fig. 8-4 The induced magnetic field produced from an electronic group *G*, and its effect at a distant nucleus *X*:
(a) $r_{G-X} \perp B_0$ (X deshielded);
(b) $r_{G-X} \parallel B_0$ (X shielded).

magnetic dipoles, the contribution to $\Delta\sigma$ from a group G is

$$\Delta\sigma = \frac{1}{3r^3} \sum_{\ell=x,y,z} \chi_\ell(1-3\cos^2\theta_\ell)/4\pi \tag{8-17}$$

where r is the distance between G and the nucleus X, and θ_ℓ is the angle between axis ℓ and r. In most cases the principal axes for bond susceptibilities will be along or perpendicular to the bond; usually the point dipole induced is considered to be at the bond centre. Only bonding electrons and lone pairs need to be considered, since inner-shell electrons associated with individual atoms will have isotropic susceptibility contributions. Equation 8-17 can be shown to reduce to

$$\Delta\sigma = \frac{1}{3r^3}[(\chi_\parallel - \chi'_\perp)(3\cos^2\theta'_\perp - 1) + (\chi_\parallel - \chi''_\perp)(3\cos^2\theta''_\perp - 1)]/4\pi \tag{8-18}$$

where $\chi_\parallel$ is the longitudinal component of susceptibility (i.e. along the bond direction), $\chi'_\perp$ and $\chi''_\perp$ are the two transverse principal susceptibilities (i.e. perpendicular to the bond), while $\theta'_\perp$ and $\theta''_\perp$ are the angles between **r** and the appropriate bond perpendicular. The general nature of the shielding is such that there is a nodal cone (of elliptical cross-section) whose axis is in the direction of one of the principal susceptibilities (see Fig. 8-5). Further simplification occurs if it is considered that the bond is cylindrically symmetrical ($\chi'_\perp = \chi''_\perp$)—in such a case the cross-section of the nodal cone is circular, all directions perpendicular to the axis of the cone are equally shielded and the half-angle of the cone is 54°44′. This will apply to a good approximation to all single and triple bonds, when the cone axis is the bond direction. In such a case

$$\Delta\sigma = \frac{1}{3r^3}(\chi_\parallel - \chi_\perp)(1-3\cos^2\theta)/4\pi \tag{8-19}$$

where θ is the angle between the bond axis and **r**.

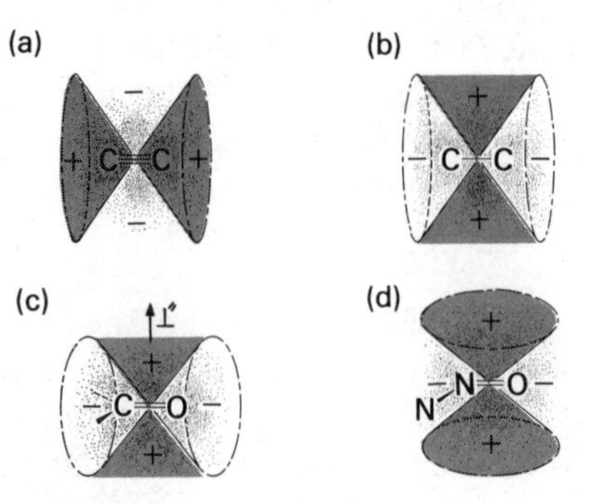

Fig. 8-5 The nodal cones of shielding arising from anisotropy of bond magnetic susceptibility for (a) the C≡C triple bond, (b) the C—C single bond, (c) the carbonyl group, and (d) the nitroso group (in *N*-nitrosamines). The effective centres of the bond magnetic dipoles are not necessarily midway along the bonds. The + and − signs refer to the effect on the shielding constant, σ, of a distant nucleus in the regions of space designated. The magnitude of the effect decreases rapidly as the distance from the bond increases.

Table 8-7 Bond magnetic susceptibility anisotropies[a] and associated shielding effects

	C—H	C—C	C═C[b]	C═O[b,f]	N═O[b,c]	C≡C
$\Delta\chi^{(d)}_{\parallel\perp'}$	90	140	150	280	370	−340
$\Delta\chi^{(d)}_{\parallel\perp''}$	90	140	150	420	1300	−340
$\Delta\sigma^{(e)}_{\parallel}$	−15	−22	−24	−55	−130	54
$\Delta\sigma^{(e)}_{\perp'}$	7·5	11	12	11	−44	−27
$\Delta\sigma^{(e)}_{\perp''}$	7·5	11	12	44	175	−27

[a] The values given are adapted from the research literature and are subject to controversy.
[b] In the cases involving double bonds the $\perp'$ direction is taken to be in the nodal plane of the π electrons.
[c] In *N*-nitroso compounds (ignoring the nitrogen lone pair).
[d] $\Delta\chi_{\parallel\perp} = \chi_{\parallel} - \chi_{\perp}$; $\Delta\chi$ is in units of 10^{-36} m^3 per molecule (alternatively $\Delta\chi$ may be expressed in m^3 per mole). It should be noted that values of χ in SI units are 4π times those in e.m.u., due to rationalization.
[e] $\Delta\sigma$ is in units of $10^{-36}/3(r/\mathrm{m})^3$; $\Delta\sigma_j$ refers to shielding in the direction of axis *j*.
[f] For ketonic C═O groups; probably not appropriate for amides.

Table 8-7 gives some published values for bond susceptibility anisotropies and Fig. 8-5 shows some of the shielding patterns diagrammatically. It may be seen that except for C≡C all the $\Delta\chi$ are positive and there is deshielding along the bond direction; there is shielding in the $\perp''$ direction for all groups involving double bonds. It is assumed that single and triple bonds have cylindrical symmetry; the shielding effect may be almost cylindrically symmetrical for C═C, but is markedly not so for C═O and N═O. In fact for N═O there appears to be deshielding in the $\perp'$ direction, and the nodal cone for shielding, in contrast to the other cases, does not have its axis along the bond direction. The magnitude of the effect is large for the N═O group and accounts for the fact that an observable chemical-shift dependence on stereochemistry is found for protons β and even γ to *N*-nitroso groups when shift contributions from other sources are probably zero due to local symmetry, e.g. in 8-[I]. In less symmetrical situations, it is more difficult to evaluate the contribution of magnetic susceptibility anisotropy. However, the important difference in shielding for axial and equatorial protons of cyclohexane (0·462 ppm, with the equatorial

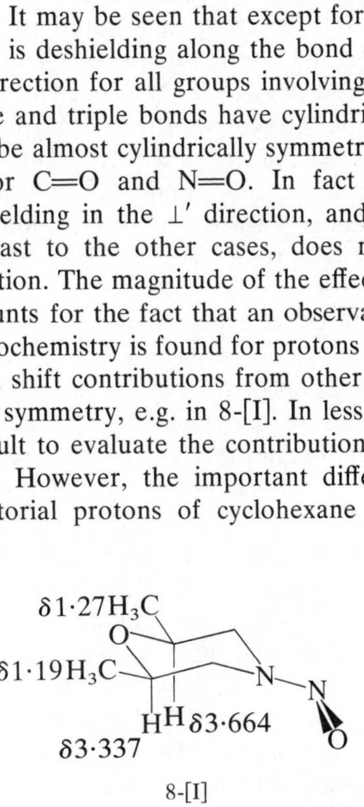

δ1·27H$_3$C
O
δ1·19H$_3$C
N
N
O
H H δ3·664
δ3·337

8-[I]

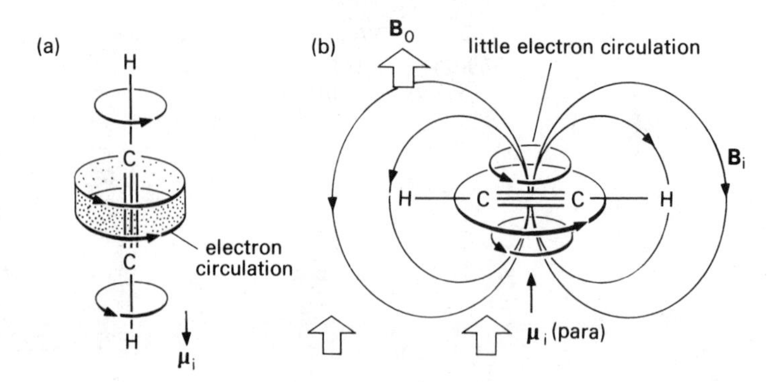

Fig. 8-6 Magnetic anisotropy effects for acetylene: (a) free electron circulation when B_0 is parallel to the C≡C bond, (b) hindered electron circulation when B_0 is perpendicular to the C≡C bond, giving a paramagnetic moment and shielding of the protons.

proton resonance to high frequency) is attributed to the combined effects of magnetic anisotropy of the C—C and C—H bonds.

The anisotropy of bond susceptibilities can be separated into anisotropy of the diamagnetic term and anisotropy of the temperature-independent paramagnetic term. A useful physical picture can be given in some cases, for instance that for acetylene, where the anisotropy is chiefly in the paramagnetic contribution. When B_0 is along the molecular axis there is no hindrance to circulation of the electrons in the C≡C bond; there is thus a zero paramagnetic effect (Fig. 8-6). When B_0 is perpendicular to the axis the nuclei hinder the circulation, and the paramagnetic effect is strong. An alternative (and more rigorous) picture is given by considerations of the mixing of excited states into the ground state by the magnetic field: mixing is possible for B_0 perpendicular to the molecular axis but not for B_0 parallel to the axis. Tumbling averages the paramagnetic term, but as far as the effect of the C≡C electrons on the proton resonance is concerned orientation (b) provides the dominant contribution, and the net result is one of shielding. Clearly this conclusion relies on the anisotropy of the susceptibility of the C≡C bond (see Table 8-7) and is in accord with Fig. 8-5.

This effect explains one apparent oddity in proton chemical shifts—for the series C_2H_2, C_2H_4, C_2H_6. One would expect δ values to decrease in that order due to the inductive effect, since the acetylene protons are most acidic and should be least shielded. However, it is found that in the gas phase the order of shielding is C_2H_4, C_2H_2, C_2H_6, with the shift between C_2H_2 and C_2H_6 (0·6 ppm) much less than that between C_2H_2 and C_2H_4 (3·8 ppm). However, the shielding pattern due to magnetic anisotropy is very different for the three molecules—even the sign of the effect is different for C_2H_2 as compared to C_2H_4 and C_2H_6 (Fig. 8-5). The difference arises because for B_0 parallel to the C—C bond there is hindrance to rotation of the electron cloud for C_2H_4 and C_2H_6 [contrast Fig. 8-6(a)]. Furthermore the protons in C_2H_2 are on a principal axis of susceptibility, which is not the case for C_2H_4 and C_2H_6. In accordance with Eqn 8-19 there is a large low-frequency shift for the acetylene protons.

It may be noticed that although the local effect of the paramagnetic term is deshielding, its long-range effect due to anisotropy may be shielding (and vice versa for the diamagnetic term). An extreme example is provided by the proton resonance of metal hydrides; δ values as low as -40 have been reported. Some representative δ values are as follows

trans-$IrHCl_2(PEt_3)_3$	$-12{\cdot}6$	$ReH_7(PEt_2PH)_2$	$-5{\cdot}8$
cis-$IrHCl_2(PEt_3)_3$	$-21{\cdot}6$	$OsH_2(CO)_4$	$-8{\cdot}7$
trans-$PtHNO_3(PEt_3)_2$	$-23{\cdot}6$	$[RhH(NH_3)_5]SO_4$	$-17{\cdot}1$

These shifts have been used as diagnostic for metal–hydrogen bonds, since there are few other examples of proton Larmor frequencies to low frequency of TMS. This feature correlates with the high-frequency shifts for resonance of the metal itself, already discussed; both effects arise from the paramagnetic shift contribution of the metal electrons. The low-lying *d* electronic levels, distorted from degeneracy by the effect of the ligands, are responsible for the size of the effect.

8-9 Ring currents

When electrons are delocalized, diamagnetic currents may have special importance since the nuclei may not hinder the motion. As the circulations are over large areas, the magnetic moment induced by the field can be quite large. In aromatic systems the delocalized π electrons give rise to a ring current when the field is perpendicular to the molecular plane, as in Fig. 8-7 (though this concept has been criticized). The induced field opposes B_0 at the middle of the molecule but reinforces it at the periphery. Thus the protons in benzene suffer a high-frequency shift—they resonate at $\delta\ 7{\cdot}27$ compared to a value of $\delta \sim 5{\cdot}7$ calculated for a corresponding ring system with alternating single and double bonds. Note that the geometry of the system is such that the diamagnetic circulation of electrons gives an observed high frequency shift. But it is possible in certain cases for protons to be situated 'above' the aromatic plane or even 'inside' the ring current. In such circumstances the ring current gives low-frequency shifts. A classic example is provided by [18]-annulene, 8-[II]. The protons are of two types—the 12 'outer' protons resonate at $\delta\ 8{\cdot}9$; the six 'inner' protons give rise to absorption at $\delta\ -1{\cdot}8$. In fact the proton spectrum was used

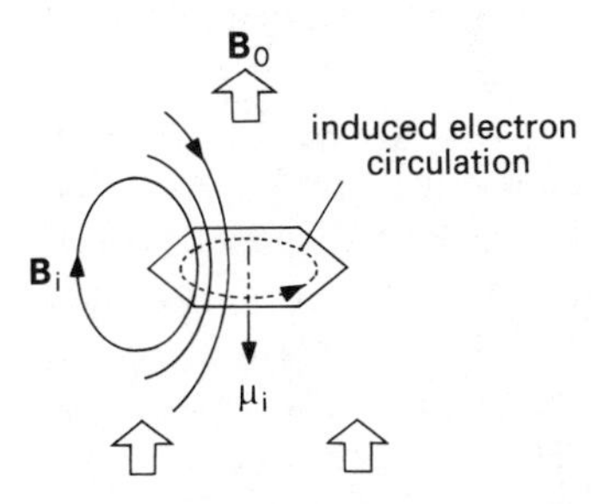

Fig. 8-7 The ring current effect for benzene

8-[II]

8-[III]

8-[IV]

as a proof of the aromaticity of [18]-annulene. There have been suggestions that proton NMR data be used to define ring current, and hence aromaticity, quantitatively. However, such a procedure is very sensitive to the choice of model non-aromatic compounds used for comparison purposes. Similar calculations of the ring currents in benzene, thiophene and furan show that resonance energy, aromatic reactivity and ring current are not related in any simple way.

The ring current does not occur when B_0 is in the plane of the aromatic molecule. Thus the chemical shift for such systems is in principle highly anisotropic; the observed shift in normal high-resolution work is an average over all orientations. Clearly the molecular susceptibility is also highly anisotropic. Indeed, it is only because of this anisotropy that the ring current leads to observable chemical shift effects at a distance; in a sense shifts due to ring currents are analogous to those arising from anisotropy of bond susceptibility, discussed in the preceding section.

The existence of the ring current effect has been used to study molecular conformations. For example, the ethylenic protons of *trans*- and *cis*-stilbene resonate at δ 6·99 and δ 6·49 respectively. This has been attributed to coplanarity of the aromatic rings (and hence a maximum high-frequency ring current shift) for the *trans*-isomer, 8-[III], but tilting of the rings for the *cis*-isomer, 8-[IV].

8-10 Neighbouring group electric dipoles

If a molecule contains strongly polar groups, there will be intramolecular electric fields. These will have the effect of distorting the electron density in the rest of the molecule, and will therefore affect screening constants. It has been shown that for shielding of, say, a proton in an axially symmetric situation there are two terms contributing to σ_e.

$$\sigma_e = -AE_z - BE^2 \tag{8-20}$$

where z is the direction of the bond to the hydrogen atom, and A and B are constants. The first term may be simply related to drift of electrons along the bond direction (see Fig. 8-8)—the minus sign in Eqn 8-20 takes account of the fact that electron drift occurs against the direction of $\mathbf{E}$. The second term can be viewed as due to an effect on σ_p(loc). Consider, for example, an atom: an electric field will distort the electronic environment of the nucleus from spherical symmetry and will therefore introduce a paramagnetic shift—the extent of this effect must depend on the square of $\mathbf{E}$ since the symmetry of the situation requires that a change in direction of $\mathbf{E}$ should not affect σ. Although at this point intramolecular electric fields are being considered, it should be pointed out that similar effects may be obtained in principle from externally applied electric fields. Normally the term in E_z is dominant for proton resonance, and it can either decrease or increase the shielding depending on the sign of E_z. However, the second term always reduces the shielding. In calculations of σ_e it is usual to use a point-dipole approximation for the bond electric moments in order to

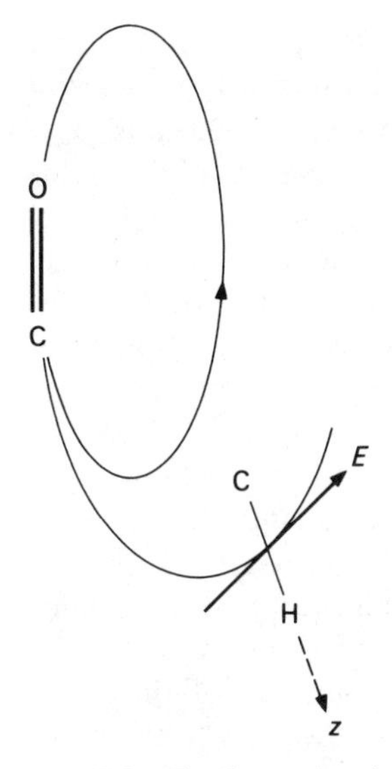

Fig. 8-8 The linear electric field effect on the chemical shift of a C—H proton due to a distant carbonyl group. Note that electrons drift towards the tail of the electric field vector **E**, and therefore, since in this case E_z is negative, the proton is shielded.

calculate E_z and E^2. For the nitrobenzenes, σ_e may be up to 0·8 ppm. Actually, if exact electronic wave functions were known there would be no need to invoke an intramolecular electric field effect, since a full direct calculation of σ_d and σ_p would suffice; however, Eqn 8-20 is necessary as soon as the approximation of separating local and long-range shielding contributions (as discussed in Section 8-4) is used.

The constants A and B differ for different nuclei; in fact for ^{19}F the second term may be more important than the first. Moreover a third term, due to time-dependent electric dipoles mutually induced between a bond X—Y and the bond to the shielded nucleus, is needed. Such dipoles give values of E_z that average to zero, but there is a non-zero average of E^2, written $\langle E^2 \rangle$. This effect has essentially the same origin as van der Waals forces between molecules, and the shielding contribution is therefore known as the van der Waals shift. It has been shown that $\langle E^2 \rangle$ depends on the polarizability and first ionization potential of the electrons in the X—Y bond; there is also an r^{-6} dependence on the distance from the centre of the time-dependent dipoles in the X—Y bond to the shielded nucleus. Chemical shifts between fluorine nuclei in closely related molecules may be dominated by differences in σ_e—for example, the shifts for chloroperfluorobenzenes from the resonance of perfluorobenzene, where the van der Waals contributions are up to 35 ppm, and greatly outweigh the E^2 effect of permanent intramolecular electric fields. The differences in this case derive from replacing F by the more polarizable Cl atom. The electric field term accounts for the so-called '*ortho* effect' (the large low-frequency shift associated with the replacement of F by Cl *ortho* to the fluorine nucleus under study). For other series of compounds, a better fit of calculated to observed parameters has been achieved by empirically using different coefficients for the E^2 and $\langle E^2 \rangle$ terms.

8-11 Hydrogen bonds

Some of the highest-frequency proton chemical shifts recorded are those of strongly hydrogen-bonded protons, both in intramolecular and in intermolecular systems. For example, the hydroxyl proton in the enol form of acetylacetone, 8-[V], resonates at $\delta = 15{\cdot}4$. The effect on the proton chemical shift of forming a hydrogen bond cannot be explained on a simple basis; naturally, all the contributions discussed in previous sections must be considered. The electron density in the vicinity of the proton is probably increased by hydrogen-bond formation (though an electrostatic model indicates the reverse)—it is well known that for an O—H · · · O bond the O, O distance is normally less

H_3C CH_3 O H O

8-[V]

Me N H CCl_3

8-[VI]

than the van der Waals diameter of oxygen, so the proton lies within the electron cloud of *both* oxygen atoms. Such an increase in electron density should lead to an increase of σ_d(loc). Other contributions must therefore predominate in order to bring about the substantial deshielding observed. The additional restraint placed on field-induced electronic circulations by the nuclei of the hydrogen bond acceptor may lead to an increase in σ_p(loc), giving a high-frequency shift. In addition, effects from susceptibility anisotropy of the acceptor group and of its electric field (including the van der Waals term) are probably substantial.

NMR has been used to measure the strength of hydrogen bonds, but considerable caution has to be used in arriving at quantitative conclusions. High-frequency shifts may, however, be used as criteria for the formation of weak hydrogen bonds. For example, the high-frequency shift for the chloroform proton when in association with *N*-methylpyrrolidine has been attributed to the formation of the weak hydrogen bond complex 8-[VI]. The resonance of the $CHCl_3$ proton of the complex was found by extrapolation to be 2·05 ppm from that of the free chloroform, and the mole-fraction equilibrium constant for complex formation was determined as 2·2.

By observing chemical shift changes between a pure liquid and a dilute solution in an inert solvent such as CCl_4, NMR provides a useful tool for distinguishing between intra- and inter-molecular hydrogen bonding. Only small changes would be observed in the case of intramolecular bonding, but changes of ca. 10 ppm may be found for intermolecular bonding. For example the hydroxyl proton resonance for liquid ethanol occurs to high frequency of the CH_2 proton resonance, but for a dilute solution in $CDCl_3$ the hydroxyl resonance is between those of the CH_2 and CH_3 protons (Fig. 8-9).

The situation for hydroxyl proton resonances (and for NH and NH_2 proton resonances) is complicated by the possible occurrence of non-hydrogen-bonded species. In principle there are frequently several different resonances for OH protons over a considerable range of δ values. However, there is usually a rapid exchange of protons between the various environments, and only a single signal is seen at an average position (see Chapter 5). The position of hydroxyl proton resonances

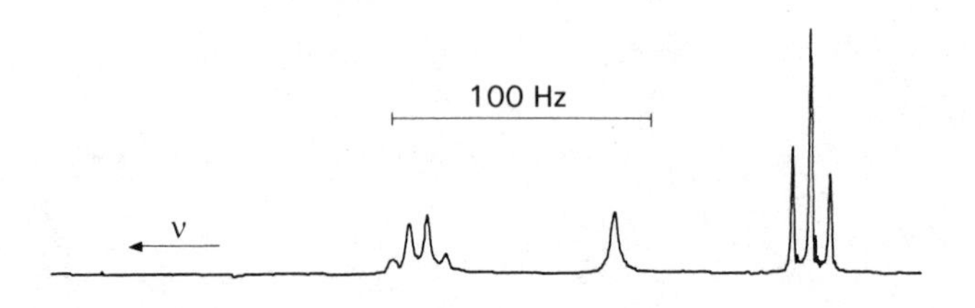

Fig. 8-9 60 MHz proton NMR spectrum of a dilute solution of ethanol in $CDCl_3$. Note that $CDCl_3$ is not a hydrogen bond acceptor. The spectrum should be compared with those of Fig. 5-1. Chloroform tends to become acidic so that 'chemical exchange decoupling' is induced, as well as break-up of the ethanol intermolecular hydrogen bonds.

will therefore vary greatly because of the variable extent of hydrogen bonding. Since the Larmor frequencies of hydrogen-bonded and non-hydrogen-bonded species may in principle be very different, the averaging process may not be sufficiently rapid to give a sharp line. Hydroxyl proton resonances are therefore frequently rather broad.

8-12 Solvent effects

Clearly, solvent molecules can affect resonance positions of a solute nucleus (and vice versa) by the same types of mechanism as the intramolecular effects already discussed. It is important at the outset to realize that when an internal reference standard such as TMS is used, measured solvent effects actually represent the difference of such effects on the solute and on the standard. Only solvent molecules near the solute or standard will make important contributions. Solvent effects (which in some situations are better referred to as *medium* effects) will be considered under three categories:

(a) The polar solvent effect This is an electric field effect. A polar solute, such as acetonitrile, will set up a reaction field in a polar solvent (Fig. 8-10). This will occur even for solute molecules with no net dipole moment if they contain polar groups [Fig. 8-10(b)]. The reaction field then causes electron drift in the solute; for polar solutes this occurs so as to reinforce the shielding or deshielding inductive effect associated with the original electric dipole. This results in changes in shielding. There is also the term in E^2 (Section 8-10) to be taken into account. Thus polar solvents usually tend to augment the inductive effect of electronegative substituents in solutes. In general, protons are at the positive end of molecular or bond dipoles and the action is one of deshielding for solute protons, as in the case of acetonitrile. An isotropic molecule, such as is usually chosen for a standard (e.g. TMS) will be unaffected, except by the van der Waals term.

(a)

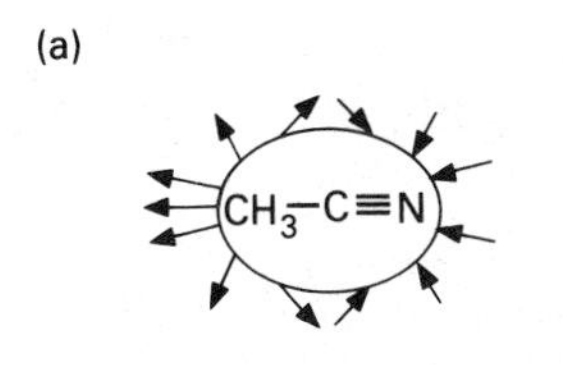

(b)

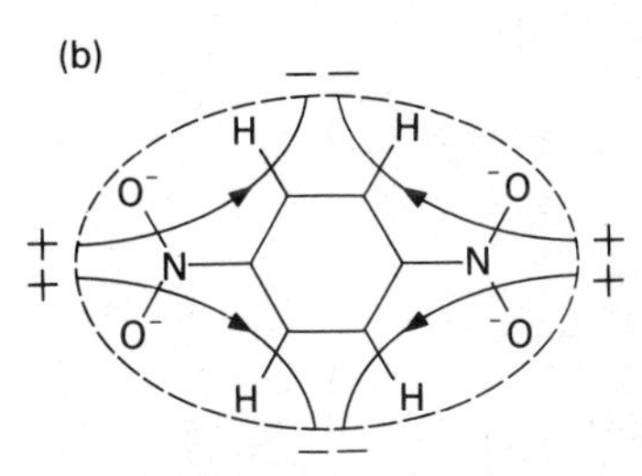

Fig. 8-10 The polar solvent effect: (a) preferred orientation of solvent dipoles around an acetonitrile solute molecule (diagrammatic). The dipoles are drawn with the positive charge at the head of the arrow. The reaction field of the solvent cage acts to augment the electron drift associated with the dipole of the acetonitrile molecule. (b) the reaction field (denoted by the lines with arrows) associated with the quadrupolar molecule *p*-dinitrobenzene.

(b) Solvent magnetic anisotropy Molecules which have a pronounced anisotropy of magnetic susceptibility will cause shifts when acting as solvents. For example, benzene, due to the ring current effect, causes shielding or deshielding at a distance. Of course, for solutes in benzene solution the effect is averaged over all possible solvent–solute molecular orientations, but because of the disk shape of benzene there is still a net effect, even for spherically symmetrical molecules such as TMS. The effect will not vary with the position in the solute molecule of the nucleus whose shielding is considered, but the magnitude of the shift will differ for different symmetrical solutes. Differential shifts will therefore be observed when TMS is used as a standard, but such effects are probably small.

However, most solutes are not spherically symmetrical, and consequently further more marked differential shifts will occur. Moreover, in this case chemically distinct protons in the same molecule will be affected differently. Note that this may imply that there is a certain preferred orientation between solute and solvent, but it does not necessarily imply any positive attraction between the molecules—there may be simply steric repulsions.

It is frequently possible to make use of such solvent effects to spread out a spectrum and make it amenable to analysis. For this purpose the best solvents are probably benzene (causing general low-frequency shifts relative to TMS due to solute protons coming above or below the benzene plane) or acetone (causing general high-frequency shifts due to solute protons being in the plane of the heavy-atom skeleton of acetone). A good example is δ-1,2,3,4,5,6-hexachlorocyclohexane (8-[VII] gives δ values in benzene solution). Comparison[7] with the chemical shifts in dioxan shows that the resonances of hydrogens A, B, and D have been shifted by ca. 1·0 ppm, but the C resonance by only ca. 0·25 ppm. Clearly, benzene molecules can approach closer to one side of 8-[VII] than to the other (this is reasonably explained

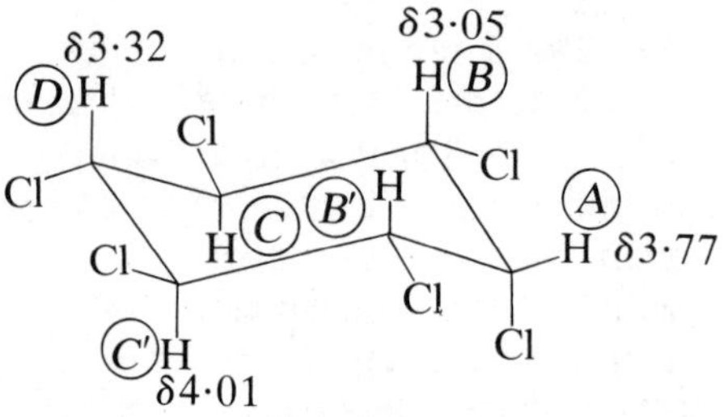

8-[VII]

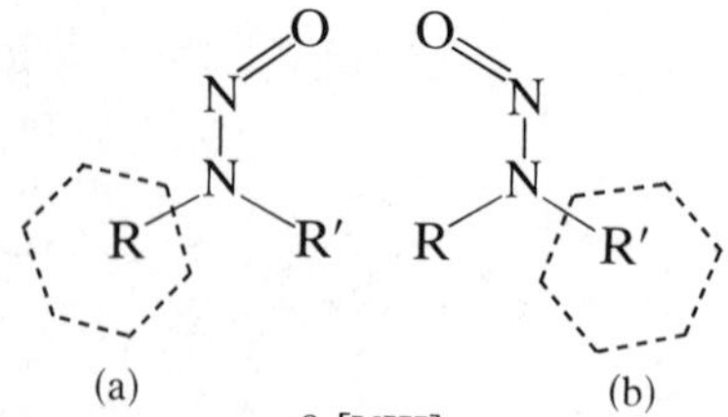

(a) (b)

8-[VIII]

on steric grounds by the effects of the chlorine atoms). This results in the unusual feature of the spectrum of the benzene solution: axial protons (*C*) resonate to high frequency of an equatorial proton (*A*). The spectrum in benzene, being 'spread out', may be analysed resonably accurately on a first-order basis.

Such preferred orientation effects (which will be enhanced if there are weak solvent–solute chemical interactions) can frequently lead to useful empirical correlations. For *N*-nitroso compounds, the benzene ring prefers to lie away from the oxygen atom of the NO group (symbolically the preferred situation is as in 8-[VIIIa]), causing greater low-frequency benzene solvent shifts for protons in groups *trans* to the oxygen atom than for those *cis* to oxygen. This effect may be used to distinguish between resonances due to isomers 8-[VIIIa] and 8-[VIIIb] for $R \neq R'$ in solutions containing mixtures of the two. The magnitude of the effect may be illustrated using the values for dimethyl-nitrosamine (8-[VIII], $R = R' = CH_3$). The solvent shift δ(in C_6H_6)$-$$\delta$(in CCl_4) is $-0{\cdot}56$ ppm for the protons of the *cis* methyl group, and $-0{\cdot}80$ ppm for those of the *trans* group.

(c) *Specific interactions* Chemical interaction between solute and solvent, or between two different solutes in the same solution, will clearly cause chemical shift changes. The intermolecularly hydrogen-bonded species discussed in Section 8-11 provide examples of such an effect. NMR evidence is frequently invoked in order to show that chemical interaction occurs. Any type of weak complex formation may be studied by NMR. In fact, there is no clear distinction between 'complex formation' on the one hand and 'preferred orientation' on the other. Measurements of an 'energy of association' become difficult to interpret when these are small (less than about 20 kJ mol^{-1}) and suggestions of a geometry for the complex are unreliable when a whole range of conformations is likely to have similar energies.

Solvent effects on the shielding of metal nuclei may be very large. They are particularly common when ligand formation is feasible, even in cases where it has not been rigorously studied. For example, the shifts of ^{205}Tl in solutions of thallium(I) nitrate are[8] (in ppm with respect to the shift in water) +1848 in *n*-butylamine, +360 in dimethylsulphoxide and −389 in pyrrole.

In many cases it is feasible to talk in terms of a complex, and it becomes possible to obtain chemical shifts of such complexes even when these cannot be isolated. Consider the formation in solution of a charge-transfer complex DA from an electron donor D and an electron acceptor A. The lifetime of an individual complex is normally so small that the NMR signal seen for a nucleus in A, for example, is the average (see Chapter 5) of those for free A and for complex DA, weighted by the respective concentrations. The chemical shift for the solution may be measured relative to that for a similar solution containing no D (i.e. that for free A). Let this relative shift be Δ. Then consideration of the weighted average shows that Eqn 8-21 applies,

$$\Delta = \Delta_0[A]/([A]+[DA]) \tag{8-21}$$

where the square brackets indicate concentrations and Δ_0 is the chemical shift of A in the pure complex (under solution conditions) relative to that for free A. The equilibrium constant, K_c, for the charge-transfer process may be written

$$K_c = [DA]/[D][A] \tag{8-22}$$

Thence it may be shown (Problem 8-4) that if D is in large excess over A for all the measurements, then

$$\Delta/[D]_0 = -\Delta K_c + \Delta_0 K_c \tag{8-23}$$

where $[D]_0$ is the *total* concentration (i.e. free plus complexed) of D.

Thus the plots of $\Delta/[D]_0$ against Δ should be linear, and will give values of both Δ_0 and K_c, even when the former cannot be directly measured because the complex is weak. Fluorine magnetic resonance studies of the acceptor 1,4-dicyano-2,3,5,6-tetrafluorobenzene with the donor hexamethylbenzene in CCl_4 solution give $K_c = 5{\cdot}2$ kg of solution per mole and $\Delta_0 = 2{\cdot}99$ ppm. The maximum value of Δ observed (2·26 ppm) was substantially less than Δ_0. NMR thus provides a powerful way of obtaining information about equilibria with high reaction rates; the values of Δ_0 may be used to speculate on the geometry of non-isolable complexes while those of K_c give information about the donor and acceptor properties of D and A respectively.

The chemical shift effects of chemical interaction will have magnetic anisotropy and electric field contributions, and, in addition, will change σ_d(loc) and σ_p(loc) because the electron distribution in D will be altered.

8-13 Empirical relationships for ¹³C chemical shifts

The discussion of chemical shifts in this book is largely aimed at giving a general understanding of the physical factors of importance. However, there is no denying that much of the value of shifts for the elucidation of chemical structure can be summarized in terms of empirical relationships. It is the purpose of this short section to illustrate this point in a simple fashion by choosing three examples.

(a) *Unstrained alkanes.* Grant and Paul [9] discussed how the ^{13}C shifts of alkanes could be fitted to a simple equation involving a limited number of empirical parameters which depend on the relationship of other carbons to the one considered. Lindeman and Adams[10] extended the concept, and there have been further refinements. At its simplest, the chemical shift from the resonance of methane may be written[11]

$$\Delta\delta = n_\alpha \mathrm{A} + n_\beta \mathrm{B} + n_\gamma \Gamma + n_\delta \Delta \qquad (8\text{-}24)$$

where n_α, n_β, n_γ and n_δ are the number of carbon atoms one, two, three and four bonds removed from the one in question. However, it is found that the parameters A, B, Γ and Δ depend on the nature (primary, secondary, tertiary or quaternary) of the carbon whose shift is desired. Moreover, term B varies if more than one β-carbon is attached to a given α-carbon. The parameters are listed in Table 8-8. It is now well-recognized that even with so many parameters Eqn 8-24 is not wholly satisfactory. In particular, the γ-effect is dependent on conformation, a relatively large shielding contribution being obtained from γ-*gauche* carbons. Influences from heteroatoms can be accounted for using a similar approach to Eqn 8-24.

(b) *Substituted benzenes.* Additivity schemes of a simpler nature are generally used for substituted benzenes. The empirical additions to the chemical shift δ_C for substituents at various positions, taken from the compilation of Wehrli and Wirthlin (see the

Table 8-8 Chemical shift contributions for alkanes[a]

n_α	A	$B_1^{(b)}$	$B_2^{(b)}$	$B_3^{(b)}$	Γ	Δ
1	6·80	9·56	8·92	8·49	−2·99	0·49
2	7·67	9·75	8·35	7·14	−2·69	0·29
3	7·82	6·60	5·57	4·90	−2·07	~0
4	6·94	2·26	1·98	2·45	0·68	~0

[a] These are the contributions in ppm *per substituent carbon atom* for the position indicated in Eqn 8-24.
[b] The subscript, j, indicates that the contribution is to be used when j β-carbons are bonded to the *same* α-carbon.

Further Reading list at the end of the book) are given in Table 8-9. Deviations occur when substituents are *ortho* to one another.

(c) *π-electron charge density.* The carbon chemical shifts of the cycloheptatrienylium cation ($C_7H_7^+$), the cyclopentadienylide anion ($C_5H_5^-$), and the cyclooctatetraenide dianion ($C_8H_8^{2-}$) are $\Delta\delta = -27{\cdot}6$, $+25{\cdot}7$ and $+42{\cdot}5$ ppm (from the resonance of benzene) respectively.[12] This suggests that the addition of an electron to a $2p\pi$ orbital causes a decrease in δ_C of ca. 160 ppm, which it is presumed is primarily due to change in the $\langle r^{-3}\rangle_{2p}$ term of Eqn 8-12.

Table 8-9 Empirical substituent effects (in ppm) on carbon chemical shifts for benzene derivations

Substituent	*ipso*	*ortho*	*meta*	*para*
CH_3	+9·3	+0·8	~0	−2·9
CO_2H	+2·1	+1·5	~0	+5·1
CHO	+8·6	+1·3	+0·6	+5·5
CN	−15·4	+3·6	+0·6	+3·9
OH	+26·9	−12·7	+1·4	−7·3
NH_2	+18·0	−13·3	+0·9	−9·8
NO_2	+20·0	−4·8	+0·9	+5·8
Cl	+6·2	+0·4	+1·3	−1·9
Br	−5·5	+3·4	+1·7	−1·6

8-14 Isotropic shifts of paramagnetic species

So far this book has been concerned almost entirely with diamagnetic species. Paramagnetic systems behave substantially differently and are often difficult to observe by NMR. This is because in the solution state special effects arising from the paramagnetism can cause both large chemical shifts and substantial line broadening. Both T_1 and T_2 can be influenced. Such phenomena occur for both intramolecular and intermolecular situations. Intermolecular effects of paramagnetics on relaxation have already been discussed (see Sections 3-16 and 4-12), including the use of such species as relaxation reagents. The present section will be concerned with effects on chemical shifts. However, the influence of paramagnetics on relaxation and on shifts have common origins in two types of interaction between unpaired electron spins and nuclear spins.

One of these types of interaction, known as the Fermi contact term, will be discussed further in Section 8-17. Suffice it to say at this point that in principle it causes splittings in NMR (and ESR) spectra of a magnitude given by the electron-nucleus hyperfine splitting constant a_N, characteristic of the given nucleus and environment. These constants are often enormous in NMR terms ($\sim 10^9$ Hz) for, say, the metal atom of paramagnetic transition metal complex, but for nuclei in the ligands of such a complex they will be reduced because unpaired electron density *at* the nucleus is required. Nevertheless, they can still be substantial (0–1 MHz). However, typical electron spin relaxation times are very short (for systems amenable to NMR T_{1e} may be 10^{-10} to 10^{-13} s), so self-decoupling (see Section 5-29) occurs and a single NMR line will be seen for a given nucleus. This line is unlikely to be sharp because the self-decoupling will not be fully efficient. Moreover, there will be a shift from the mid-point of the spectrum because the Boltzmann populations of the different electronic spin states deviate significantly from equality. Treatment as a multi-site exchange situation (Problem 8-8) shows that the *contact shift* is given by

$$\Delta\sigma = -2\pi a_N g \mu_B S(S+1)/3kT\gamma \qquad (8\text{-}25)$$

where S is the electron spin quantum number, a_N is in frequency units (Hz), and the electronic g-factor has been assumed to be isotropic. The direction of this shift depends on the *sign* of a_N.

The magnitude of the isotropic shift for a given compound is determined by the *magnitude* of a_N, which will vary from nucleus to nucleus in an organic ligand of a metal complex. Thus measurements of isotropic contact shifts for such systems give information about the distribution of unpaired spin density. As an example, the proton contact shifts and π-spin densities at the associated carbon atoms in the γ-substituted nickel(II) N,N'-diethylaminotroponeiminate 8-[IX] are given in Table 8-10. Clearly, conjugation extends to some extent

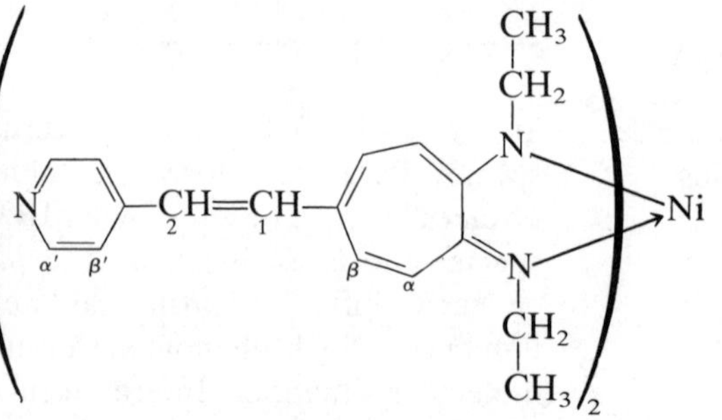

8-[IX]

Table 8-10 Proton contact shifts and carbon spin densities in compound 8-[IX][(a)]

Position	α	β	CH_2	CH_3
$\Delta\sigma$/ppm	−95·4	+54·0	+155·4	+12·7
Spin density	+0·0390	−0·0220	—	—
Position	CH(1)	CH(2)	α'	β'
$\Delta\sigma$/ppm	+44·8	−70·8	+2·5	−15·2
Spin density	−0·0183	+0·0288	−0·0012	+0·0062

[(a)] D. R. Eaton, R. E. Benson, C. J. Bottomley & A. D. Josey, *J. Amer. Chem. Soc.*, **94,** 5996 (1972).

throughout the ligand. The contact shifts of several of the protons are well outside the shift range for diamagnetic systems.

The second type of interaction causing shifts in paramagnetic systems is the electron–nucleus dipolar term. In this context it is often called the pseudo-contact interaction. The theoretical treatment for the shift depends on the relationship between the correlation time for molecular tumbling, τ_c, the electron spin relaxation time (or exchange lifetime), T_{1e}, and the anisotropy of the electron spin g-tensor. For axial symmetry of the g-tensor, and for the case $\tau_c^{-1} \ll T_{1e}^{-1}$ and $|g_{\|} - g_{\perp}|\ \mu_B B_0/\hbar$ it can be shown that:

$$\Delta\sigma = \mu_B^2 S(S+1)(g_{\|}^2 - g_{\perp}^2)(1 - 3\cos^2\theta)/9kTr^3 \qquad (8\text{-}26)$$

where r is the distance from the metal to the ligand nucleus and θ is the angle between **r** and the symmetry axis of the g-factor. Equation 8-26 has similar features to Eqn 8-19 since both are direct magnetic effects—Eqn 8-26 may be couched in a form involving susceptibility anisotropy. The pseudo-contact term 8-26 clearly contains information relating to molecular geometry. If the electronic g-factor is isotropic, Eqn 8-26 shows that there will be no shielding effect arising from the dipolar interactions with the unpaired electron.

8-15 Shift reagents

Strictly speaking, any chemical entity which causes differential chemical shifts for the nuclei of a substrate may be called a shift reagent. Indeed, aromatic solvent induced shifts (ASIS) were in popular use for a time.[13] However, in general the term 'shift reagent' is now commonly limited to complexes of the lanthanides, which operate by means of the mechanisms discussed in the preceding section. Such complexes were discovered by Hinckley[14] to be particularly effective in causing differential shifts. Those of Eu(III) and Pr(III) are generally preferred because the electron spin relaxation times are usually very short, so that sharp NMR lines are observed for substrate nuclei even when substantial shifts are induced. The favoured ligands are 2,2,6,6-tetramethylheptane-3,5-dionato (8-[X]), abbreviated to tmhd (but also known as dipivaloylmethanato, dpm), and 1,1,1,2,2,3,3-heptafluoro-7,7-dimethyl-4,6-octadienato (8-[XI]), referred to as fod.

$CF_3CF_2CF_2$

8-[X] 8-[XI]

The lanthanide complexes act by expanding the valence shell of the metal by interaction with functional groups (such as —OH and —NH_2) on the substrate, thus forming an overall paramagnetic entity. Exchange between the free substrate and the complexed species is generally fast on the NMR timescale so that only an average spectrum is observed. It is found that Eu shift reagents generally cause high-frequency shifts, whereas the reverse is usually true for Pr shift reagents.

The differential shifts can be quite dramatic, as shown in Fig. 8-11, even for relatively small amounts of the shift reagents (e.g. a molar ratio of shift reagent to substrate of ca. 0·1). In the case of proton NMR it is generally accepted that the contact contribution to the shifts is negligible, but in the case of other nuclei such as ^{13}C and ^{31}P this is not always true, particularly for substrate sites near to the active functional group. Of course, the dipolar effect is strongly dependent (as r^{-3}) on the distance from the nucleus in question to the paramagnetic centre, and it is this feature that is largely responsible for the differential nature of the shifts.

Shift reagents may be used in one of two ways, viz:

(i) To increase the frequency-spread of a spectrum to make it amenable to interpretation. This is particularly valuable in ^1H NMR, enabling first-order analysis to be made. Shift reagents

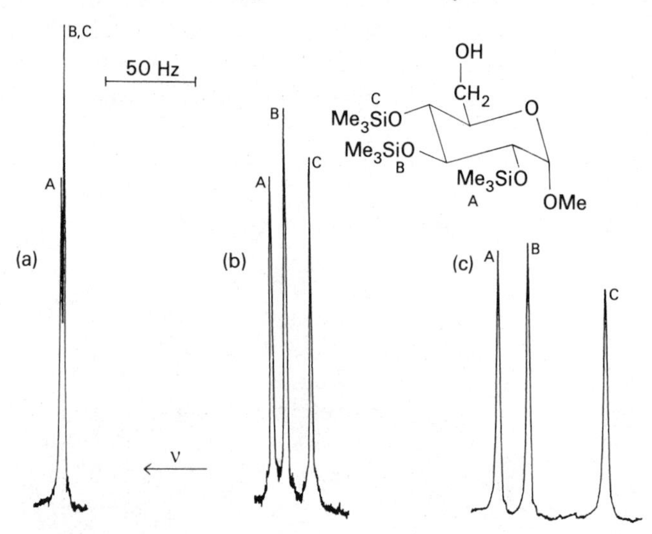

Fig. 8-11 The differential effect of a shift reagent on resonance frequencies for the ^1H NMR spectrum (at 100 MHz) of methyl 2,3,4-tri-O-trimethylsilyl-α-D-glucopyranoside. Spectra of the trimethylsilyl region are shown (a) in the absence of shift reagent; (b) and (c) with increasing concentration of $Pr(dpm)_3$. [Adapted from R. C. Rao, Ph.D. thesis, University of East Anglia (1978)].

are therefore occasionally referred to as the poor man's alternative to a superconducting magnet! Such a use requires no theoretical interpretation of the factors affecting the shift, and the only major problems to arise are the lack of a suitable functional group in the substrate or an excessively large number of functional groups.

(ii) It has been stated that Eqn 8-26 contains information on molecular geometry. Detailed use of theory can therefore enable such information to be evaluated provided chemical shifts at several sites in the substrate are measured and that the number of such sites exceeds the number of unknown geometrical factors. Many studies have purported to obtain accurate information in this way, but there are a number of severe problems which can be encountered. For example, the equilibrium constant for formation of the shift reagent/substrate complex is one of the unknown parameters. Unfortunately, if there are several active functional groups on the substrate or if a variety of complexes (1:1, 2:1, etc.) are formed for other reasons, the matter becomes complex. Moreover, the question of the correct equation to use is not clear-cut, particularly if axial symmetry of the electronic g-factor cannot be assumed. Finally, the question arises as to whether the derived geometry is properly that of the substrate alone or whether it reflects a different situation for the complex. For these reasons, great caution needs to be exercized in interpreting shift data quantitatively. However, in suitable cases there is no doubt that valid information on molecular geometry is accessible.

8-16 Shielding anisotropy

As discussed in Section 6-6, shielding is actually a tensor property and so in general data on three principal components, σ_{11}, σ_{22} and σ_{33}, are required, together with data on the orientation of the principal axes in the molecular framework. So far in this Chapter, discussion has concentrated on the isotropic average of the shielding constant, i.e. $\bar{\sigma}=\frac{1}{3}(\sigma_{11}+\sigma_{22}+\sigma_{33})$, since this is the quantity most commonly obtained in experimental studies. However, work on liquid crystal solutions or on solids yields the separate values of σ_{11}, σ_{22} and σ_{33}, providing yet more detailed information about the molecules. Some results for ^{13}C are shown in Table 8-11. It appears that methyl groups are frequently axially symmetrical, possibly arising from rapid internal rotation, and in any case have rather small anisotropies, whereas carboxyl and carbonyl groups are generally asymmetric with large anisotropies ($\sigma_{33}-\sigma_{11}\sim 200$ ppm). Acetylenic and aromatic anisotropies are also relatively large. For dimethylacetylene the *sp* carbon shows axial symmetry as expected, and, in the case of hexamethylbenzene the situation changes to one of axial symmetry as the temperature is increased from 87 K to 296 K due to the onset of rapid rotation about the six-fold axis. This identifies σ_{33} as the shielding about the

Table 8-11 Carbon-13 shielding anisotropies[(a)]

Carbon type	*Compound*	$\sigma_{11}-\bar{\sigma}$ /ppm	$\sigma_{22}-\bar{\sigma}$ /ppm	$\sigma_{33}-\bar{\sigma}$ /ppm
Methyl	$CH_3C{\equiv}CCH_3$	−5	−5	9
Methyl	acetone	−17	−17	33
Methyl	C_6Me_6	−12	−3	16
Carboxyl	acetic acid	−81	4	78
Carbonyl	acetone	−71	−57	129
Carbonyl	$CH_3CO_2CH_3$	−85	22	62
Acetylenic	$CH_3C{\equiv}CCH_3$	−67	−67	135
Aromatic	C_6Me_6 (at 87 K)	−95	−17	113
Aromatic	C_6Me_6 (at 296 K)	−56	−56	112

[(a)] From the compilation in M. Mehring, *NMR Basic Principles & Prog.*, **11** (1976).

direction perpendicular to the aromatic ring (unaffected by the molecular rotation), which is of interest in terms of ring-current ideas.

Relating the principal components of shielding to molecular directions requires a study of single crystals. Obviously shielding can be affected by intermolecular interactions so that the principal axes of shielding do not necessarily follow molecular symmetry. For example, Haeberlen[15] has shown that for proton shielding in pyromellitic dianhydride (8-[XII], showing shielding components in ppm with respect

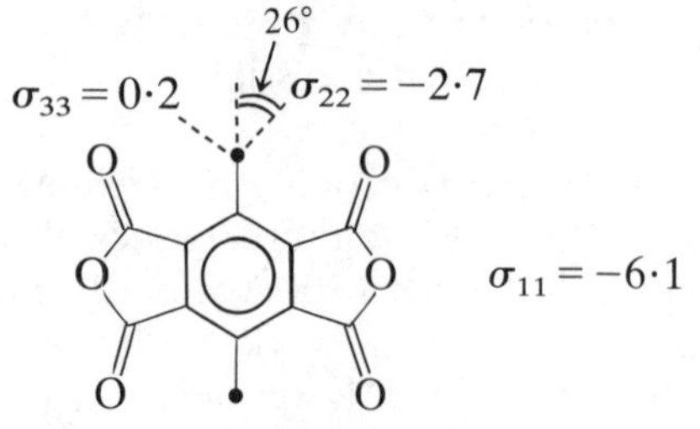

26°

$\sigma_{33}=0{\cdot}2$ $\sigma_{22}=-2{\cdot}7$

$\sigma_{11}=-6{\cdot}1$

8-[XII]

to that of liquid water), the in-plane shielding components deviate by 26° from the C—H bond direction and its perpendicular. Moreover, as is clear from Fig. 6-11(d), different crystallographic sites differ in $\bar{\sigma}$. This arises from differences in magnitude in the individual shielding components for different sites. The components also vary in direction with respect to the molecular framework for different sites.

The magnitude of shift anisotropies depends on the nuclide considered. Those with large shielding ranges tend to have large anisotropies. For instance, it has been shown[16] that $\sigma_{\parallel}-\sigma_{\perp}$ for ^{199}Hg in Me_2Hg is 7475 ppm. Shielding anisotropies are significantly larger than the isotropic shielding differences between relevant differing chemical sites. Thus, the ^{13}C anisotropies for carbonyl groups (ca. 200 ppm) exceed the normal variation in $\bar{\sigma}$ for such groups (~50 ppm as observed in solution state studies of organic compounds). Clearly the information content contained in the tensor is significantly higher than that in the isotropic average. So far, however, little use has been made of shielding components for chemical structure purposes. However, they are invaluable as tests for theories of shielding. In fact there is little point in attempting only to fit isotropic average shifts by theory, since the equations outlined earlier in this chapter either provide values of shielding components directly or can be readily modified to do so. Unfortunately, corrections to experimental data to eliminate intermolecular effects are not easy to make.

PART B COUPLING CONSTANTS

8-17 Coupling between electron and nuclear spins

Direct (dipolar) coupling between nuclear spins has been shown in Section 4-2 to be averaged to zero by molecular tumbling in isotropic liquids. However spin–spin coupling still occurs, as is shown by the observations of fine splitting in NMR spectra (Section 1-11). Such coupling must be *indirect*, i.e. it must be transmitted via the electrons in the system. In fact, this implies the existence of coupling between nuclei and electrons. Coupling involving electrons occurs by more than one mechanism. First, there is the interaction between the magnetic moments due to nuclear spin and electron orbital motion (the electron-orbital term). This requires a term in the Hamiltonian of the form of Eqn. 8-27.

$$\hat{\mathcal{H}} = \frac{\mu_0}{4\pi} g_L \mu_B \gamma_N \hbar \sum_e 2\hat{\mathbf{L}}_e \cdot \hat{\mathbf{I}}_N r_{eN}^{-3} \tag{8-27}$$

where e and N refer to electron and nucleus respectively, $\hat{\mathbf{L}}$ is the orbital angular momentum operator, and g_L is the electron orbital g factor (=1). Secondly there is the dipolar interaction between the electron and nuclear spins (the electron spin-dipole term). The classical energy of interaction is that given by Eqn (4-3) and therefore the term in the Hamiltonian is

$$\hat{\mathcal{H}} = -\frac{\mu_0}{4\pi} g_S \mu_B \gamma_N \hbar \sum_e \{r_{eN}^{-3} \hat{\mathbf{I}}_N \cdot \hat{\mathbf{S}}_e - 3r_{eN}^{-5}(\hat{\mathbf{I}}_N \cdot \mathbf{r}_{eN})(\hat{\mathbf{S}}_e \cdot \mathbf{r}_{eN})\} \tag{8-28}$$

where r_{eN} is the distance between electron e and nucleus N. However, this equation does not completely account for this type of interaction because it assumes point dipoles and therefore is not adequate as $r \rightarrow 0$. It does not properly allow for the possibility that the electron may be in the same region of space as the nucleus. Quantum-mechanically this *is* possible so a further energy term is required. This is known as the Fermi contact term. It can be shown to give a contribution 8-29 to the Hamiltonian.

$$\hat{\mathcal{H}}_{\text{contact}} = \tfrac{2}{3}\mu_0 g_S \mu_B \gamma_N \hbar \sum_e \delta(r_{eN}) \hat{\mathbf{I}}_N \cdot \hat{\mathbf{S}}_e \tag{8-29}$$

where $\delta(r_{eN})$ is the Dirac delta function (1 if $r_{eN} = 0$ but zero otherwise) and the other symbols have their usual meanings. The use of the delta function gives meaning to the word 'contact' used for this term. In both Eqns 8-28 and 8-29 the electron spin g-factor, g_S, may be taken as 2.

8-18 The Dirac vector model

Before further discussion of the theory of coupling is given, it may be helpful to take a simple physical picture to show how coupling can occur. However, it must be emphasized that the Dirac vector model mentioned here is a gross oversimplification of the real situation, and many conclusions based on it are found to be incorrect. Equation 8-29 shows that the contact term (which is usually dominant—see the next section) stabilizes the anti-parallel orientation of an electron spin *at* a nuclear spin if γ_N is positive; the electron and nuclear magnetic moments are therefore stable when parallel. Now pairs of electrons in the same orbital have opposed spins according to the Pauli principle, so for HD the stable situation is as in Fig. 8-12(a). This figure assumes that when one of the electrons in a bonding pair is near one of the nuclei there is a tendency for the second electron to be near the second nucleus. As noted in Section 1-13 the NMR energy Eqn 1-32 is written in such a way that a coupling constant is defined as positive if coupling stabilizes anti-parallel spins. The Dirac vector model of coupling (Fig. 8-12) therefore indicates that the coupling constant should be positive for the directly-bonded nuclei of HD, and, indeed, for any directly-bonded nuclei with positive magnetogyric ratios (in practice not all such coupling constants are positive).

If another atom intervenes between the coupled nuclei, and the bonds are all localized, the interaction between electrons occupying different orbitals in the region of space around this atom must be considered. Hund's rule states that such electrons should have their spins parallel, and Fig. 8-12(b) shows that this implies a negative coupling constant between nuclei separated by two bonds. A logical extension indicates that the sign of coupling constants should alternate as the number of bonds separating the nuclei increases, being positive for an odd number of intervening bonds. This rule is generally obeyed for (H, H) coupling, though there are exceptions. Notice that the model only considers coupling 'through bonds' via σ electrons when bonds are localized. Since there must be a Hund-type interaction at each intervening atom for the transmission of the spin information to occur, the Dirac vector model predicts a rapid attenuation in the magnitude of J as the number of bonds separating the nuclei increases. In fact for saturated compounds (H, H) coupling is only rarely observed when more than four bonds intervene, but the attenuation is less rapid than the Dirac model would suggest.

When comparing coupling constants in different molecules (e.g. in a study of substituent effects) it is useful to classify the values according to the number of intervening bonds; this number may be conveniently indicated by means of a superscript prefix to the symbol J. Thus

(a)

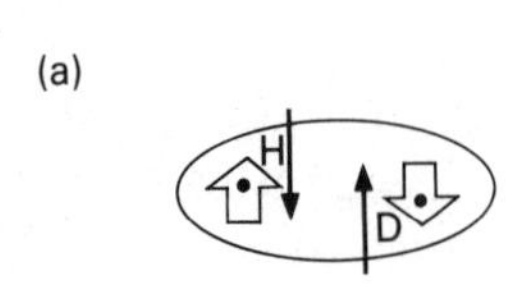

(b)

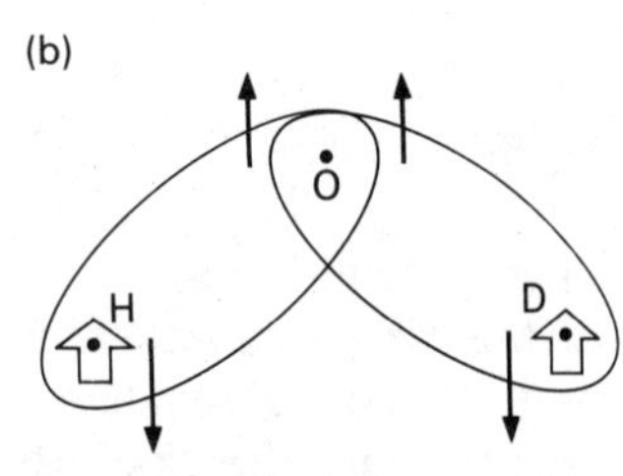

Fig. 8-12 Dirac vector model for (H, H) coupling in (a) HD, (b) HDO. The most favoured mutual arrangement of spin components is indicated. The contact term actually involves electrons in the same region of space as the nuclei.

coupling between directly-bonded nuclei is often indicated by ^1J. This classification is occasionally ambiguous for cyclic systems. Coupling through two and three bonds (^2J and ^3J respectively) is often known as *geminal* and *vicinal* coupling respectively.

The (H, H) coupling constant for the hydrogen molecule has been shown to be (see Section 8-20) $^1J_{HH} = 284$ Hz. It is clear that coupling through more than one bond must involve delocalization of electrons, though this is not necessarily best described by the model of Fig. 8-12(b). *Geminal* (H, H) coupling constants via an intervening carbon atom are negative in saturated systems, as predicted by the Dirac vector model, but are over an order of magnitude smaller than $^1J_{HH}$—they range from -9 to -21 Hz. When the intervening carbon atom is sp^2 hybridized, values of $^2J_{HH}$ are normally positive (the highest value is ca. $+41$ Hz for formaldehyde); this is clearly in contradiction to the vector model. *Vicinal* (H, H) coupling constants (see Section 8-23) are of the same order of magnitude as *geminal* values, but are invariably positive. In fact frequently $|^2J_{HH}| < |^3J_{HH}|$, and most theories show that there are important contributions to $^2J_{HH}$ that differ in sign and thus tend to cancel. Coupling between protons separated by more than three bonds is only appreciable in special circumstances (see Sections 8-24 and 8-25).

Coupling constants involving nuclei other than hydrogen may be substantially larger than the values mentioned above for J_{HH}. Thus, typical values for $^1J(^{205}Tl, ^{13}C)$ are in the range 2–10 kHz, and even *three*-bond ($^{205}Tl, ^{13}C$) coupling constants are ca. 1000 Hz. Generally speaking, the level of understanding of (H, H), (C, H) and (C, C) coupling constants is higher than that of other types of coupling. This is partly because of the large amount of data available for coupling involving only H or C, and partly because of the relatively simple electronic situations for carbon and hydrogen, especially the absence of lone pairs of electrons.

After the theory of coupling is explored in more detail in the next section, a few selected classes of coupling will be discussed, while others will be completely ignored.

8-19 Electronic coupling of nuclear spins

In order to develop the theory of nuclear spin coupling any further it must be recognized that it is necessary to explain an energy term involving the scalar product $\hat{\mathbf{I}}_N \cdot \hat{\mathbf{I}}_{N'}$ for nuclei N and N'. This occurs in an empirical Hamiltonian, as in Eqn. 2-38, which is normally evaluated using nuclear spin wave functions alone. However, it must be possible to equate such a term to others appearing in the theoretical total Hamiltonian which includes terms for all possible physical interactions. Now the physical interactions under consideration are of very small energy compared to the total electronic energy of the system. Thus their effect may be evaluated using perturbation theory. Moreover, direct interactions between nuclear spins average to zero for a liquid, so it is necessary to consider the electron–nuclear interactions discussed in Section 8-17. First-order perturbation terms are of the type $\langle\bar{\Psi}|\,\hat{\mathscr{H}}_{eN}\,|\bar{\Psi}\rangle$, where $\hat{\mathscr{H}}_{eN}$ is the perturbation in question and $\bar{\Psi}$ is the molecular wave function; such a term involves only one nucleus and cannot be equated to the coupling term from the empirical Hamiltonian. Therefore, one must turn to second-order perturbation theory, which gives a contribution to the ground-state energy which is of the form 8-30.

$$-\sum_{n\neq 0}\langle\bar{\Psi}_0|\,\hat{\mathscr{H}}_{eN}\,|\bar{\Psi}_n\rangle\langle\bar{\Psi}_n|\,\hat{\mathscr{H}}_{eN}\,|\bar{\Psi}_0\rangle/(U_n-U_0) \qquad (8\text{-}30)$$

In this term $\bar{\Psi}_0$ and $\bar{\Psi}_n$ refer to the ground-state and nth excited-state total wave functions, calculated before including the electron–nuclear perturbation $\hat{\mathscr{H}}_{eN}$; U_0 and U_n are the energies of $\bar{\Psi}_0$ and $\bar{\Psi}_n$. Since this term involves $\hat{\mathscr{H}}_{eN}$ *twice* it will contain contributions involving $\hat{\mathbf{I}}_N\,.\,\hat{\mathbf{I}}_{N'}$. The problem of evaluating the coupling constant $J_{NN'}$ therefore reduces to equating corresponding terms in 8-30 and in $\langle\Psi_{\text{nuclear}}|\,\hat{\mathscr{H}}_{NMR}\,|\Psi_{\text{nuclear}}\rangle$, where $\hat{\mathscr{H}}_{\text{NMR}}$ is given by Eqn 2-38. Like σ, J is in principle an anisotropic (tensor) quantity; the zz component, arising from the contact term alone, will be considered in detail. The treatment discussed above leads to

$$h\langle\Psi_{\text{nuclear}}|\,J_{NN'zz}\hat{I}_{Nz}\hat{I}_{N'z}\,|\Psi_{\text{nuclear}}\rangle=-2(\tfrac{2}{3}\mu_0 g_S\mu_B\hbar)^2\gamma_N\gamma_{N'}$$
$$\sum_{n\neq 0}\langle\bar{\Psi}_0|\sum_e\delta(r_{eN})\hat{I}_{Nz}\hat{S}_{ez}\,|\bar{\Psi}_n\rangle\,\langle\bar{\Psi}_n|\sum_{e'}\delta(r_{e'N'})\hat{I}_{N'z}\hat{S}_{e'z}\,|\bar{\Psi}_0\rangle/(U_n-U_0)$$
$$(8\text{-}31)$$

The factor 2 appears on the right-hand side because nucleus N may be in either the first or the second bracket of 8-30. Factorization of $\bar{\Psi}$ into nuclear and electronic parts shows that when $\bar{\Psi}_n$ and $\bar{\Psi}_0$ contain the same nuclear part there is a factor on the right-hand side of $\langle\Psi_{\text{nuclear}}|\,\hat{I}_{Nz}\hat{I}_{N'z}\,|\Psi_{\text{nuclear}}\rangle$; this factor also occurs on the left-hand side, giving Eqn (8–32):

$$J_{NN'zz}=-2h(\mu_0 g_S\mu_B/3\pi)^2\gamma_N\gamma_{N'}$$
$$\sum_{n\neq 0}\langle\bar{\Psi}_0|\sum_e\delta(r_{eN})\hat{S}_{ez}\,|\bar{\Psi}_n\rangle\,\langle\bar{\Psi}_n|\sum_{e'}\delta(r_{e'N'})\hat{S}_{e'z}\,|\bar{\Psi}_0\rangle/(U_n-U_0) \quad (8.32)$$

where $\bar{\Psi}$ is now purely electronic. As for σ, isotropic tumbling averages J, so that the observed value is $\frac{1}{3}$ the sum of Eqn 8-32 and corresponding expressions for $J_{NN'xx}$ and $J_{NN'yy}$. The resulting equation is

$$J_{NN'}=-\tfrac{2}{3}h(\mu_0 g_S\mu_B/3\pi)^2\gamma_N\gamma_{N'}\sum_{n\neq 0}\langle\bar{\Psi}_0|\sum_e\delta(r_{eN})\hat{S}_e\,|\bar{\Psi}_n\rangle$$
$$\times\langle\bar{\Psi}_n|\sum_{e'}\delta(r_{e'N'})\hat{S}_{e'}\,|\bar{\Psi}_0\rangle/(U_n-U_0) \quad (8\text{-}33)$$

It may be noted that the nature of the electron spin operators shows that only triplet excited states need be considered in the summation. The above treatment has been carried out to give J in frequency units (Hz).

Table 8-12 Some calculated[a] contributions to coupling

Coupling constant	*Molecule*	*Contributions*/Hz		
		contact	*orbital*	*dipolar*
$^1J_{CC}$	C_2H_6	11·99	−0·55	0·30
	C_2H_4	29·75	−3·76	0·93
	C_2H_2	69·43	4·13	2·84
$^1J_{FC}$	CH_3F	−126·30	−4·15	6·92
	CF_4	−85·33	−20·17	2·40
$^2J_{FF}$	CF_4	−1·09	6·20	16·73

[a] A. D. C. Towl & K. Schaumburg, *Mol. Phys.*, **22,** 49 (1971).

Further evaluation of Eqn 8-33 is possible if details of electronic wave functions are available; a similar treatment of the electron orbital and electron spin–dipole terms may be made. Theoretical calculations of coupling constants have, in fact, been carried out using both valence bond and molecular orbital theory, and qualitative agreement with experimental values is obtained. For H_2 it has been estimated that the contact, electron spin–dipole and electron orbital terms make contributions to the coupling constant approximately in the ratio 1:0·0058:0·0028. Table 8-12 lists some calculated contributions to coupling constants. The contact term generally seems to be dominant for (C, H) and (C, C) coupling. However, orbital and dipolar contributions become significant for $^1J_{CC}$ involving multiply-bonded carbon atoms, and are frequently important for coupling involving fluorine. For other elements the contact term has often been *assumed* to dominate, even when there is little experimental or theoretical evidence for such a conclusion.

The physical meaning of the use of second-order perturbation theory is clear. The coupling involves two electron–nucleus interactions, one for each of the coupled nuclei. The coupling information is carried by the electrons. In ESR spectroscopy, hyperfine splitting involves a single electron–nucleus interaction. It is therefore not surprising that ESR hyperfine splitting constants are several orders of magnitude larger than NMR coupling constants, nor that the two types of parameter are related quantities. One further note of explanation is necessary regarding the electron spin-dipole term: the first-order contribution of this term to the Hamiltonian averages to zero when molecular tumbling occurs, but the second-order contribution does not. Therefore this mechanism contributes to NMR coupling constants but *not* to ESR splitting constants. Under solution conditions the latter are entirely determined by the contact term.

8-20 The effect of magnetogyric ratios

Equation 8-33 shows that the contact contribution to coupling constants is proportional to the product of the magnetogyric ratios of the nuclei; the same is true of the electron orbital and electron spin–dipole terms. This fact arises because all the interactions are of magnetic origin and involve products of nuclear magnetic moments. Use may be made of this feature in order to evaluate coupling constants between magnetically equivalent nuclei. Such interactions do not normally affect NMR spectra, but isotopic substitution will immediately render nuclei non-equivalent and the analogous coupling constants appear as direct first-order splittings in the spectra. Since isotopic substitution does not affect electronic wave functions, the effect on J is entirely due to the change in magnetogyric ratios. Thus measurement of the spectrum of HD shows that $|^1J_{DH}| = 42{\cdot}94$ Hz; the corresponding value of $|^1J_{HH}|$ is obtained from Eqn 8-34

$$J_{H_2}/J_{DH} = \gamma_H/\gamma_D = 6{\cdot}514 \qquad (8\text{-}34)$$

The numerical value of this ratio applies to any H → D substitution. This method is in quite common use—for example, for $^2J_{HH}$ of CH_3 groups the corresponding molecule containing CH_2D is studied. To a first approximation, shielding constants (in ppm) are unaffected by isotopic substitution, so that direct deuterium resonance often gives first-order spectra for fully deuterated molecules even when the corresponding proton resonance shows second-order effects.

It has been suggested that comparison between coupling constants of different elements is only useful when the effect of magnetogyric ratios has been eliminated. Thus it is possible to define *reduced* coupling

Table 8-13 Coupling constants between directly bonded nuclei

Nuclei	*Molecule*	$K/10^{20}$ N A^{-2} m^{-3} [(a)]	^1J/Hz
H, H	H_2	(+)2·3	±276
H, C	CH_4	+4·1	+125
H, O	H_2O	±4·5	±73
H, F	HF	(−)4·6	(−)521
H, Si	SiH_4	+8·5	−202
H, P	PH_3	+3·7	+182
H, Sn	SnH_4	+43·0	(−)1931 (^{1}H, ^{119}Sn)
C, C	CH_3CH_3	+4·6	+35
C, F	CF_4	−9·1	−259
C, Si	Me_4Si	+8·7	−52
C, Se	Me_2Se	−10·8	−62
C, Te	Me_2Te	−17·0	+162
C, Hg	Me_2Hg	+128·0	+689
P, P	P_2H_4	−5·5	−108
Sn, Sn	Sn_2Me_6	+261·1	+4264 (^{117}Sn, ^{119}Sn)

[(a)] In electromagnetic units (with μ_0 considered dimensionless) K is often quoted in cm^{-3}. Eqns 8-35 and 8-36 remain as originally defined (not involving μ_0); this means the unit of K in SI is *not* m^{-3}.

constants, usually given the symbol K, as follows

$$K_{jk} = \frac{J_{jk}}{h}\left(\frac{2\pi}{\gamma_j}\right)\left(\frac{2\pi}{\gamma_k}\right) \tag{8-35}$$

This definition is used so that the coupling energy is expressed in terms of nuclear magnetic moments as

$$U_{jk} = K_{jk}\mu_j\mu_k \tag{8-36}$$

On the SI system the magnetogyric ratios are in rad T^{-1} s^{-1} and K appears in N A^{-2} m^{-3}. Table 8-13 gives examples for directly-bonded nuclei. There is a general tendency for the magnitude of ^1K to increase as the atomic numbers of the coupled nuclei increase. However, because γ for the proton is high, coupling constants involving protons are anomalously large if quoted in Hz. Since magnetogyric ratios may be either positive or negative, the sign of K is not necessarily the same as that of the corresponding J. It has been shown that when the hybridizations are constant, values of $(^1K_{XH})^{1/2}$ and $(^2K_{XCH})^{1/2}$ are proportional to Z_X, the atomic number of X, except for $Z_X < 10$. This occurs because of the dependence of reduced coupling constants on the square of the atomic s-electron charge density at the nucleus (see Section 8-22).

8-21 Coupling between directly bonded nuclei

The contact contribution to coupling depends on the probability of an electron being *at* the nucleus. However, for atoms only electrons in s orbitals have any electron density at the nucleus. Thus any development of Eqn 8-33 using molecular orbital theory needs to consider only the contributions of the valence atomic s orbitals for the coupled nuclei. The theory of Pople and Santry[17] shows that if the molecular wave functions Ψ are expressed in terms of molecular orbitals ψ and if the latter are expanded as linear combinations of atomic orbitals ϕ according to Eqn 8-37

$$\psi_i = \sum_{\mu} c_{i\mu}\phi_{\mu} \tag{8-37}$$

then the space part (involving δ) and the spin part (involving $\hat{S}$) of Eqn 8-33 may be evaluated independently and the reduced coupling constant between nuclei j and k may be written

$$K_{jk} = (\tfrac{2}{3}\mu_0\mu_B)^2\langle s_j|\delta(r_j)|s_j\rangle\langle s_k|\delta(r_k)|s_k\rangle\Pi_{s_j,s_k} \tag{8-38}$$

where s_j and s_k are the valence shell s functions for j and k respectively, the integrals $\langle s|\delta|s\rangle$ represent s-electron density at a nucleus and Π_{s_j,s_k} is known as the mutual polarizability of the orbitals, given by Eqn (8-39).

$$\Pi_{s_j,s_k} = 4\sum_i^{\text{occ}}\sum_{\ell}^{\text{unocc}}(\varepsilon_i - \varepsilon_\ell)^{-1}c_{is_j}c_{is_k}c_{\ell s_j}c_{\ell s_k} \tag{8-39}$$

Only excited states involving the promotion of one electron are considered, and the excitation energy is therefore written as the difference of one-electron energies $(\varepsilon_i - \varepsilon_\ell)$; the coefficients c are those of Eqn 8-37. In comparing Eqn 8-38 with Eqn 8-33 the g-factor has been set at 2 and a further factor of $\frac{3}{2}$ has appeared. The 3 arises from the separate and equal contributions of the three excited triplet states for each excitation energy $(\varepsilon_i - \varepsilon_\ell)$; the factor $\frac{1}{2}$ appears from the combined effects of evaluating the electron spin part of Eqn 8-33 and of the necessary use of anti-symmetrized molecular wave functions Ψ. Clearly, Eqn 8-38 describes the mixing of excited states into the ground-state wave function by the magnetic field, as discussed in Section 8-19. Each 'excited state' can be seen to make its own contribution to K; however, the contribution may be positive or negative, depending on the signs of the coefficients in Eqn 8-39. The value of K itself may be positive or negative—both situations are found experimentally for directly-bonded coupling—in contrast to the predictions of the simple Dirac vector model. The most important 'excitations' are those with small values of $(\varepsilon_i - \varepsilon_\ell)$.

Equation 8-38 is valid for coupling through any number of bonds; we will discuss at this point its application to coupling between directly-bonded nuclei. For tetrahedral molecules or ions such as BH_4^-, CH_4, and NH_4^+, only the two totally symmetrical orbitals (one bonding and one anti-bonding) involve the s wave function of the central atom, and therefore only one 'excitation' contributes to K. The product of coefficients for Eqn 8-39 is positive, and therefore so is K; the magnitude of K is principally determined by the s-electron densities at the nuclei for the valence shell atomic orbitals. These densities increase in the same order as atomic number for the elements considered here (H, B, C, N, O, F). Calculated and observed values of K for the three tetrahedral species are listed in Table 8-14.

For (H, X) coupling in NH_3, H_2O and HF there are two 'excitations' to consider since three molecular orbitals (two occupied) involve the s function of X. These two excitations give contributions to K of opposite sign (Problem 8-6); when the energy difference between s and p electrons is large, the negative term dominates. Now this energy difference increases in the series C, N, O, F and the calculated sign of K changes in going from K_{CH} to K_{FH}. For the corresponding series of (C, X) coupling constants it has been established experimentally that $^1K_{FC}$ is negative whereas $^1K_{CH}$ is positive. In fact, all signs so far

Table 8-14 Coupling constants for $[BH_4]^-$ CH_4 and $[NH_4]^+$

	$K_{XH}/10^{20}$ N A^{-2} m^{-3} *Calculated*	*Found*
$[BH_4]^-$	+2·2	(+)2·13
CH_4	+4·4	+4·18
$[NH_4]^+$	+7·5	(+)5·9

determined for ^1K involving fluorine are negative. Other one-bond reduced coupling constants, such as $^1K_{PC}$ and $^1K_{PP}$, actually exhibit both positive and negative signs, which in the cases mentioned depend primarily on the valency of the phosphorus atom or atoms involved and are related to the presence or absence of lone pairs of electrons. Thus $^1J_{PP}$ is -180 Hz for Me_2PPMe_2 but $+19$ Hz for the corresponding disulphide.

8-22 Hybridization and related effects

Attention can be focused to some of the physical concepts behind Eqn 8-38 if the bond between the coupled nuclei is considered to be of the localized σ type. Then it can be shown that the mutual polarizability reduces to $\Pi_{s_j,s_k} = \alpha_j^2\alpha_k^2/\Delta U$ where α_j^2 is the s character of the hybrid orbital used by j in the j, k bond, and ΔU is the (positive) energy of excitation from the bonding to the anti-bonding j, k molecular orbital. The coupling constant may be written:

$$^1J_{ik} \propto \gamma_j\gamma_k\alpha_j^2\alpha_k^2\langle s_j|\,\delta(\mathbf{r}_j)\,|s_j\rangle\langle s_k|\,\delta(\mathbf{r}_k)\,|s_k\rangle/\Delta U \tag{8-40}$$

Eqn. 8-40 is useful for predicting certain variations in J. In particular it shows that coupling constants should have a pronounced dependence on the s character of the σ-bond between the nuclei, i.e. on hybridization. This is well illustrated by $^1J_{CH}$ (see Table 8-15), though the accuracy of the proportionality may be fortuitous. Equations such as 8–40 are less successful when one or both of the coupled nuclei possess lone pairs of electrons having appreciable s-character. Contributions to the coupling (as expressed in Eqn 8-38) from such orbitals can be significant because the relevant excitation energy tends to be relatively small. These contributions generally have the opposite sign to those arising from s-electrons in the j, k bond.

Equation 8-40 may be viewed as an example of an average energy approximation treatment. This may be valid in certain circumstances, but general use of such an approximation is not to be recommended. In fact, only positive one-bond reduced coupling constants would be predicted, whereas several negative values are known (see the preceding section). However, an empirical relationship of coupling constants to hybridization may have a wider validity. For instance, for PF_5 derivatives it is well-known that $^1J_{PF}$ is invariably negative and that its *magnitude* is greater for equatorial fluorine atoms than for axial fluorines. It has been argued that this is because equatorial bonds have

Table 8-15 Dependence of $^1J_{CH}$ on carbon hybridization

Molecule	*Hybridization*	α_C^2	J/Hz
Ethane	sp^3	$\frac{1}{4}$	124·9
Ethene	sp^2	$\frac{1}{3}$	156·4
Ethyne	sp	$\frac{1}{2}$	248·7

greater s-character than axial bonds, as is reflected in the fact that the former are shorter than the latter,

Similarly it has been noted that for methane derivatives, CHXYZ, there are additive substituent contributions, ξ, such that the (C, H) coupling constants may be expressed as:

$$^{1}J_{CH} = \xi_X + \xi_Y + \xi_Z \tag{8-41}$$

There are deviations for multiple substitution by highly electronegative groups and extra terms must be introduced into Eqn 8-41 to account for pairwise interactions between substituents. The parameters ξ correlate with the electronegativity of the substituent, but this affects the s-electron density at the carbon atom in at least two ways. First, the hybridization of the carbon atom used in the various bonds (i.e. α^2) is affected: electronegative substituents prefer to make use of carbon p character, since the electron density is at a maximum further from the carbon nucleus than for the carbon s orbital. This results in an increase in the s character to the remaining substituents (a lone pair will occupy an orbital of high s character)[18] and the coupling constant will therefore increase. Secondly, the bond to an electronegative substituent is polarized. The electron density around carbon is reduced, but this results in an increased effective nuclear charge for the carbon nucleus so that the radial electron distribution is modified. There is a contraction of the electron cloud of carbon; $\langle s_C | \delta(r_C) | s_C \rangle$ would increase if α^2 remained constant and therefore coupling constants increase (see Eqn 8-40). Both effects act in the same direction and it is not clear which is dominant. It is certainly unwise to derive s characters from values of J_{CH}. However, in the case of the platinum complex *cis*-$[PtClMe(Et_3P)_2]$, the two chemically shifted phosphorus resonances have very different coupling constants to platinum ($^{1}J_{PtP} = 4179$ Hz and 1719 Hz). In this situation $\langle 6s_{Pt} | \delta(r_{Pt}) | 6s_{Pt} \rangle$ is common to the two cases and the observed differences appear to be due to different values of α^2_{Pt} (provided the contact term is dominant and π-type electrons can be ignored).

8-23 *Vicinal* (H, H) coupling

Valence bond theory has been of great success in qualitatively describing trends in $^{3}J_{HH}$. The situation is clearly rather complex because of the considerable number of electrons and geometrical parameters involved, so we will give no details of the theory. The most important feature is that, when other factors are constant, the *vicinal* coupling constant through carbon is predicted to depend on the dihedral angle, ϕ, between the C—H bonds as in Eqn 8-42

$$^{3}J = A + B \cos \phi + C \cos 2\phi \tag{8-42}$$

where A, B, and C are constants with approximate values 4·0 Hz, −0·5 Hz, and 4·5 Hz respectively.[19] However, use of these values usually underestimates ^{3}J, and a better empirical set for hydrocarbons is found to be $A = 7$ Hz, $B = -1$ Hz, $C = 5$ Hz. This variation with ϕ is

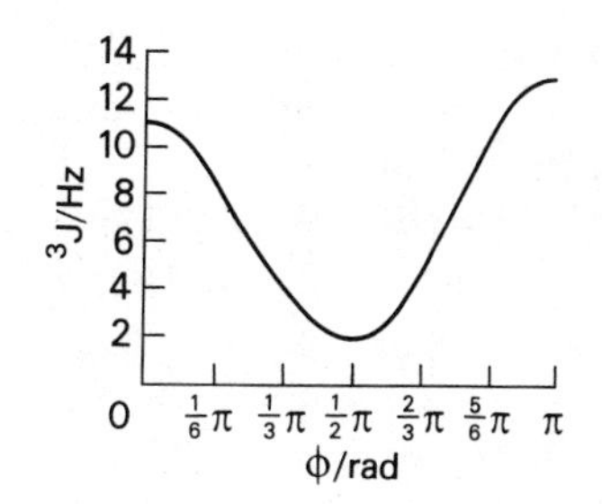

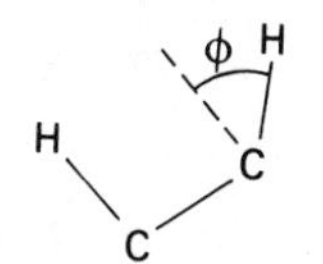

Fig. 8-13 The Karplus curve: variation of *vicinal* (H, H) coupling constants in hydrocarbons with dihedral angle.

shown in Fig. 8-13. Equation 8-42 is known as the Karplus equation,[20] and has been used to predict dihedral angles in compounds of unknown conformation. However, there are considerable dangers in this procedure since the Karplus model is only valid in the absence of electronegative substituents and of departure from tetrahedral angles at carbon. Thus qualitative use of Eqn 8-42 is useful, particularly when values of A, B, and C are determined from model compounds closely related to the unknown, whereas quantitative use is usually to be avoided. Experiment shows that *vicinal* (H, H) coupling constants increase in the order *gauche* ($\phi = \pi/3$) < *cis* ($\phi = 0$) < *trans* ($\phi = \pi$) in the absence of other influences than ϕ. Some typical values are given for schemes 8-[XIII], 8-[XIV] and 8[XV].

8-[XIII]	8-[XIV]	8-[XV]
$J_{cis} = 11{\cdot}6$ Hz ($\phi = 0$)	$J_{ab} = 4{\cdot}31$ Hz ($\phi \sim \pi/3$)	$J_{aa'} = 11{\cdot}0$ Hz ($\phi \sim 0$)
$J_{trans} = 19{\cdot}1$ Hz ($\phi = \pi$)	$J_{ac} = 11{\cdot}07$ Hz ($\phi \sim \pi$)	$J_{ab'} = 7{\cdot}5$ Hz ($\phi \sim 2\pi/3$)

Only values averaged over internal rotation about the C—C bond are normally obtained for acyclic ethane derivatives; for compounds of the type CH_3CHXY the observed value is $\frac{2}{3}J_g + \frac{1}{3}J_t$ where J_g and J_t are the *gauche* and *trans* coupling constants respectively. The situation is more clear-cut in the case of cyclohexane derivatives in the chair form. There are five possible mutual orientations of the protons, each of which may be equatorial (e) or axial (a)—see Table 8-16. The first three rows of this table are for 'fixed' chair conformations, which occur when the two possible chair forms are non-equivalent (or at low temperatures in cases where they are equivalent); the last two rows apply when rapid ring inversion occurs, as is true at room temperature when the inversion isomers are equivalent. In practice, $^3J_{ee}$ is generally found to be less than $^3J_{ae}$, particularly when electronegative substituents are present. As an example of this type of information (typical

Table 8-16 Orientations and *vicinal* coupling constants for cyclohexane derivatives

Position of protons	ϕ	*Expected* ^3J/Hz[(a)]
aa	π	13
ae	$\pi/3$	4
ee	$\pi/3$	4
aa ⇌ *ee*	$\pi \rightleftharpoons \pi/3$	8·5
ae ⇌ *ea*	$\pi/3 \rightleftharpoons \pi/3$	4

(a) Using Eqn 8-42 with $A = 7$ Hz, $B = -1$ Hz, $C = 5$ Hz.

of the use of NMR in organic chemistry), consider 2,3,5,6-tetramethylpiperazine, of which there are five configurational isomers. The NMR spectrum for the dihydrochloride of one of these, the α-form, shows there is only one type of CH hydrogen, and the vicinal coupling constant between two such ring hydrogens is 11·1 Hz. Therefore these hydrogens must all be axial, and the structure must be as in 8-[XVI].

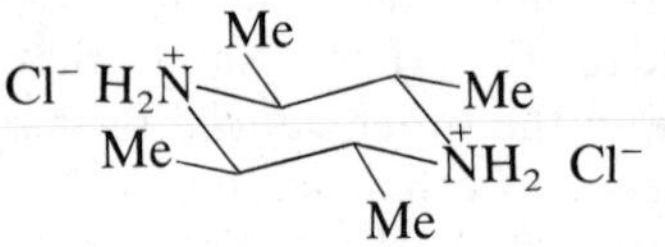

α isomer
8-[XVI]

Valence bond calculations also make the following predictions, which agree with experimental observation—though the effects are not independent:

(a) Electronegative substituents decrease ^3J.
(b) Increase of HCC bond angles decreases ^3J.
(c) Increase of C—C bond length decreases ^3J.

Effect (a) is well illustrated by ethyl compounds and may be expressed by Eqn 8-43.

$$^3J(CH_3CH_2X) = {}^3J^0[1 - 0{\cdot}09(E_X - E_H)] \qquad (8\text{-}43)$$

The quantity $^3J^0$ is the value for ethane itself (8·0 Hz), while E_X and E_H are the electronegativities of the substituent X and hydrogen respectively. However, the magnitude of the electronegativity effect depends on the dihedral angle, and empirically it is found that the effect is greatest (i.e. $^3J_{HH}$ least) when X is *trans* to one of the coupled protons. Some examples will make this clear:

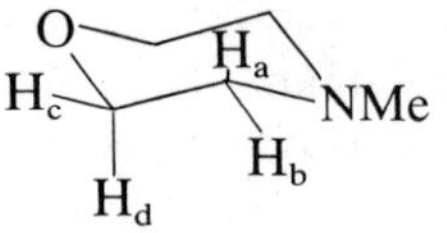

$J_{ac} = 2{\cdot}96$ Hz
$J_{bc} < 1{\cdot}0$ Hz
8-[XVII]

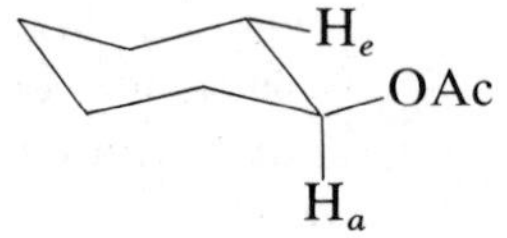

$J_{ae} \sim 5{\cdot}5$ Hz
8-[XVIIIa]

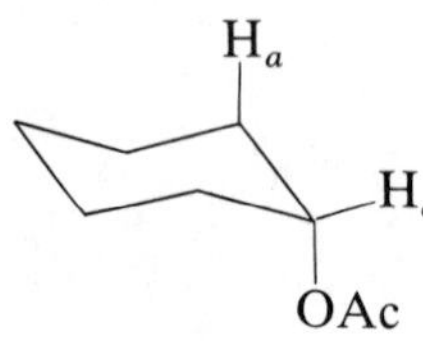

$J_{ae} \sim 3{\cdot}0$ Hz
8-[XVIIIb]

The sign of $^3J_{HH}$ is invariably found to be positive, in agreement with the simple Dirac vector model. Indeed, relative signs of other coupling constants are often based on an assumption that $^3J_{HH}$ is positive. *Vicinal* coupling constants are also normally positive for other nuclei coupled to hydrogen, although negative values of $^3J_{FF}$ are well-known. However, if the Dirac vector model were entirely valid, no dependence of the coupling constant on dihedral angle would be

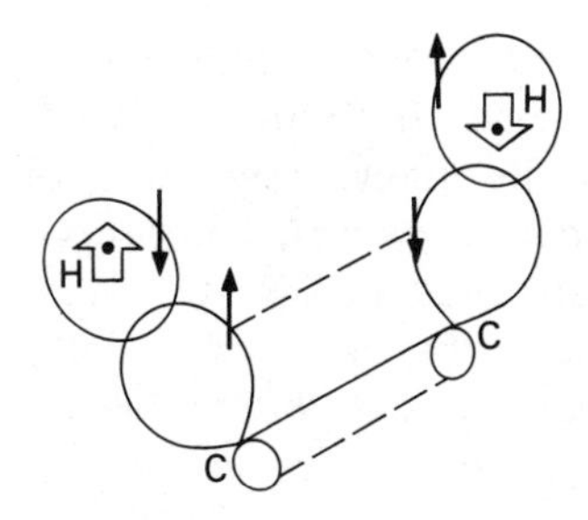

Fig. 8-14 Spin correlation for *vicinal* (H, H) coupling.

expected. In fact, the most important exchange integral determining the electron spin correlation between electrons in different bonds is that between *vicinal* orbitals (Fig. 8-14). This causes the electrons to be stable when anti-parallel and thus leads to a positive *vicinal* coupling constant (as does the Dirac vector model). Figure 8-12 makes it clear that the electron correlation depends greatly on dihedral angle (since overlap of the *vicinal* orbitals does), and indicates that it will be optimum when the bonds are *cis* or *trans*. In fact the overlap (and hence ^3J) is slightly greater for *trans* bonds than for *cis* bonds.

Karplus-type equations are not general for nuclei other than hydrogen, but they have been developed for several classes of coupling, e.g. (^{13}C, ^{1}H), (^{13}C, ^{13}C) and (^{31}P, ^{13}C), and have reasonable validity provided closely similar compounds are treated, multiply-bonded atoms are excluded, and the coupled nuclei do not possess lone pairs.

8-24 Long-range (H, H) coupling

Coupling over more than three chemical bonds is normally termed 'long-range'. For protons in saturated compounds such coupling is normally small (often negligible). Moreover, it is often highly stereospecific; the optimum path for such coupling appears to be a planar zig-zag atomic chain (schemes 8-[XIX] and 8-[XX]). Some examples are given in schemes 8-[XXI] to [XXIV]. Note that $|^4J|$ increases

^4J

8-[XIX]

^5J

8-[XX]

$|^4J_{2e,4e}| = 1{\cdot}0$ Hz $|^4J_{2e,4a}| = 0{\cdot}45$ Hz
$|^5J_{2e,5e}| = 0{\cdot}9$ Hz $|^5J_{2e,5a}| < 0{\cdot}3$ Hz

8-[XXI]

$^4J_{ac} = +7{\cdot}4$ Hz
$^4J_{bc} = -0{\cdot}2$ Hz

8-[XXII]

$|^4J| = 18$ Hz

8-[XXIII]

$|^5J| = 2{\cdot}3$ Hz

8-[XXIV]

rapidly as the number of 'coupling paths' increases. One of the commonest favourable situations is that of coupling over four bonds between equatorial protons in six-membered saturated rings. Another

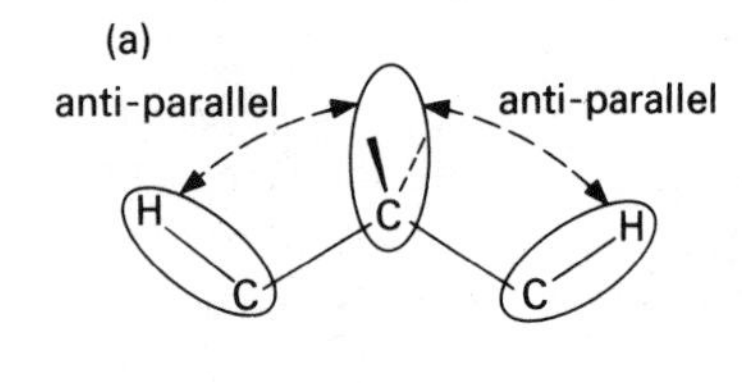

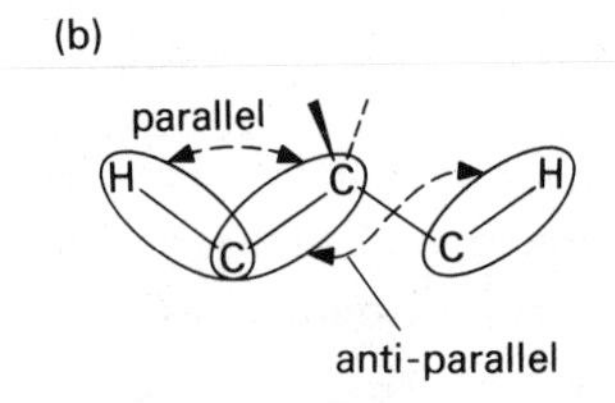

Fig. 8-15 Coupling paths for $^4J_{HH}$

favourable situation for ^4J occurs when the intervening atom is sp^2-hybridized, as in acetone and dimethylsulphoxide ($|^4J|=0{\cdot}55$ Hz and 0·45 Hz, respectively, for the neat liquids). Normally, however, the coupling for acyclic compounds is averaged over several conformations, few of which have the optimum geometry, so the observed value is usually very small or undetectable. It may be noted that both positive and negative values of $^4J_{HH}$ have been reported, but the value for the planar zig-zag relationship (when π effects are absent) is probably always positive.

Both the stereospecificity and the frequent occurrence of positive signs indicate that the Dirac vector model cannot adequately account for the observed values of $^4J_{HH}$. The stereospecificity indicates that *vicinal* exchange interactions (i.e. direct spin correlation between electrons in *vicinal* orbitals) are again of importance. It has been suggested that at least two types of coupling path must be considered. These may be described in terms of spin correlation between electrons in different bonds (Fig. 8-15). Path (a) involves two *vicinal* orbital interactions (see Fig. 8-14) and gives a negative contribution to ^4J, whereas path (b) involves one *vicinal* orbital interaction and one *geminal* interaction [spins parallel as for Hund coupling—see Fig. 8-12(b)] leading to a positive contribution to ^4J. The total coupling constant may therefore be positive or negative; since the *vicinal* orbital interactions are maximum when the orbitals are *trans*, path (b) dominates for the zig-zag conformation, giving a positive sign (and, in fact, maximum magnitude) in this case. The Dirac vector model provides a further coupling path which gives a negative contribution to $^4J_{HH}$, independent of conformation.

8-25 Coupling through π electrons

The π electrons appear to be involved in the nuclear spin–spin coupling for unsaturated compounds; in fact the observed coupling constant may be written as the sum of two parts, J^σ and J^π. Of course, the π-electron distributions have nodes at the nuclei and so cannot contribute to the contact term directly. Consequently it is necessary to introduce spin correlation between σ and π electrons. This same interaction accounts for hyperfine coupling between protons and unpaired electrons in the ESR of related free radicals. It is possible to obtain a relation between J and the hyperfine splitting constants. Three types of σ–π exchange repulsion may be considered:

(a) acetylenic—the σ electron on H is stabilized anti-parallel to the π electron on C for a H—C≡group.
(b) ethylenic—again the anti-parallel orientation is stabilized.
(c) allylic—in this case the parallel orientation is stable.

Figure 8-16 shows the stable situation for propene, involving both ethylenic and allylic interactions. It may be seen that a negative value of ^4J is predicted. In fact the allylic σ–π exchange repulsion depends on the dihedral angle about the C—C single bond, being approximately zero when all the nuclei involved are coplanar, and maximum

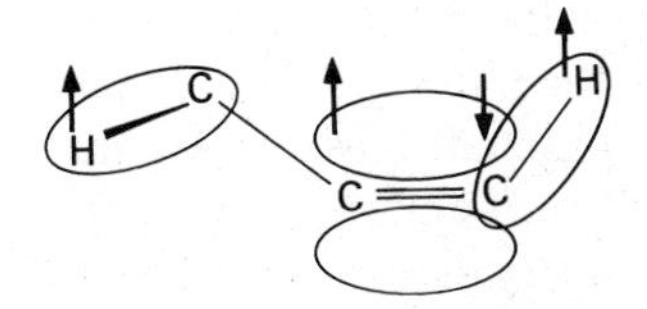

Fig. 8-16 Spin correlation for allylic $^4J^{\pi}_{HH}$

when the allylic C—H bond is perpendicular to the C=C plane. For allyl compounds the observed coupling constant is a weighted average over all conformations. Further conclusions of the theory are:

(i) ^3J (ethylenic) should be positive, ^4J (allylic) should be negative, and ^5J (known as homoallylic for the group HC—C=C—CH) should be positive.

(ii) Orientation about a double bond should not affect J; e.g. the *cisoid* ^4J illustrated in Fig. 8-16 should equal the corresponding *transoid* ^4J.

(iii) Since π-electron systems are often delocalized, spin correlation may occur for electrons at atoms separated by many chemical bonds, i.e. π-coupling should be long-ranged. The magnitude of coupling constants in conjugated systems may be evaluated using linear combinations of atomic orbitals in a similar fashion to that developed in Section 8-21.

These conclusions agree in general with experimental observation, though it is often difficult to estimate J^{π} because of contributions of unknown magnitude from J^{σ}. Values of $^3J_{HH}$ are dominated by the σ contribution; both σ and π effects may be important for ^4J and ^5J. Table 8-17 lists some values of (H, H) coupling constants in unsaturated systems. The signs, where known, are in agreement with prediction (i) above. Orientation about the double bond seems to affect ^4J and ^5J in *cis*- and *trans*-2-butene to only a small extent. The values of ^5J for butatriene are large due to the existence of several paths for π-coupling; $^5J_{cis}$ and $^5J_{trans}$ are equal within experimental error, thus indicating that the π contribution is dominant. The case of 1,3-butadiene is of considerable interest; of the three different coupling constants over five bonds, the values for *cis-cis* and *cis-trans* orientations are approximately equal (0·69 Hz and 0·60 Hz respectively) and are therefore suggested to be dominated by J^{π}, whereas the *trans-trans* value is considerably larger (1·30 Hz) indicating a contribution from J^{σ}. This is not unexpected since the coupling path is of the planar zig-zag type known to be favourable for σ coupling (see the preceding section).

Table 8-17 (H, H) coupling constants in some unsaturated compounds

Compound	^4J/Hz	^5J/Hz
cis-2-Butene	−1·79	+1·18
trans-2-Butene	−1·73	+1·60
Allene	−7·0	—
Butatriene	—	±8·95 (*cis*)
		±8·95 (*trans*)
1,3-Butadiene	−0·83 (*cis*)	+0·69 (*cis, cis*)
	−0·86 (*trans*)	+0·60 (*cis, trans*)
		+1·30 (*trans, trans*)

The ESR hyperfine splitting constant of the ethylenic type is in fact somewhat smaller in magnitude (−65 MHz) than the allylic constant averaged over all the positions of a methyl group, (+75 MHz). Consequently, replacement of an ethylenic proton by a methyl group should affect π coupling to any other proton by (a) a slight increase in magnitude, and (b) a change of sign. This prediction has been used to test whether coupling is of the π-type or not. Data for the butadiene series shows that such a test confirms that $^5J_{cis,cis}$ and $^5J_{cis,trans}$ are π-dominated whereas $^5J_{trans,trans}$ has roughly equal contributions from σ and π mechanisms.

The long-range nature of J^{π} is well illustrated by the compound $CH_3C{\equiv}CC{\equiv}CC{\equiv}CCH_2OH$, where an observable (H, H) coupling (0·4 Hz) is obtained through no fewer than nine chemical bonds.

Similar considerations to the above apply to (C, H) and (C, C) coupling constants for conjugated compounds.

8-26 'Through-space' coupling

In certain instances, anomalously large coupling constants are observed for nuclei many bonds apart but physically close in space. This occurs particularly for (F, F) coupling but has been found in other cases also. Some representative examples are shown below:

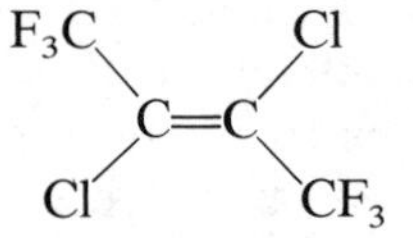

$F_3C(Cl)C{=}C(Cl)CF_3$ (trans)

$^5J_{FF} = 1{\cdot}44$ Hz

8-[XXV]

$F_3C(Cl)C{=}C(Cl)CF_3$ (cis)

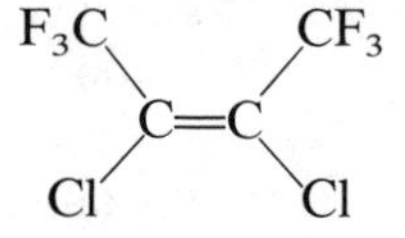

$^5J_{FF} = 13{\cdot}39$ Hz

8-[XXVI]

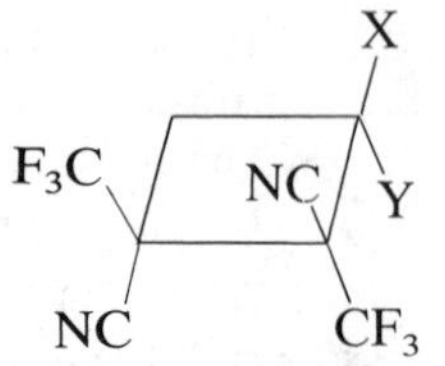

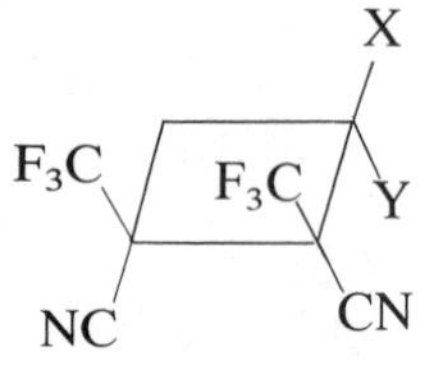

X = H, Y = OEt and vice versa

J_{FF} < ca. 1 Hz

8-[XXVII]

$J_{FF} = 10$ to 12 Hz

8-[XXVIII]

It should be noted that there is some dispute about the existence and origin of this effect. However, it is not necessary to postulate any novel coupling mechanism; all that is implied is that there is direct electron spin correlation between electrons of fluorine nuclei (or of bonds to such nuclei) close in space. The name 'through-space' is therefore rather misleading. 'Through-space' coupling may be the dominant effect for (F, F) interactions in saturated compounds when the nuclei are separated by five or more chemical bonds.

It has been suggested, from both theoretical and experimental standpoints, that 'through-space' effects may also be noticeable for

(H, F) and (H, H) coupling constants. The contribution to J_{HH} from this effect is usually positive. Such an effect provides a further coupling path for $^4J_{HH}$ which clearly is highly dependent on conformation.

Notes and references

[1] The expression 'paramagnetic' is somewhat misleading—it should be emphasized that this term (sometimes called the temperature-independent paramagnetic term) is to be sharply distinguished from the temperature-dependent paramagnetism associated with systems having unpaired electrons.

[2] A. A. Cheremisin & P. V. Schastnev, *J. Magn. Reson.*, **40,** 459 (1980).

[3] See, for example, K. A. K. Ebraheem, G. A. Webb & M. Witanowski, *Org. Magn. Reson.*, **8,** 317 (1976).

[4] C. J. Jameson and H. S. Gutowsky, *J. Chem. Phys.*, **40,** 1714 (1964).

[5] (a) J. M. Letcher & J. R. Van Wazer in *Topics in Phosphorus Chemistry*, Vol. 5, Eds. M. Grayson & E. J. Griffiths, Wiley & Sons, New York (1967).
(b) G. Engelhardt, R. Radeglia, H. Jancke, E. Lippmaa & M. Magi, *Org. Magn. Reson.*, **5,** 561 (1973).

[6] H. C. E. McFarlane & W. McFarlane, *J.C.S. Dalton*, 2416 (1973).

[7] R. K. Harris & N. Sheppard, *Mol. Phys.*, **7,** 595 (1964).

[8] J. F. Hinton & R. W. Briggs, *J. Magn. Reson.*, **19,** 393 (1975); **25,** 379 (1977).

[9] D. M. Grant & E. G. Paul, *J. Amer. Chem. Soc.*, **86,** 2984 (1964).

[10] L. P. Lindeman & J. Q. Adams, *Anal. Chem.*, **43,** 1245 (1971).

[11] The method of presentation used here differs from that of Lindeman and Adams but appears to the present author to provide pedagogical advantages. The numerical values in Table 8-8 are, however, adapted from the work of Lindeman and Adams.

[12] H. Spiesecke & W. G. Schneider, *Tetrahedron Lett.* 468 (1961);
G. A. Olah & G. D. Mateșcu, *J. Amer. Chem. Soc.*, **92,** 1430 (1970).

[13] P. Laszlo, *Prog. NMR Spectrosc.*, **3,** 231 (1967).

[14] C. C. Hinckley, *J. Amer. Chem. Soc.*, **91,** 5160 (1969).

[15] U. Haeberlen, *Faraday Symposium* **13,** 31 (1978).

[16] J. D. Kennedy & W. McFarlane, *J.C.S. Faraday II*, **72,** 1653 (1976).

[17] J. A. Pople & D. P. Santry, *Mol. Phys.*, **8,** 1 (1964).

[18] H. A. Bent, *Chem. Rev.* **61,** 275 (1961).

[19] This equation is sometimes written in the form $^3J = D\cos^2\phi + E$, where D is somewhat larger for $\pi/2 \leq \phi \leq \pi$ than for $0 \leq \phi \leq \pi/2$.

[20] M. Karplus, *J. Chem. Phys.*, **30,** 11 (1959).

Further reading

Applications of NMR Spectroscopy in Organic Chemistry (second edition), L. M. Jackman & S. Sternhell, Pergamon Press (1969).

'Van der Waals forces and shielding effects', F. H. A. Rummens, *NMR Basic Principles & Prog.*, **10,** 1 (1975).

'Semi-empirical calculations of the chemical shifts of nuclei other than protons', K. A. K. Ebraheem & G. A. Webb, *Prog. NMR Spectrosc.* **11,** 149 (1978).

'Solvent effects and NMR', P. Laszlo, *Prog. NMR Spectrosc.*, **3,** 231 (1967).

'The theory of nuclear spin-spin coupling in high-resolution NMR spectroscopy', J. N. Murrell, *Prog. NMR Spectrosc.*, **6,** 1 (1971).

'Nuclear shielding of the transition metals', R. G. Kidd, *Ann. Repts. NMR Spectrosc.*, **10A,** 2 (1980).

'NMR spectroscopy of paramagnetic species', K. G. Orrell, *Ann. Repts. NMR Spectrosc.*, **9,** 1 (1979).

'Long-range proton spin-spin coupling', M. Barfield & B. Chakrabarti, *Chem. Rev.*, **69,** 757 (1969).

'The 'through-space' mechanism in spin-spin coupling', J. Hilton & L. H. Sutcliffe, *Prog. NMR Spectrosc.*, **10,** 27 (1975).

'Calculations of nuclear spin–spin coupling', J. Kowalewski, *Ann. Repts. NMR Spectrosc.* **12,** 81 (1982).

Problems

8-1 By transformation to polar coordinates, show that Eqn 8-7 reduces to Eqn 8-8 for a spherical charge distribution.

8-2 From the data of Table 8-7 calculate the angle which the nodal surface of shielding (due to magnetic anisotropy) makes with the bond direction in the nodal plane of the π electrons for the carbonyl group.

8-3 Show that Eqn 8-17 reduces to 8-18 in all cases, and becomes 8-19 if $\chi'_\perp = \chi''_\perp$.

8-4 Show that if $[D]_0 \gg [A]_0$, Eqn 8-23 applies to NMR measurements for a charge-transfer equilibrium.

8-5 The hydrogen molecule has molecular orbitals $\psi_g = (s_A + s_B)/\sqrt{2}$ and $\psi_u = (s_A - s_B)/\sqrt{2}$, where s_A and s_B denote $1s$ atomic orbitals on the two hydrogen atoms A and B. The ground and the first excited electronic states, Ψ_0 and Ψ_1 respectively are:

$$\Psi_0 = \psi_g(\mathbf{r}_1)\psi_g(\mathbf{r}_2)[\alpha_1\beta_2 - \beta_1\alpha_2]/\sqrt{2}$$

$$\Psi_1\ (m_s = 0 \text{ component}) = \tfrac{1}{2}[\psi_g(\mathbf{r}_1)\psi_u(\mathbf{r}_2) - \psi_u(\mathbf{r}_1)\psi_g(\mathbf{r}_2)][\alpha_1\beta_2 + \beta_1\alpha_2]$$

Evaluate the contact contribution to the (H, H) coupling constant using Eqn 8-33 and the following information:

$$U_1 - U_0 = 1{\cdot}6 \times 10^{-18}\ \text{J}$$

$$\langle s_j|\ \delta(r_j)\ |s_j\rangle = 2{\cdot}15 \times 10^{30}\ \text{electrons m}^{-3}$$

(Evaluate the S_z component and multiply by 3 to take account of S_x and S_y contributions.)

8-6 Consider the three molecular orbitals of HF obtained from the hydrogen $1s$ and the fluorine $2s$ and $2p_z$ atomic orbitals (z being the bond direction). Take in turn the hypothetical cases where the energies of the orbitals are (a) $H_{1s} \sim F_{2p_z} \gg F_{2s}$, and (b) $H_{1s} \gg F_{2p_z} \gg F_{2s}$ and estimate which 'excitation' is most important for (H, F) coupling. Give the predicted sign of J_{HF} in these two cases.

8-7 The resonances of the central methylene protons of 8-[XXIX] and 8-[XXX] occur at very different frequencies. Which has the higher δ value? Why?

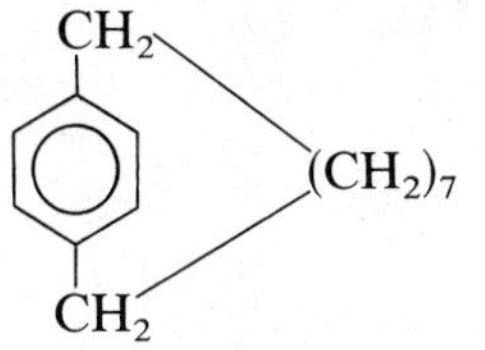

CH_2 $(CH_2)_7$ CH_2 $Ph(CH_2)_9Ph$

8-[XXIX] 8-[XXX]

8-8 Show that, when the situation is treated as rapid exchange between unequally-populated sites, the contact shift for the NMR resonance of a nucleus in a paramagnetic system with $S=\frac{1}{2}$ is given by Eqn 8-25.

What is the significance of the following results?

8-9 In dilute solution in CCl_4 the hydroxyl protons of phenol and methyl salicylate give NMR signals at $\delta=4{\cdot}24$ and $\delta=10{\cdot}58$ respectively.

8-10 The ^{19}F chemical shifts (from the resonance of $CFCl_3$, with positive values to high frequency) of CF_3Cl and CF_3I are $-28{\cdot}1$ and -4.8 ppm respectively.

8-11 $Me_3SnSnMe_3$ has $^2J(^{117}Sn, ^1H)=47{\cdot}0$ Hz, $^2J(^{119}Sn, ^1H)=49{\cdot}6$ Hz.

8-12 *Vicinal* coupling constants for ring protons of 8-[XXXI] and 8-[XXXII] are as below (in Hz):

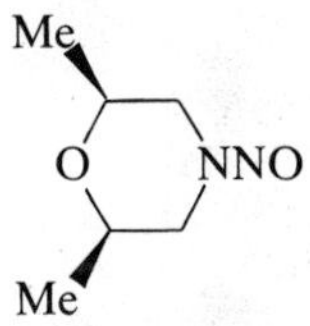

Me O NNO Me Me MeN NNO Me

10·7 and 2·8 4·6 and 1·2

8-[XXXI] 8-[XXXII]

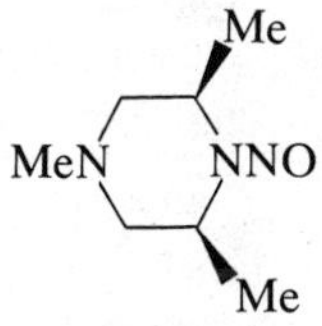

8-13 2,4-disubstituted benzaldehydes 8-[XXXIII] have long-range (H, H) coupling constants (in Hz) as follows:

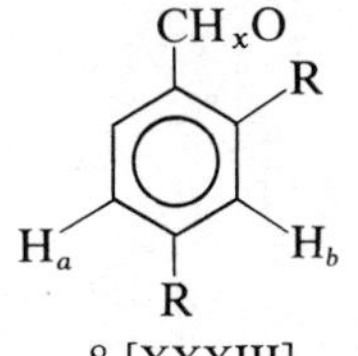

CH_xO R H_a H_b R

8-[XXXIII]

R = OH	$J_{ax}<0{\cdot}1$	$J_{bx}=0{\cdot}51$
R = NO_2	$J_{ax}=0{\cdot}57$	$J_{bx}=0{\cdot}21$

Appendix 1 NMR properties of the spin-½ nuclei[a]

Isotope[b]	*Natural abundance*[c] C/%	*Magnetic moment*[d] μ/μ_N	*Magnetogyric ratio*[d] $\gamma/10^7$ rad T^{-1} s^{-1}	*NMR frequency*[e] Ξ/MHz	*Standard*	*Relative receptivity*[f] D^P	D^C
^{1}H	99·985	4·83724	26·7519	100·000 000	Me_4Si	1·000	$5{\cdot}67\times10^3$
^{3}H[g]	—	5·1596	28·535	(106·664)	Me_4Si-t	—	—
^{3}He	$1{\cdot}3\times10^{-4}$	−3·6851	−20·380	(76·182)	—	$5{\cdot}75\times10^{-7}$	$3{\cdot}26\times10^{-3}$
^{13}C	1·108	1·2166	6·7283	25·145 004	Me_4Si	$1{\cdot}76\times10^{-4}$	1·00
^{15}N	0·37	−0·4903	−2·712	10·136 783	$MeNO_2$ or $[NO_3]^-$	$3{\cdot}85\times10^{-6}$	$2{\cdot}19\times10^{-2}$
^{19}F	100	4·5532	25·181	94·094 003[i]	CCl_3F	0·834	$4{\cdot}73\times10^3$
^{29}Si	4·70	−0·96174	−5·3188	19·867 184	Me_4Si	$3{\cdot}69\times10^{-4}$	2·10
^{31}P	100	1·9602	10·841	40·480 737	85% H_3PO_4	0·0665	$3{\cdot}77\times10^2$
^{57}Fe	2·19	0·1566	0·8661	(3·238)	$Fe(CO)_5$	$7{\cdot}43\times10^{-7}$	$4{\cdot}22\times10^{-3}$
^{77}Se	7·58	0·925	5·12	19·071 523	Me_2Se	$5{\cdot}30\times10^{-4}$	3·01
^{89}Y	100	−0·23786	−1·3155	(4·917)	$Y(NO_3)_3$aq.	$1{\cdot}19\times10^{-4}$	0·675
^{103}Rh	100	−0·153	−0·846	3·172 310	mer-$[RhCl_3(SMe_2)_3]$	$3{\cdot}16\times10^{-5}$	0·179
(^{107}Ag)	51·82	−0·1966	−1·087	4·047 897	Ag^+aq.	$3{\cdot}48\times10^{-5}$	0·197
^{109}Ag	48·18	−0·2260	−1·250	4·653 623		$4{\cdot}92\times10^{-5}$	0·279
(^{111}Cd)	12·75	−1·0293	−5·6926	21·215 478	$CdMe_2$	$1{\cdot}23\times10^{-3}$	6·97
^{113}Cd[h]	12·26	−1·0768	−5·9550	22·193 173		$1{\cdot}35\times10^{-3}$	7·67
(^{115}Sn)	0·35	−1·590	−8·792	(32·86)		$1{\cdot}24\times10^{-4}$	0·705
(^{117}Sn)	7·61	−1·732	−9·578	35·632 295	Me_4Sn	$3{\cdot}49\times10^{-3}$	19·8
^{119}Sn	8·58	−1·8119	−10·021	37·290 662		$4{\cdot}51\times10^{-3}$	25·6
(^{123}Te[h])	0·87	−1·275	−7·049	(26·35)	Me_2Te	$1{\cdot}59\times10^{-4}$	0·903
^{125}Te	6·99	−1·537	−8·498	31·549 802		$2{\cdot}24\times10^{-3}$	12·7
^{129}Xe	26·44	−1·345	−7·441	(27·81)	$XeOF_4$	$5{\cdot}69\times10^{-3}$	32·3
^{169}Tm	100	−0·400	−2·21	(8·27)	—	$5{\cdot}66\times10^{-4}$	3·21
^{171}Yb	14·31	0·8520	4·712	(17·61)	—	$7{\cdot}82\times10^{-4}$	4·44
^{183}W	14·40	0·2025	1·120	4·161 733	WF_6	$1{\cdot}06\times10^{-5}$	$5{\cdot}99\times10^{-2}$
^{187}Os	1·64	0·111	0·616	2·282 343	OsO_4	$2{\cdot}00\times10^{-7}$	$1{\cdot}14\times10^{-3}$
^{195}Pt	33·8	1·043	5·768	21·414 376	$[Pt(CN)_6]^{2-}$	$3{\cdot}39\times10^{-3}$	19·2
^{199}Hg	16·84	0·87072	4·8154	17·910 841	Me_2Hg	$9{\cdot}82\times10^{-4}$	5·57
(^{203}Tl)	29·50	2·7912	15·436	(57·70)	$TlNO_3$aq.	0·0567	$3{\cdot}22\times10^2$
^{205}Tl	70·50	2·8187	15·589	57·633 833		0·140	$7{\cdot}91\times10^2$
^{207}Pb	22·6	1·002	5·540	20·920 597	Me_4Pb	$2{\cdot}01\times10^{-3}$	11·4

[a] A complete list, excluding most radioactive nuclei.

[b] Nuclei in brackets are considered to be not the most favourable for the element concerned.

[c] Data from *Handbook of Chemistry and Physics*, 55th edition, CRC Press (1974–5), pages B248-332, except for the value for ^{13}C (which is taken from page E69).

[d] Data derived from the compilation of G. H. Fuller, *J. Phys. Chem. Ref. Data* **5,** 835 (1976), which lists values of $\mu_{max} = \gamma\hbar I$, corrected for diamagnetic shielding.

[e] Resonance frequency in a magnetic field such that the protons of TMS resonate at exactly 100 MHz. The values quoted are for the resonances of the standards listed in the next column and are taken (except where otherwise stated) from *NMR and the Periodic Table*, Eds R. K. Harris & B. E. Mann, Academic Press (1978). Values in brackets are calculated from the magnetogyric ratios given in the preceding column, and are therefore relative to the resonant frequency of bare protons.

[f] D^p is the receptivity relative to that of ^1H whereas D^C is relative to ^{13}C (see Section 3-6 and Eqn 5-30).

[g] Radioactive (half-life 12 y)

[h] Long-lived radioactive isotope

[i] S. Brownstein & J. Bornais, *J. Magn. Reson.* **38,** 131 (1980).

Appendix 2 The spin properties of quadrupolar nuclei[(a)]

Isotope[(b)]	*Spin*[(c)]	*Natural abundance*[(d)] C/%	*Magnetic moment*[(c)] μ/μ_N	*Magnetogyric ratio*[(c)] $\gamma/10^7$ rad T^{-1} s^{-1}	*Quadruple moment*[(c,e)] 10^{28} Q/m^2	*NMR frequency*[(f)] Ξ/MHz	*Linewidth factor*[(g)] $10^{56}\ell/m^4$	*Relative receptivity*[(b)] D^p	D^C
^{2}H[(i)]	1	0·015	1·2126	4·1066	2·8 × 10^{-3}	15·351	3·9 × 10^{-5}	1·45 × 10^{-6}	8·21 × 10^{-3}
^{6}Li	1	7·42	1·1625	3·9371	−8 × 10^{-4}	14·717	3·2 × 10^{-6}	6·31 × 10^{-4}	3·58
^{7}Li	$\frac{3}{2}$	92·58	4·20394	10·3975	−4 × 10^{-2}	38·866	2·1 × 10^{-3}	0·272	1·54 × 10^{3}
^{9}Be	$\frac{3}{2}$	100	−1·52008	−3·75958	5 × 10^{-2}	14·054	3·3 × 10^{-3}	1·39 × 10^{-2}	78·7
^{10}B	3	19·58	2·0792	2·8746	8·5 × 10^{-2}	10·746	1·4 × 10^{-3}	3·93 × 10^{-3}	22·3
^{11}B	$\frac{3}{2}$	80·42	3·4708	8·5843	4·1 × 10^{-2}	32·089	2·2 × 10^{-3}	0·133	7·52 × 10^{2}
^{14}N[(i)]	1	99·63	0·57099	1·9338	1 × 10^{-2}	7·228	5·0 × 10^{-4}	1·00 × 10^{-3}	5·69
^{17}O	$\frac{5}{2}$	0·037	−2·2407	−3·6279	−2·6 × 10^{-2}	13·561	2·2 × 10^{-4}	1·08 × 10^{-5}	6·11 × 10^{-2}
^{21}Ne	$\frac{3}{2}$	0·257	−0·85433	−2·1130	9 × 10^{-2}	7·899	1·1 × 10^{-2}	6·33 × 10^{-6}	3·59 × 10^{-2}
^{23}Na	$\frac{3}{2}$	100	2·86265	7·08013	0·10	26·466	1·3 × 10^{-2}	9·27 × 10^{-2}	5·26 × 10^{2}
^{25}Mg	$\frac{5}{2}$	10·13	−1·012	−1·639	0·22	6·126	1·5 × 10^{-2}	2·72 × 10^{-4}	1·54
^{27}Al	$\frac{5}{2}$	100	4·3084	6·9760	0·15	26·077	7·2 × 10^{-3}	0·207	1·17 × 10^{3}
^{33}S	$\frac{3}{2}$	0·76	0·8308	2·055	−5·5 × 10^{-2}	7·681	4·0 × 10^{-3}	1·72 × 10^{-5}	9·77 × 10^{-2}
^{35}Cl	$\frac{3}{2}$	75·53	1·0610	2·6240	−0·10	9·809	1·3 × 10^{-2}	3·56 × 10^{-3}	20·2
^{37}Cl	$\frac{3}{2}$	24·47	0·88313	2·1842	−7·9 × 10^{-2}	8·165	8·3 × 10^{-3}	6·66 × 10^{-4}	3·78
^{39}K	$\frac{3}{2}$	93·1	0·50533	1·2498	4·9 × 10^{-2}	4·672	3·2 × 10^{-3}	4·75 × 10^{-4}	2·69
(^{41}K)	$\frac{3}{2}$	6·88	0·27740	0·68608	6·0 × 10^{-2}	2·565	4·8 × 10^{-3}	5·80 × 10^{-6}	3·29 × 10^{-2}
^{43}Ca	$\frac{7}{2}$	0·145	−1·4936	−1·8025	0·2[(j)]	6·738	5·4 × 10^{-3}	8·67 × 10^{-6}	4·92 × 10^{-2}
^{45}Sc	$\frac{7}{2}$	100	5·3927	6·5081	−0·22	24·328	6·6 × 10^{-3}	0·302	1·72 × 10^{3}
^{47}Ti	$\frac{5}{2}$	7·28	−0·93292	−1·5105	0·29	5·646	2·7 × 10^{-2}	1·53 × 10^{-4}	0·867
^{49}Ti	$\frac{7}{2}$	5·51	−1·25198	−1·51093	0·24	5·648	7·8 × 10^{-3}	2·08 × 10^{-4}	1·18
(^{50}V)[(k)]	6	0·24	3·6152	2·6717	±6 × 10^{-2}	9·987	1·4 × 10^{-4}	1·34 × 10^{-4}	0·759
^{51}V	$\frac{7}{2}$	99·76	5·8379	7·0453	−5 × 10^{-2}	26·336	3·4 × 10^{-4}	0·383	2·17 × 10^{3}
^{53}Cr	$\frac{3}{2}$	9·55	−0·6113	−1·512	3 × 10^{-2}	5·651	1·2 × 10^{-3}	8·62 × 10^{-5}	0·489
^{55}Mn	$\frac{5}{2}$	100	4·081	6·608	0·4	24·70	5·1 × 10^{-2}	0·176	9·97 × 10^{2}
^{59}Co	$\frac{7}{2}$	100	5·234	6·317	0·38	23·61	2·0 × 10^{-2}	0·277	1·57 × 10^{3}
^{61}Ni	$\frac{3}{2}$	1·19	−0·9680	−2·394	0·16	8·949	3·4 × 10^{-2}	4·06 × 10^{-5}	0·231
^{63}Cu	$\frac{3}{2}$	69·09	2·8696	7·0974	−0·211	26·530	5·94 × 10^{-2}	6·45 × 10^{-2}	3·66 × 10^{2}
^{65}Cu	$\frac{3}{2}$	30·91	3·0741	7·6031	−0·195	28·421	5·07 × 10^{-2}	3·55 × 10^{-2}	2·01 × 10^{2}
^{67}Zn	$\frac{5}{2}$	4·11	1·0356	1·6768	0·16	6·2679	8·2 × 10^{-3}	1·18 × 10^{-4}	0·670
(^{69}Ga)	$\frac{3}{2}$	60·4	2·6007	6·4323	0·19	24·044	4·8 × 10^{-2}	4·19 × 10^{-2}	2·38 × 10^{2}
^{71}Ga	$\frac{3}{2}$	39·6	3·3046	8·1731	0·12	30·551	1·9 × 10^{-2}	5·65 × 10^{-2}	3·20 × 10^{2}
^{73}Ge	$\frac{9}{2}$	7·76	−0·97197	−0·93574	−0·18	3·498	2·4 × 10^{-3}	1·10 × 10^{-4}	0·622
^{75}As	$\frac{3}{2}$	100	1·858	4·595	0·29	17·18	0·11	2·53 × 10^{-2}	1·44 × 10^{2}
(^{79}Br)	$\frac{3}{2}$	50·54	2·7182	6·7228	0·37	25·130	0·18	4·01 × 10^{-2}	2·28 × 10^{2}
^{81}Br	$\frac{3}{2}$	49·46	2·9300	7·2468	0·31	27·089	0·13	4·92 × 10^{-2}	2·79 × 10^{2}
^{83}Kr	$\frac{9}{2}$	11·55	−1·073	−1·033	0·26	3·860	5·0 × 10^{-3}	2·19 × 10^{-4}	1·24
(^{85}Rb)	$\frac{5}{2}$	72·15	1·6002	2·5909	0·26	9·685	2·2 × 10^{-2}	7·65 × 10^{-3}	43·4

Isotope[b]	Spin[c]	Natural abundance[d] C/%	Magnetic moment[c] μ/μ_N	Magnetogyric ratio[c] $\gamma/10^7$ rad T^{-1} s^{-1}	Quadruple moment[c,e] 10^{28} Q/m²	NMR frequency[f] Ξ/MHz	Linewidth factor[g] $10^{56}\ell/m^4$	Relative receptivity[b]	
								D^p	D^C
^{87}Rb[k]	$\frac{3}{2}$	27·85	3·5502	8·7807	0·13	32·823	$2{\cdot}3\times10^{-2}$	$4{\cdot}92\times10^{-2}$	$2{\cdot}79\times10^{2}$
^{87}Sr	$\frac{9}{2}$	7·02	−1·208	−1·163	0·3	4·349	$6{\cdot}7\times10^{-3}$	$1{\cdot}91\times10^{-4}$	1·08
^{91}Zr	$\frac{5}{2}$	11·23	−1·5415	−2·4959	−0·21[l]	9·3298	$1{\cdot}4\times10^{-2}$	$1{\cdot}06\times10^{-3}$	6·04
^{93}Nb	$\frac{9}{2}$	100	6·818	6·564	−0·22	24·54	$3{\cdot}6\times10^{-3}$	0·487	$2{\cdot}77\times10^{3}$
^{95}Mo	$\frac{5}{2}$	15·72	1·081	1·750	±0·12	6·542	$4{\cdot}6\times10^{-3}$	$5{\cdot}14\times10^{-4}$	2·92
(^{97}Mo)	$\frac{5}{2}$	9·46	−1·104	−1·787	±1·1	6·679	0·39	$3{\cdot}29\times10^{-4}$	1·87
^{99}Tc[k]	$\frac{9}{2}$	—	6·281	6·046	0·3	22·60	$6{\cdot}7\times10^{-3}$	—	—
^{99}Ru	$\frac{5}{2}$	12·72	−0·7623[m]	−1·234[m]	$7{\cdot}6\times10^{-2}$	4·614[m]	$1{\cdot}8\times10^{-3}$	$1{\cdot}46\times10^{-4}$	0·827
^{101}Ru	$\frac{5}{2}$	17·07	−0·8544[m]	−1·383[m]	0·44	5·171[m]	$6{\cdot}2\times10^{-2}$	$2{\cdot}75\times10^{-4}$	1·56
^{105}Pd	$\frac{5}{2}$	22·23	−0·760	−1·23	0·8	4·60	0·20	$2{\cdot}52\times10^{-4}$	1·43
(^{113}In)	$\frac{9}{2}$	4·28	6·1058	5·8782	0·82	21·973	$5{\cdot}0\times10^{-2}$	$1{\cdot}50\times10^{-2}$	85·0
^{115}In[k]	$\frac{9}{2}$	95·72	6·1190	5·8908	0·83	22·020	$5{\cdot}1\times10^{-2}$	0·337	$1{\cdot}91\times10^{3}$
^{121}Sb	$\frac{5}{2}$	57·25	3·9747	6·4355	−0·28	24·056	$2{\cdot}5\times10^{-2}$	$9{\cdot}30\times10^{-2}$	$5{\cdot}27\times10^{2}$
(^{123}Sb)	$\frac{7}{2}$	42·75	2·8876	3·4848	−0·36	13·026	$1{\cdot}8\times10^{-2}$	$1{\cdot}98\times10^{-2}$	$1{\cdot}13\times10^{2}$
^{127}I	$\frac{5}{2}$	100	3·3238	5·3817	−0·79	20·117	0·20	$9{\cdot}50\times10^{-2}$	$5{\cdot}39\times10^{2}$
^{131}X[i]	$\frac{3}{2}$	21·18	0·8918	2·206	−0·12	8·245	$1{\cdot}9\times10^{-2}$	$5{\cdot}94\times10^{-4}$	3·37
^{133}Cs	$\frac{7}{2}$	100	2·9231	3·5277	-3×10^{-3}	13·187	$1{\cdot}2\times10^{-6}$	$4{\cdot}82\times10^{-2}$	$2{\cdot}73\times10^{2}$
(^{135}Ba)	$\frac{3}{2}$	6·59	1·080	2·671	0·18	9·984	$4{\cdot}3\times10^{-2}$	$3{\cdot}28\times10^{-4}$	1·86
^{137}Ba	$\frac{3}{2}$	11·32	1·208	2·988	0·28	11·17	0·10	$7{\cdot}89\times10^{-4}$	4·47
^{139}La	$\frac{7}{2}$	99·911	3·150	3·801	0·22	14·210	$6{\cdot}6\times10^{-3}$	$6{\cdot}02\times10^{-2}$	$3{\cdot}42\times10^{2}$
^{177}Hf	$\frac{7}{2}$	18·50	0·8960	1·081	4·5	4·042	2·8	$2{\cdot}57\times10^{-4}$	1·46
^{179}Hf	$\frac{9}{2}$	13·75	−0·705	−0·679	5·1	2·54	1·9	$7{\cdot}42\times10^{-5}$	0·421
^{181}Ta	$\frac{7}{2}$	99·988	2·66	3·22	3	12·0	1·2	$3{\cdot}65\times10^{-2}$	$2{\cdot}07\times10^{2}$
(^{185}Re)	$\frac{5}{2}$	37·07	3·753	6·077	2·3	22·72	1·7	$5{\cdot}13\times10^{-2}$	$2{\cdot}91\times10^{2}$
^{187}Re[k]	$\frac{5}{2}$	62·93	3·791	6·138	2·2	22·94	1·5	$8{\cdot}81\times10^{-2}$	$5{\cdot}00\times10^{2}$
^{189}Os[i]	$\frac{3}{2}$	16·1	0·8475	2·096	0·8	7·836	0·85	$3{\cdot}87\times10^{-4}$	2·20
(^{191}Ir)	$\frac{3}{2}$	37·3	0·1877	0·4643	1·1	1·735	1·6	$9{\cdot}77\times10^{-6}$	$5{\cdot}54\times10^{-2}$
^{193}Ir	$\frac{3}{2}$	62·7	0·2044	0·5054	1·0	1·889	1·3	$2{\cdot}11\times10^{-5}$	0·120
^{197}Au	$\frac{3}{2}$	100	0·18701	0·46254	0·59	1·729	0·46	$2{\cdot}58\times10^{-5}$	0·147
^{201}Hg[i]	$\frac{3}{2}$	13·22	−0·71871	−1·7776	0·44	6·645	0·26	$1{\cdot}94\times10^{-4}$	1·10
^{209}Bi[k]	$\frac{9}{2}$	100	4·511	4·342	−0·38	16·23	$1{\cdot}1\times10^{-2}$	0·141	$8{\cdot}01\times10^{2}$

[a] Excluding the lanthanides.

[b] See footnote b to Appendix 1.

[c] Loc. cit. in footnote d to Appendix 1, except where otherwise stated.

[d] Loc. cit. in footnote c to Appendix 1.

[e] It should be noted that reported values of Q may be in error by as much as 20–30%.

[f] Calculated from the quoted value of γ (therefore diamagnetic corrections are included and the frequency quoted is not with respect to TMS).

[g] $\ell=(2I+3)Q^2/I^2(2I-1)$.

[h] See footnote f to Appendix 1.

[i] A useful isotope of $I=\frac{1}{2}$ exists.

[j] R. Neumann, F. Träger, J. Kowalski & G. zu Putlitz, *Z. Physik* **A279,** 249 (1976).

[k] Radioactive, with a long half-life. Other radioactive nuclei (e.g. ^{40}K, ^{41}Ca) have been examined by NMR but are unimportant for chemical studies.

[l] S. Büttgenbach, R. Dicke, H. Gebauer, R. Kuhnen & F. Träber, *Z. Physik* **A286,** 125 (1978).

[m] Derived from data in C. Brevard & P. Granger, *J. Chem. Phys.* **75,** 4175 (1981).

[n] S. Büttgenbach, R. Dicke, H. Gebauer & M. Herschel, *Z. Physik* **A280,** 217 (1977).

Appendix 3 Time-dependent perturbation theory and NMR transitions

Quantum mechanics is greatly concerned with solving the Schrödinger Equation. This is expressed in two forms, one independent of time (A3-1), the other time-dependent (A3-2).

$$\hat{\mathcal{H}}\,|\psi_r\rangle = U_r\,|\psi_r\rangle \tag{A3-1}$$

$$i\hbar\frac{\partial}{\partial t}|\Psi\rangle = \hat{\mathcal{H}}\,|\Psi\rangle \tag{A3-2}$$

The former equation is used in several places in this book, notably in Chapter 2. The time-dependent wave functions which are the solutions of A3-2 will be denoted in capital Greek letters to distinguish them from the eigenfunctions of A3-1. In fact, by substitution it can be shown that if ψ_r is an eigenfunction of A3-1, then a solution of A3-2 is given by:

$$|\Psi\rangle = |\psi_r\rangle\exp(-iU_rt/\hbar) \tag{A3-3}$$

The corresponding bra is:

$$\langle\Psi| = \exp(iU_rt/\hbar)\,\langle\psi_r| \tag{A3-4}$$

A general solution to A3-2 is a linear expansion of the type:

$$|\Psi\rangle = \sum_r |\psi_r\rangle\, a_r(t)\exp(-iU_rt/\hbar) \tag{A3-5}$$

Suppose one is dealing with a Hamiltonian with a large static component $\hat{\mathcal{H}}_0$ and a small time-dependent one $\lambda\hat{\mathcal{H}}(t)$, i.e.

$$\hat{\mathcal{H}} = \hat{\mathcal{H}}_0 + \lambda\hat{\mathcal{H}}(t) \tag{A3-6}$$

where λ is introduced for mathematical convenience. The eigenfunctions $|\psi_r\rangle$ are established under the influence of $\hat{\mathcal{H}}_0$ alone, and one wants to investigate the influence of $\hat{\mathcal{H}}(t)$. Substitution of A3-5 and A3-6 into A3-2, followed by differentiation of each term in A3-5 as a product yields:

$$i\hbar\sum_r \exp(-iU_rt/\hbar)\,|\psi_r\rangle\frac{\partial a_r(t)}{\partial t} = \lambda\sum_r \exp(-iU_rt/\hbar)\hat{\mathcal{H}}(t)\,|\psi_r\rangle\, a_r(t) \tag{A3-7}$$

where the summations run separately. Integration with the complex conjugate function $\exp(iU_s t/\hbar)\langle\psi_s|$, with use of the orthonormality condition, gives

$$i\hbar\frac{\partial a_s(t)}{\partial t}=\lambda\sum_r\langle\psi_s|\,\hat{\mathscr{H}}(t)\,|\psi_r\rangle a_r(t)\exp(-i\omega_{rs}t) \qquad \text{(A3-8)}$$

where $\omega_{rs}=(U_r-U_s)/\hbar$—i.e. it is the transition angular frequency $s\leftrightarrow r$.

The coefficients $a_r(t)$ may be expanded as a Taylor series in powers of λ:

$$a_r(t)=a_r^{(0)}(t)+a_r^{(1)}(t)+a_r^{(2)}(t)+\ldots. \qquad \text{(A3-9)}$$

The only concern here is with the zero- and first-order terms. The expansion A3-9 may be inserted into Eqn A3-8 and terms in successive powers of λ collected to give

$$i\hbar\frac{\partial a_s^{(0)}(t)}{\partial t}=0 \qquad \text{(A3-10)}$$

$$i\hbar\frac{\partial a_s^{(1)}(t)}{\partial t}=\sum_r\langle\psi_s|\,\hat{\mathscr{H}}(t)\,|\psi_r\rangle\,a_r^{(0)}(t)\exp(-i\omega_{rs}t) \qquad \text{(A3-11)}$$

In fact λ can then be set as unity without loss of generality.

Equation A3-11 expresses the fact that $\hat{\mathscr{H}}(t)$ causes transitions. If the system is initially in the state r, then $a_r^{(0)}(0)=1$. Thus:

$$a_s^{(1)}(t)=\frac{1}{i\hbar}\int_0^t\langle\psi_s|\,\hat{\mathscr{H}}(t')\,|\psi_r\rangle\exp(-i\omega_{rs}t')\,\mathrm{d}t' \qquad \text{(A3-12)}$$

The probability that the system has made a transition from state r to state s at time t is $|a_s^{(1)}(t)|^2$. Note that evaluation of this probability involves a complex conjugate.

Two cases are of interest in this book. The first concerns transitions simulated by coherent radiofrequency radiation. In this case, for a single type of nucleus, $\hat{\mathscr{H}}(t)$ is given by:

$$\hat{\mathscr{H}}(t)=2\gamma\hbar B_1\,\hat{I}_x\cos\omega t \qquad \text{(A3-13)}$$

where it should be recognized that $2\cos\omega t$ has a r.m.s value of unity. Eqn A3-12 may be integrated directly to yield:

$$a_s^{(1)}(t)=-\gamma B_1\langle\psi_s|\,\hat{I}_x\,|\psi_r\rangle\left[\frac{\exp\{i(\omega_{rs}-\omega)t'}{\omega_{rs}-\omega}+\frac{\exp\{i(\omega_{rs}+\omega)t'\}}{\omega_{rs}+\omega}\right]_0^t \qquad \text{(A3-14)}$$

$$=-\gamma B_1\langle\psi_s|\,\hat{I}_x\,|\psi_r\rangle\left[\frac{\exp\{i(\omega_{rs}-\omega)t\}-1}{\omega_{rs}-\omega}+\frac{\exp\{i(\omega_{rs}+\omega)t\}-1}{\omega_{rs}+\omega}\right] \qquad \text{(A3-15)}$$

When $\omega_{rs}\sim\omega$ (i.e. near resonance) the second term in A3-15 is much smaller than the first, and so may be omitted. Multiplying the first term of A3-15 by its complex conjugate gives the transition probability as:

$$|a_s^{(1)}(t)|^2 = 4\gamma^2 B_1^2 \langle\psi_s|\, \hat{I}_x \,|\psi_r\rangle^2 \sin^2\{\tfrac{1}{2}i(\omega_{rs}-\omega)t\}/(\omega_{rs}-\omega)^2 \quad \text{(A3-16)}$$

The probability of transition per unit time can therefore be written:

$$P_{rs} = 2\pi\gamma^2 B_1^2 \langle\psi_s|\, \hat{I}_x \,|\psi_r\rangle^2 \,[\sin^2\{\tfrac{1}{2}i(\omega_{rs}-\omega)t\}/2\pi t\tfrac{1}{4}(\omega_{rs}-\omega)^2] \quad \text{(A3-17)}$$

The expression in square brackets behaves like the δ-function—it has a sharp maximum at $\omega = \omega_{rs}$ and its area is unity, so P_{rs} becomes:

$$P_{rs} = 2\pi\gamma^2 B_1^2 \langle\psi_s|\, \hat{I}_x \,|\psi_r\rangle^2 \,\delta(\omega_{rs}-\omega) \quad \text{(A3-18)}$$

This is identical to the equation given in Chapter 2, i.e. Eqn 2-44, provided it is realized that, since the delta function has dimensions which are the reciprocal of those of its argument, $\delta(\omega) \equiv \hbar\delta(U)$.

The second case of interest concerns transitions occurring during spin-lattice relaxation. In this case the Hamiltonian $\hat{\mathcal{H}}(t)$ represents a randomly-varying interaction with average value zero, which fluctuates many times in the interval t. For the purposes of evaluation, the equation for $|a^{(1)}(t)|^2$ is best cast in the form:

$$|a^{(1)}(t)|^2 = \frac{1}{\hbar^2}\int_0^t \mathrm{d}t' \int_0^t \overline{\langle\psi_r|\, \hat{\mathcal{H}}(t') \,|\psi_s\rangle\langle\psi_s|\, \hat{\mathcal{H}}(t'') \,|\psi_r\rangle} \exp[-i\omega_{rs}(t'-t'')]\,\mathrm{d}t'' \quad \text{(A3-19)}$$

where the bar indicates that an average is taken over a statistical ensemble of spins. This may be put into a form involving a time difference, $\tau = t' - t''$, as follows:

$$|a^{(1)}(t)|^2 = \frac{1}{\hbar^2}\int_0^t \mathrm{d}t'' \int_{-t''}^{t-t''} \overline{\langle\psi_r|\, \hat{\mathcal{H}}(t''+\tau) \,|\psi_s\rangle\langle\psi_s|\, \hat{\mathcal{H}}(t'') \,|\psi_r\rangle} \exp(-i\omega_{rs}\tau)\,\mathrm{d}t \quad \text{(A3-20)}$$

The Hamiltonian elements causing relaxation may generally be written in terms of products of constants, time-dependent spin factors, and time-dependent factors. The case of the dipolar Hamiltonian (see Eqn 4-5) will be considered here. For a heteronuclear two-spin (AX) system, this may be written:

$$\hat{\mathcal{H}}_{\mathrm{dd}}(t) = \sum_{m=-2}^{2} Y_{2,m} A_m \hbar^2 \gamma_{\mathrm{A}}\gamma_{\mathrm{X}} r_{\mathrm{AX}}^{-3} \frac{\mu_0}{4\pi} \quad \text{(A3-21)}$$

where $Y_{2,m}$ is the angular (space) part, which is time-dependent due to molecular tumbling, and is expressed as second-order spherical harmonics:

$$Y_{2,0} = \sqrt{\tfrac{5}{4}}(1 - 3\cos^2\theta) \quad \text{(A3-22)}$$

$$Y_{2,\pm 1} = \sqrt{\tfrac{15}{2}}\cos\theta\sin\theta\exp(\pm i\phi) \quad \text{(A3-23)}$$

$$Y_{2,\pm 2} = -\sqrt{\tfrac{15}{8}}\sin^2\theta\exp(\pm 2i\phi) \quad \text{(A3-24)}$$

and A_m is the spin part:

$$A_0 = \sqrt{\tfrac{4}{5}}[\hat{I}_{Az}\hat{I}_{Xz} - \tfrac{1}{4}(\hat{I}_{A+}\hat{I}_{X-} + \hat{I}_{A-}\hat{I}_{X+})] \qquad \text{(A3-25)}$$

$$A_{\pm 1} = -\sqrt{\tfrac{3}{10}}(\hat{I}_{A\pm}\hat{I}_{Xz} + \hat{I}_{Az}\hat{I}_{X\pm}) \qquad \text{(A3-26)}$$

$$A_{\pm 2} = -\sqrt{\tfrac{3}{10}}\hat{I}_{A\pm}\hat{I}_{X\pm} \qquad \text{(A3-27)}$$

It is assumed that r_{AX} is time-independent, i.e. vibrations are ignored. The numerical coefficients in Eqns A3-22 to A3-27 have been deliberately adjusted to normalize the angular functions to unit r.m.s. average.[1] Equation 3-20 therefore becomes:

$$|a^{(1)}(t)|^2 = (2\pi R)^2 \sum_m \overline{\langle\psi_r|\, A_m\, |\psi_s\rangle^2} \int_0^t \mathrm{d}t''$$

$$\times \int_{-t''}^{t-t''} \overline{Y_{2,m}(t+\tau)\, Y_{2,m}(t)} \exp(-i\omega_{rs}\tau)\, \mathrm{d}\tau \qquad \text{(A3-28)}$$

where R is the dipolar interaction constant, $(\mu_0/4\pi)\gamma_A\gamma_X r_{AX}^{-3}\hbar/2\pi$ (in frequency units). The factor $\overline{Y_{2,m}(t+\tau)\, Y_{2,m}(t)}$ depends only on τ, and is the autocorrelation function $G(\tau)$ given in Chapter 3, Eqn 3-41. For fixed τ it has a definite value which does not depend on t. The function $G(\tau)$ is normalized to unity, i.e. $G(0) = \overline{Y_{2,m}^2(t)} = 1$, and it is usually a rapidly-decreasing function of τ. The second integral in A3-28 becomes almost independent of t when t is large, so that the limits may be extended to $\pm\infty$. The first integral then gives a factor of t. The transition rate per unit time, for which the symbol W_{rs} has been used earlier, is therefore:

$$W_{rs} = (2\pi R)^2 \sum_m \overline{\langle\psi_r|\, A_m\, |\psi_s\rangle^2} \int_{-\infty}^{\infty} G(\tau)\exp(-i\omega_{rs}\tau)\, \mathrm{d}\tau \qquad \text{(A3-29)}$$

The integral will be recognized as the spectral density, $J(\omega)$, first used in Eqn 3-43. In principle it depends on m, but this complication is ignored here. However, W_{rs} definitely depends on m through the spin term. In accordance with Fig. 4-3 and the usual rules of spin operators, the spin term yields the following for the two-spin case:

$\langle\psi_r|\, A_0\, |\psi_s\rangle^2 = \frac{4}{5} \cdot \frac{1}{16}$ for W_0 (from the flip-flop term only)

$= \frac{3}{10} \cdot \frac{1}{4}$ for W_1

$= \frac{3}{10}$ for W_2

Substitution in Eqn A3-29 using the appropriate resonance frequencies gives:

$$W_0 = \tfrac{1}{20}(2\pi R)^2 J(\omega_X - \omega_A) \qquad \text{(A3-30)}$$

$$W_1 = \tfrac{3}{40}(2\pi R)^2 J(\omega_A) \qquad \text{(A3-31)}$$

$$W_2 = \tfrac{3}{10}(2\pi R)^2 J(\omega_X + \omega_A) \qquad \text{(A3-32)}$$

which appear in the main text as Eqns 4-16 to 4-18. The theory may be developed similarly for other relaxation mechanisms.

[1] For isotropic tumbling $\langle(1-3\cos^2\theta)^2\rangle=\frac{4}{5}$, $\langle\sin^2\theta\cos^2\theta\rangle=\frac{2}{15}$, and $\langle\sin^4\theta\rangle=\frac{8}{15}$. Other authors include these factors in $J(\omega)$, thus making it necessary to distinguish between spectral densities corresponding to transitions differing in Δm_T. The values of $J(\omega)$ as used in this book are sometimes referred to as *reduced* spectral densities.

Appendix 4 The effects of rapid sample rotation

Several interactions of interest for NMR of solids contain the term $(3\cos^2\theta-1)$, where θ is the angle between a particular direction $\mathbf{r}$ (such as the distance between two nuclei) and the static magnetic field $\mathbf{B}_0$. An example is the dipolar interaction between two nuclei, terms A and B, as expressed in Eqns 4-5 to 4-7. It is of importance to know how such functions are affected by rapid sample rotation of $\mathbf{r}$ about an axis fixed in space relative to $\mathbf{B}_0$.

Consider, as defined in Fig. A1, a unit vector OU in direction $\mathbf{r}$ rotating about axis $\mathbf{s}$, which is inclined at angle β to the direction $\mathbf{B}_0$. Let the angle between $\mathbf{r}$ and $\mathbf{s}$ be χ and the angle between $\mathbf{r}$ and $\mathbf{B}_0$ be θ. Then the required cosine is given by OY.

Now $\mathrm{OV}=\cos\chi$

$$\therefore \mathrm{OX}=\mathrm{OV}\cos\beta=\cos\chi\cos\beta$$

Also $\mathrm{VU}=\sin\chi$

$$\therefore \mathrm{VZ}=\sin\chi\cos\phi$$

$$\therefore \mathrm{XY}=\sin\chi\sin\beta\cos\phi \text{ since } \widehat{\mathrm{XVZ}}=90^\circ-\widehat{\mathrm{XVO}}=\widehat{\mathrm{VOX}}=\beta$$

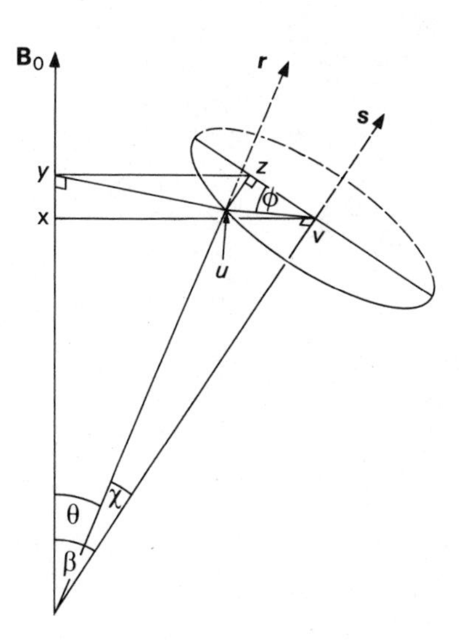

Fig. A1 Geometry definitions for the effect of rotation.

[N.B. the UVZ plane is perpendicular to the plane containing $\mathbf{B}_0$ and $\mathbf{s}$]

$$\therefore \text{OY} = \text{OX} + \text{XY} = \cos\chi \cos\beta + \sin\chi \sin\beta \cos\phi$$

In the case of rotation about $\mathbf{s}$ at rate ω, the angle $\phi = \omega t$ if $\mathbf{r}$ is in the plane of $\mathbf{B}_0$ and $\mathbf{s}$ at time $t=0$. Thus $\cos\theta = \cos\chi\cos\beta + \sin\chi\sin\beta \cos\omega t$ at time t. Thence

$$\langle\cos^2\theta\rangle = \cos^2\beta\cos^2\chi + 2\sin\beta\cos\beta\sin\chi\cos\chi\langle\cos\omega t\rangle + \sin^2\beta\sin^2\chi\langle\cos^2\omega t\rangle$$

Now $\langle\cos\omega t\rangle = 0$ and $\langle\cos^2\omega t\rangle = \frac{1}{2}$

$$\begin{aligned}\therefore \langle\cos^2\theta\rangle &= \cos^2\beta\cos^2\chi + \tfrac{1}{2}\sin^2\beta\sin^2\chi \\ &= \cos^2\beta\cos^2\chi + \tfrac{1}{2}(1-\cos^2\beta)(1-\cos^2\chi) \\ &= \tfrac{1}{3} + \tfrac{1}{6}(3\cos^2\beta - 1)(3\cos^2\chi - 1)\end{aligned}$$

$$\therefore \langle 3\cos^2\theta - 1\rangle = \tfrac{1}{2}(3\cos^2\beta - 1)(3\cos^2\chi - 1)$$

[N.B. the extrema of the angle θ are clearly $\beta - \chi$ and $\beta + \chi$]

Further reading: advanced texts

Review series

Progress in Nuclear Magnetic Resonance Spectroscopy (Eds. J. W. Emsley, J. Feeney & L. H. Sutcliffe), Pergamon Press, Oxford.

Advances in Magnetic Resonance (Ed. J. S. Waugh), Academic Press, New York.

Annual Review of NMR Spectroscopy [Eds. E. F. Mooney (vols. 1–6) and G. A. Webb (vol. 6 on)] (renamed *Annual Reports on NMR Spectroscopy* from Vol. 3 onwards), Academic Press, New York.

Specialist Periodical Reports on NMR [Eds. R. K. Harris (vols. 1–5), R. J. Abraham (vols. 6–8) & G. A. Webb (vol. 9 on)], Royal Society of Chemistry, London.

NMR Basic Principles and Progress (Eds. P. Diehl, E. Fluck & R. Kosfeld), Springer-Verlag, Berlin.

Topics in Carbon-13 NMR Spectroscopy (Ed. G. C. Levy), Wiley & Sons, New York.

Biological Magnetic Resonance (Eds. L. J. Berliner & J. Reuben), Plenum Press.

Books and articles on topics not covered in the present work

Magnetic Resonance of Biomolecules, P. F. Knowles, D. Marsh & H. W. E. Rattle, Wiley & Sons (1976).

NMR in Biochemistry: Principles and Applications, T. L. James, Academic Press (1975).

'The theory of chemically-induced dynamic spin polarization', J. H. Freed & J. B. Pedersen, *Adv. Magn. Reson.*, **8**, 2 (1976).

'Chemically induced dynamic nuclear and electron polarizations', C. Richard & P. Granger, *NMR Basic Principles & Prog.*, **8,** 1 (1974).

NMR Spectroscopy using Liquid Crystal Solvents, J. W. Emsley & J. C. Lindon, Pergamon Press (1975).

'NMR in medicine' Ed. R. Damadian, *NMR Basic Principles & Prog.*, **19,** 1 (1981).

'NMR imaging in biomedicine', P. Mansfield & P. G. Morris, *Adv. Magn. Reson.* Suppl. 2 (1982).

NMR and its Application to Living Systems, D. G. Gadian, Oxford Univ. Press (1982).

NMR in Molecular Biology, O. Jardetzky & G. C. K. Roberts, Academic Press (1981).

Books and articles on specific nuclei

NMR and the Periodic Table, Eds. R. K. Harris & B. E. Mann, Academic Press (1978).

Handbook of High-resolution Multinuclear NMR, C. Brevard & P. Granger, John Wiley & Sons (1981).

'NMR spectroscopy of boron compounds', H. Nöth & B. Wrackmeyer, *NMR Basic Principles & Prog.*, **14,** 1 (1978).

Carbon-13 NMR Spectroscopy (second edition), G. C. Levy, R. L. Lichter & G. L. Nelson, Wiley & Sons (1980).

Interpretation of Carbon-13 NMR Spectroscopy, F. W. Wehrli & T. Wirthlin, Heyden & Son Ltd. (1976).

Nitrogen-15 NMR Spectroscopy, G. C. Levy & R. L. Lichter, Wiley & Sons (1979).

Nitrogen NMR, Eds. M. Witanowski & G. A. Webb, Plenum Press (1973).

'Fluorine chemical shifts', J. W. Emsley & L. Phillips, *Prog. NMR Spectrosc.*, **7,** 1 (1971).

Phosphorus-31 NMR, M. M. Crutchfield, C. H. Dungan, J. H. Letcher, V. Mark & J. R. Van Wazer, *Topics in Phosphorus Chemistry*, Vol. 5, Wiley & Sons (1967).

General texts

High-resolution Nuclear Magnetic Resonance, J. A. Pople, W. G. Schneider & H. J. Bernstein, McGraw-Hill (1959).

High-resolution Nuclear Magnetic Resonance Spectroscopy (2 volumes), J. W. Emsley, J. Feeney & L. H. Sutcliffe, Pergamon Press (1965).

Principles of Nuclear Magnetism, A. Abragam, Clarendon Press (1961).

Principles of Magnetic Resonance (second edition), C. P. Slichter, Springer-Verlag (1978).

Practical NMR Spectroscopy, M. L. Martin, J.-J. Delpuech & G. J. Martin, Heyden & Son Ltd. (1980).

NMR of Paramagnetic Molecules, Eds. G. N. La Mar, W. DeW. Horrocks & R. H. Holm, Academic Press (1973).

Answers to problems

Chapter 1

1-1 $\delta 4{\cdot}33$; 433 Hz
1-2 (a) 12·69 MHz (b) 11·42 MHz (c) 6·26 MHz
1-3 1·0 T, 2·47 T
1-6 0·799 and 9·836; 45·6, 50·3, 587·8 and 592·5 Hz.
1-7 520 Hz; −5·20 ppm; 12·22 μT
1-17 204

Chapter 2

2-9 (a) 0 (b) $i/2[I(I+1)-m(m-1)]^{1/2}$ (c) 2
(d) $\frac{1}{2}[I(I+1)-m(m+1)]^{1/2}$ (e) $\sqrt{3}$ (f) 2
2-10 0; 0; $-\frac{1}{2}$; 0; $(c_1+c_2+c_3)$
2-12 $9A+9B$. There will be lines at ν_A and ν_B (from the TS and ST states respectively).
2-13 $J=4{\cdot}5$ Hz; $\delta=6{\cdot}910$ and 6·999 ppm; intensities 0·358 : 1·642; frequencies 688·16, 692·66, 698·17 and 702·67 Hz.

Chapter 3

3-2 $2\sqrt{2}/T_2$ angular frequency units
3-3 No signal
3-4 $T_1=20{\cdot}8$ s
3-5 $T_1=26.5$ s

Chapter 4

4-2 2
4-3 $1-\gamma_S/\gamma_I$
4-8 0·0164 s^{-1}

Chapter 8

8-2 61°31′
8-5 209 Hz

Index

Northern Michigan University
3 1854 003 466 789
EZNO
QD96 N8 H37 1983
Nuclear magnetic resonance spectroscopy